Tribology in Machine Design

T. A. STOLARSKI

MSc, PhD, DSc, DIC, CEng, MIMechE

BUTTERWORTH
HEINEMANN

OXFORD AUCKLAND BOSTON JOHANNESBURG MELBOURNE NEW DELHI

Butterworth-Heinemann
Linacre House, Jordan Hill, Oxford OX2 8DP
225 Wildwood Avenue, Woburn, MA 01801-2041
A division of Reed Educational and Professional Publishing Ltd

A member of the Reed Elsevier plc group

First published 1990
Reprinted 2000

British Library Cataloguing in Publication Data
A catalogue record for this book is available from the British Library

Library of Congress Cataloguing in Publication Data
A catalogue record for this book is available from the Library of Congress

ISBN 0 7506 3623 8

Printed and bound in Great Britain

Contents

Preface   xi

1.   Introduction to the concept of tribodesign   **1**
1.1.   Specific principles of tribodesign   4
1.2.   Tribological problems in machine design   6
    1.2.1.   Plain sliding bearings   6
    1.2.2.   Rolling contact bearings   7
    1.2.3.   Piston, piston rings and cylinder liners   8
    1.2.4.   Cam and cam followers   9
    1.2.5.   Friction drives   10
    1.2.6.   Involute gears   10
    1.2.7.   Hypoid gears   11
    1.2.8.   Worm gears   12

2.   Basic principles of tribology   **13**
2.1.   Origins of sliding friction   13
2.2   Contact between bodies in relative motion   14
2.3   Friction due to adhesion   15
2.4.   Friction due to ploughing   16
2.5.   Friction due to deformation   17
2.6   Energy dissipation during friction   18
2.7   Friction under complex motion conditions   18
2.8.   Types of wear and their mechanisms   19
    2.8.1.   Adhesive wear   19
    2.8.2.   Abrasive wear   20
    2.8.3.   Wear due to surface fatigue   21
    2.8.4.   Wear due to chemical reactions   22
2.9.   Sliding contact between surface asperities   23
2.10. The probability of surface asperity contact   26
2.11. Wear in lubricated contacts   31
    2.11.1. Rheological lubrication regime   33
    2.11.2. Functional lubrication regime   33
    2.11.3. Fractional film defect   34
    2.11.4. Load sharing in lubricated contacts   37
    2.11.5. Adhesive wear equation   39
    2.11.6. Fatigue wear equation   40
    2.11.7. Numerical example   41

2.12		Relation between fracture mechanics and wear	45
	2.12.1.	Estimation of stress intensity under non-uniform applied loads	47
2.13.		Film lubrication	48
	2.13.1	Coefficient of viscosity	48
	2.13.2.	Fluid film in simple shear	49
	2.13.3.	Viscous flow between very close parallel surfaces	50
	2.13.4.	Shear stress variations within the film	51
	2.13.5.	Lubrication theory by Osborne Reynolds	51
	2.13.6.	High-speed unloaded journal	53
	2.13.7.	Equilibrium conditions in a loaded bearing	53
	2.13.8.	Loaded high-speed journal	54
	2.13.9.	Equilibrium equations for loaded high-speed journal	57
	2.13.10.	Reaction torque acting on the bearing	59
	2.13.11.	The virtual coefficient of friction	59
	2.13.12.	The Sommerfeld diagram	60
		References	63
3.		**Elements of contact mechanics**	**64**
3.1.		Introduction	64
3.2.		Concentrated and distributed forces on plane surfaces	65
3.3.		Contact between two elastic bodies in the form of spheres	67
3.4.		Contact between cylinders and between bodies of general shape	70
3.5.		Failures of contacting surfaces	71
3.6.		Design values and procedures	73
3.7.		Thermal effects in surface contacts	74
	3.7.1	Analysis of line contacts	75
	3.7.2.	Refinement for unequal bulk temperatures	79
	3.7.3.	Refinement for thermal bulging in the conjunction zone	80
	3.7.4.	The effect of surface layers and lubricant films	80
	3.7.5.	Critical temperature for lubricated contacts	82
	3.7.6.	The case of circular contact	83
	3.7.7.	Contacts for which size is determined by load	85
	3.7.8.	Maximum attainable flash temperature	86
3.8.		Contact between rough surfaces	87
	3.8.1.	Characteristics of random rough surfaces	87
	3.8.2.	Contact of nominally flat rough surfaces	90
3.9.		Representation of machine element contacts	94
		References	96
4.		**Friction, lubrication and wear in lower kinematic pairs**	**97**
4.1.		Introduction	97
4.2.		The concept of friction angle	98
	4.2.1.	Friction in slideways	98
	4.2.2.	Friction stability	100

4.3.	Friction in screws with a square thread	103
	4.3.1. Application of a threaded screw in a jack	105
4.4.	Friction in screws with a triangular thread	109
4.5.	Plate clutch – mechanism of operation	111
4.6.	Cone clutch – mechanism of operation	114
	4.6.1. Driving torque	115
4.7.	Rim clutch – mechanism of operation	116
	4.7.1. Equilibrium conditions	117
	4.7.2. Auxiliary mechanisms	119
	4.7.3. Power transmission rating	120
4.8.	Centrifugal clutch – mechanism of operation	120
4.9.	Boundary lubricated sliding bearings	121
	4.9.1. Axially loaded bearings	123
	4.9.2. Pivot and collar bearings	124
4.10.	Drives utilizing friction force	127
	4.10.1. Belt drive	128
	4.10.2. Mechanism of action	129
	4.10.3. Power transmission rating	132
	4.10.4. Relationship between belt tension and modulus	133
	4.10.5. V-belt and rope drives	134
4.11.	Frictional aspects of brake design	136
	4.11.1. The band brake	136
	4.11.2. The curved brake block	138
	4.11.3. The band and block brake	144
4.12.	The role of friction in the propulsion and the braking of vehicles	145
4.13.	Tractive resistance	150
4.14.	Pneumatic tyres	151
	4.14.1. Creep of an automobile tyre	152
	4.14.2. Transverse tangential forces	152
	4.14.3. Functions of the tyre in vehicle application	154
	4.14.4. Design features of the tyre surface	154
	4.14.5. The mechanism of rolling and sliding	155
	4.14.6. Tyre performance on a wet road surface	157
	4.14.7. The development of tyres with improved performance	159
4.15.	Tribodesign aspects of mechanical seals	160
	4.15.1. Operation fundamentals	161
	4.15.2. Utilization of surface tension	162
	4.15.3. Utilization of viscosity	162
	4.15.4. Utilization of hydrodynamic action	163
	4.15.5. Labyrinth seals	164
	4.15.6. Wear in mechanical seals	164
	4.15.7. Parameters affecting wear	168
	4.15.8. Analytical models of wear	169
	4.15.9. Parameters defining performance limits	170
	4.15.10. Material aspects of seal design	170

4.15.11. Lubrication of seals   172
References   173

5.   Sliding-element bearings   **174**
5.1.   Derivation of the Reynolds equation   174
5.2.   Hydrostatic bearings   178
5.3.   Squeeze-film lubrication bearings   181
5.4.   Thrust bearings   183
       5.4.1.   Flat pivot   184
       5.4.2.   The effect of the pressure gradient in the direction
                of motion   186
       5.4.3.   Equilibrium conditions   188
       5.4.4.   The coefficient of friction and critical slope   188
5.5.   Journal bearings   189
       5.5.1.   Geometrical configuration and pressure
                generation   189
       5.5.2.   Mechanism of load transmission   192
       5.5.3.   Thermoflow considerations   194
       5.5.4.   Design for load-bearing capacity   196
       5.5.5.   Unconventional cases of loading   197
       5.5.6.   Numerical example   199
       5.5.7.   Short bearing theory – CAD approach   201
5.6.   Journal bearings for specialized applications   204
       5.6.1.   Journal bearings with fixed non-preloaded pads   205
       5.6.2.   Journal bearings with fixed preloaded pads   205
       5.6.3.   Journal bearings with special geometric features   207
       5.6.4.   Journal bearings with movable pads   207
5.7.   Gas bearings   210
5.8.   Dynamically loaded journal bearings   212
       5.8.1.   Connecting-rod big-end bearing   213
       5.8.2.   Loads acting on main crankshaft bearing   213
       5.8.3.   Minimum oil film thickness   214
5.9.   Modern developments in journal bearing design   217
       5.9.1.   Bearing fit   218
       5.9.2.   Grooving   219
       5.9.3.   Clearance   219
       5.9.4.   Bearing materials   220
5.10. Selection and design of thrust bearings   221
       5.10.1. Tilting-pad bearing characteristics   223
       5.10.2. Design features of hydrostatic thrust bearings   225
5.11. Self-lubricating bearings   226
       5.11.1. Classification of self-lubricating bearings   226
       5.11.2. Design considerations   228
       References   230

6.   Friction, lubrication and wear in higher kinematic pairs   **232**
6.1.   Introduction   232
6.2.   Loads acting on contact area   233

6.3.	Traction in the contact zone	233
6.4.	Hysteresis losses	234
6.5.	Rolling friction	235
6.6.	Lubrication of cylinders	238
6.7.	Analysis of line contact lubrication	242
6.8.	Heating at the inlet to the contact	244
6.9.	Analysis of point contact lubrication	245
6.10.	Cam-follower system	246
	References	247
7.	**Rolling-contact bearings**	**248**
7.1.	Introduction	248
7.2.	Analysis of friction in rolling-contact bearings	248
	7.2.1. Friction torque due to differential sliding	249
	7.2.2. Friction torque due to gyroscopic spin	250
	7.2.3. Friction torque due to elastic hysteresis	251
	7.2.4. Friction torque due to geometric errors	252
	7.2.5. Friction torque due to the effect of the raceway	252
	7.2.6. Friction torque due to shearing of the lubricant	252
	7.2.7. Friction torque caused by the working medium	253
	7.2.8. Friction torque caused by temperature increase	254
7.3.	Deformations in rolling-contact bearings	254
7.4.	Kinematics of rolling-contact bearings	256
	7.4.1. Normal speeds	256
	7.4.2. High speeds	258
7.5.	Lubrication of rolling-contact bearings	259
	7.5.1. Function of a lubricant	259
	7.5.2. Solid film lubrication	260
	7.5.3. Grease lubrication	261
	7.5.4. Jet lubrication	262
	7.5.5. Lubrication utilizing under-race passages	263
	7.5.6. Mist lubrication	264
	7.5.7. Surface failure modes related to lubrication	265
	7.5.8. Lubrication effects on fatigue life	265
	7.5.9. Lubricant contamination and filtration	266
	7.5.10. Elastohydrodynamic lubrication in design practice	266
7.6.	Acoustic emission in rolling-contact bearings	268
	7.6.1. Inherent source of noise	268
	7.6.2. Distributed defects on rolling surfaces	269
	7.6.3. Surface geometry and roughness	269
	7.6.4. External influences on noise generation	270
	7.6.5. Noise reduction and vibration control methods	271
	References	272
8.	**Lubrication and efficiency of involute gears**	**273**
8.1.	Introduction	273
8.2.	Generalities of gear tribodesign	273
8.3.	Lubrication regimes	275

8.4.	Gear failure due to scuffing	278
	8.4.1. Critical temperature factor	280
	8.4.2. Minimum film thickness factor	281
8.5.	Gear pitting	282
	8.5.1. Surface originated pitting	283
	8.5.2. Evaluation of surface pitting risk	283
	8.5.3. Subsurface originated pitting	284
	8.5.4. Evaluation of subsurface pitting risk	284
8.6.	Assessment of gear wear risk	285
8.7.	Design aspect of gear lubrication	286
8.8.	Efficiency of gears	288
	8.8.1. Analysis of friction losses	289
	8.8.2. Summary of efficiency formulae	293
	References	294
Index		295

Preface

The main purpose of this book is to promote a better appreciation of the increasingly important role played by tribology at the design stage in engineering. It shows how algorithms developed from the basic principles of tribology can be used in a range of practical applications.

The book is planned as a comprehensive reference and source book that will not only be useful to practising designers, researchers and postgraduate students, but will also find an essential place in libraries catering for engineering students on degree courses in universities and polytechnics. It is rather surprising that, in most mechanical engineering courses, tribology – or at least the application of tribology to machine design – is not a compulsory subject. This may be regarded as a major cause of the time-lag between the publication of new findings in tribology and their application in industry. A further reason for this time-lag is the fact that too many tribologists fail to present their results and ideas in terms of principles and concepts that are directly accessible and appealing to the design engineer.

It is hoped that the procedures and techniques of analysis explained in this book will be found helpful in applying the principles of tribology to the design of the machine elements commonly found in mechanical devices and systems. It is designed to supplement the Engineering Science Data Unit (ESDU) series in tribology (well known to practising engineers), emphasizing the basic principles, giving the background and explaining the rationale of the practical procedures that are recommended. On a number of occasions the reader is referred to the appropriate ESDU item number, for data characterizing a material or a tribological system, for more detailed guidance in solving a particular problem or for an alternative method of solution. The text advocates and demonstrates the use of the computer as a design tool where long, laborious solution procedures are needed.

The material is grouped according to applications: elements of contact mechanics, tribology of lower kinematic pairs, tribology of higher kinematic pairs, rolling contact bearings and surface damage of machine elements. The concept of tribodesign is introduced in Chapter 1. Chapter 2 is devoted to a brief discussion of the basic principles of tribology, including some new concepts and models of lubricated wear and friction under complex kinematic conditions. Elements of contact mechanics, presented in Chapter 3, are confined to the most technically important topics. Tribology of lower kinematic pairs, sliding element bearings and higher kinematic pairs are discussed in Chapters 4, 5 and 6, respectively. Chapter 7 contains a discussion of rolling contact bearings with particular emphasis on contact problems, surface fatigue and lubrication techniques. Finally, Chapter 8 concentrates on lubrication and surface failures of involute gears.

At the end of Chapters 2–8 there is a list of books and selected papers providing further reading on matters discussed in the particular chapter. The choice of reference is rather personal and is not intended as a comprehensive literature survey.

The book is based largely on the notes for a course of lectures on friction, wear and lubrication application to machine design given to students in the Department of Mechanical Engineering, Technical University of Gdansk and in the Mechanical Engineering Department, Brunel University.

I would like to express my sincere appreciation to some of my former colleagues from the Technical University of Gdansk where my own study of tribology started. I owe a particular debt of gratitude to Dr B. J. Briscoe of the Imperial College of Science and Technology, who helped me in many different ways to continue my research in this subject. Finally, special thanks are due to my wife Alicja for her patience and understanding during the preparation of the manuscript.

Brunel University   T.A.S.

1 Introduction to the concept of tribodesign

The behaviour and influence of forces within materials is a recognized basic subject in engineering design. This subject, and indeed the concept of transferring forces from one surface to another when the two surfaces are moving relative to one another, is neither properly recognized as such nor taught, except as a special subject under the heading *friction and lubrication*. The interaction of contacting surfaces in relative motion should not be regarded as a specialist subject because, like strength of materials, it is basic to every engineering design. It can be said that there is no machine or mechanism which does not depend on it.

Tribology, the collective name given to the science and technology of interacting surfaces in relative motion, is indeed one of the most basic concepts of engineering, especially of engineering design. The term tribology, apart from its conveniently collective character describing the field of friction, lubrication and wear, could also be used to coin a new word – *tribodesign*. It should not be overlooked, however, that the term tribology is not all-inclusive. In fact, it does not include various kinds of mechanical wear such as erosion, cavitation and other forms of wear caused by the flow of matter.

It is an obvious but fundamental fact that the ultimate practical aim of tribology lies in its successful application to machine design. The most appropriate form of this application is tribodesign, which is regarded here as a branch of machine design concerning all machine elements where friction, lubrication and wear play a significant part.

In its most advanced form, tribodesign can be integrated into machine design to the extent of leading to novel and more efficient layouts for various kinds of machinery. For example, the magnetic gap between the rotor and stator in an electric motor could be designed to serve a dual purpose, that is, to perform as a load-carrying film of ambient air eliminating the two conventional bearings. The use of the process fluid as a lubricant in the bearings of pumps and turbo-compressors, or the utilization of high-pressure steam as a lubricant for the bearings of a steam turbine are further examples in this respect. Thus, it can be safely concluded that tribodesign is an obvious, and even indispensible, branch of machine design and, therefore, of mechanical engineering in general.

In any attempt to integrate tribology and tribodesign into mechanical engineering and machine design, it is advantageous to start by visualizing the engineering task of mechanical engineers in general, and of machine designers in particular. The task of a mechanical engineer consists of the control, by any suitable means, of flows of force, energy and matter, including any combination and interaction of these different kinds of flow. Conversion from one form of energy to another may also result in kinetic energy, which in turn involves motion. Motion also comes into play when one aims not so much at kinetic energy as at a controlled time-variation of the position of some element. Motion is also essential in converting mechanical energy into thermal energy in the form of frictional heat.

Certain similar operations are also important in tribology, and particularly in tribodesign. For instance, from the present point of view, wear may be regarded as an undesirable flow of matter that is to be kept within bounds by controlling the flows of force and energy (primarily frictional heat), particularly where the force and energy have to pass through the contact area affected by the wear.

In order to provide further examples illustrating the operations in mechanical engineering, let us consider the transmission of load from one rubbing surface to its mating surface under conditions of dry contact or boundary lubrication. In general, the transmission of load is associated with concentration of the contact pressure, irrespective of whether the surfaces are conformal, like a lathe support or a journal in a sleeve bearing, or whether they are counterformal, like two mating convex gear teeth, cams and tappets or rolling elements on their raceways. With conformal surfaces, contact will, owing to the surface roughness, confine itself primarily to, or near to, the summits of the highest asperities and thus be of a dispersed nature. With counterformal surfaces, even if they are perfectly smooth, the contact will still tend to concentrate itself. This area of contact is called *Hertzian* because, in an elastic regime, it may be calculated from the Hertz theory of elastic contact. Because of surface roughness, contact will not in general be obtained throughout this area, particularly at or near its boundaries. Therefore, the areas of real contact tend to be dispersed over the Hertzian area. This Hertzian area may be called a conjunction area as it is the area of closest approach between the two rubbing surfaces.

It is clearly seen that, with both conformal and counterformal contacting surfaces, the cross-sectional area presented to the flow of force (where it is to be transmitted through the rubbing surfaces themselves) is much smaller than in the bulk of the two contacting bodies. In fact, the areas of real contact present passages or inlets to the flow of force that are invariably throttled to a severe extent. In other words, in the transmission of a flow of force by means of dry contact a rather severe constriction of this flow cannot, as a rule, be avoided. This is, in a way, synonymous with a concentration of stress. Thus, unless the load to be transmitted is unusually small, with any degree of conformity contact pressures are bound to be high under such dry conditions. Nothing much can be done by boundary lubricating layers when it comes to protecting (by means of smoothing of the flow of force in such layers), the surface material of the rubbing bodies from constrictional overstressing, that is, from wear caused by mechanical factors. Such protection must be sought by other expedients. In fact, even when compared with the small size of the dispersed contact areas on conformal surfaces, the thickness of boundary lubricating layers is negligibly small from the viewpoint of diffusion.

On the one hand, if only by conformal rubbing surfaces, the constrictional overstressing can be reduced very effectively by a full fluid film. Such a film keeps the two surfaces fully separated and offers excellent opportunities for diffusion of the flow of force, since all of the conjunction area is covered by the film and is thus entirely utilized for the diffusion concerned. The result is that again, with the conformal rubbing surfaces with which we are concerned here, the risk of overstressing the surface material will be much diminished whenever full fluid film can be established. This means that a full fluid film will eliminate all those kinds of mechanical wear that might otherwise be caused by contact between rubbing surfaces. The only possible kind of mechanical wear under these conditions is erosion, exemplified by the cavitation erosion that may occur in severely dynamically loaded journal bearings.

On the other hand, the opportunities to create similar conditions in cases of counterformal surfaces are far less probable. It is now known from the theory of elastohydrodynamic lubrication of such surfaces that, owing to the elastic deformation caused by the film pressures in the conjunction area between the two surfaces, the distribution of these pressures can only be very similar to the Hertzian distribution for elastic and dry contact. This means that with counterformal surfaces very little can be gained by interposing a fluid film. The situation may even be worsened by the occurrence of the narrow pressure spike which may occur near the outlet to the fluid film, and which may be much higher than Hertz's maximum pressure, and may thus result in severe local stress concentration which, in turn, may aggravate surface fatigue or pitting. Having once conceived the idea of constriction of the flow of force, it is not difficult to recognize that, in conjunction, a similar constriction must occur with the flow of thermal energy generated as frictional heat at the area of real contact. In fact, this area acts simultaneously as a heat source and might now, in a double sense, be called a constrictional area. Accordingly, contact areas on either conformal or counterformal rubbing surfaces are stress raisers and temperature raisers.

The above distinction, regarding the differences between conformal and counterformal rubbing surfaces, provides a significant and fairly sharp line of demarcation and runs as a characteristic feature through tribology and tribodesign. It has proved to be a valuable concept, not only in education, but also in research, development and in promoting sound design. It relates to the nature of contact, including short-duration temperatures called flash temperatures, and being indicative of the conditions to which both the rubbing materials and lubricant are exposed, is also important to the materials engineer and the lubricant technologist. Further, this distinction is helpful in recognizing why full fluid film lubrication between counterformal rubbing surfaces is normally of the elastohydrodynamic type. It also results in a rational classification of boundary lubrication.

From the very start of the design process the designer should keep his eye constantly upon the ultimate goal, that is, the satisfactory, or rather the optimum, fulfilment of all the functions required. Since many machine designers are not sufficiently aware of all the really essential functions required in the various stages of tribodesign, on many occasions, they simply miss the optimum conceivable design. For instance, in the case of self-acting hydrodynamic journal bearings, the two functions to be fulfilled, i.e. guidance and support of the journal, were recognized a long time ago. But the view that the hydrodynamic generation of pressure required for these two functions is associated with a journal-bearing system serving as its own pump is far from common. The awareness of this concept of pumping action should have led machine designers to conceive at least one layout for a self-acting bearing that is different from the more conventional one based on the hydrodynamic wedging and/or squeezing effect. For example, the pumping action could be achieved through suitable grooving of the bearing surface, or of the opposite rubbing surface of the journal, or collar, of a journal of thrust bearing.

1.1. Specific principles of tribodesign

Two principles, specific to tribodesign, that is, the principle of preventing contact between rubbing surfaces, and the equally important principle of regarding lubricant films as machine elements and, accordingly, lubricants as engineering materials, can be distinguished.

In its most general form the principle of contact prevention is also taken to embody inhibiting, not so much the contact itself as certain consequences of the contact such as the risk of constrictional overstressing of the surface material of a rubbing body, i.e. the risk of mechanical wear. This principle, which is all-important in tribodesign, may be executed in a number of ways. When it is combined with yet another principle of the optimal grouping of functions, it leads to the expediency of the protective layer. Such a layer, covering the rubbing surface, is frequently used in protecting its substrate from wear. The protective action may, for example, be aimed at lowering the contact pressure by using a relatively soft solid for the layer, and thereby reducing the risk of constrictional overstressing of the mating surface.

The protective layer, in a variety of forms, is indeed the most frequently used embodiment of the principle of contact prevention. At the same time, the principle of optimal grouping is usually involved, as the protective layer and the substrate of the rubbing surface each has its own function. The protective function is assigned to the layer and the structural strength is provided by the substrate material. In fact, the substrate serves, quite often, as support for the weaker material of the layer and thus enables the further transmission of the external load. Since the protective layer is an element interposed in the flow of force, it must be designed so as not to fail in transmitting the load towards the substrate. From this point of view, a distinction should be made between protective layers made of some solid material (achieved by surface treatment or coating) and those consisting of a fluid, which will be either a liquid or a gaseous lubricant.

Solid protective layers should be considered first. With conformal rubbing surfaces, particularly, it is often profitable to use a protective layer consisting of a material that is much softer and weaker than both the substrate material and the material of the mating surface. Such a layer can be utilized without incurring too great a risk of structural failure of the relatively weak material of the protective layer considered here. In the case of conformal surfaces this may be explained by a very shallow penetration of the protective layer by surface asperities. In fact, the depth of penetration is comparable to the size of the micro-contacts formed by the contacting asperities. This is a characteristic feature of the nature of contact between conformal surfaces. Unless the material of the protective layer is exceedingly soft, and the layer very thick indeed, the contact areas, and thus the depth of penetration, will never become quite as large as those on counterformal rubbing surfaces.

Other factors to be considered are the strengthening and stiffening effects exerted on the protective layer by the substrate. It is true that the soft material of the protective layer would be structurally weak if it were to be used in bulk. But with the protective layer thin enough, the support by the comparatively strong substrate material, particularly when bonding to the substrate is firm, will considerably strengthen the layer. The thinner the protective layer, the greater is the stiffening effect exerted by the substrate. But the stiffening effect sets a lower bound to the thickness of the layer. For the layer to be really protective its thickness should not be reduced to anywhere near the depth of penetration. The reason is that the stiffening effect would become so pronounced that the contact pressures would, more or less, approach those of the comparatively hard substrate material. Other requirements, like the ability to accommodate misalignment or deformations of at least one of the two rubbing bodies under loading, and also the need for embedding abrasive particles that may be trapped between the two rubbing surfaces, set the permissible lower bound to thicknesses much higher than the depth of penetration. In fact, in many cases, as in heavily loaded bearings of high-speed internal combustion engines, a compromise has to be struck between the various requirements, including the fatigue endurance of the protective layer. The situation on solid protective layers formed on counterformal rubbing surfaces, such as gear teeth, is quite different, in that there is a much greater depth of penetration down to which the detrimental effects of the constriction of the flow of force are still perceptible. The reason lies in the fact that the size of the Hertzian contact area is much greater than that of the tiny micro-contact areas on conformal surfaces. Thus, if they are to be durable, protective layers on counterformal surfaces cannot be thin, as is possible on conformal surfaces. Moreover, the material of the protective layer on a counterformal surface should be at least as strong in bulk, or preferably even stronger, as that of the substrate. These two requirements are indeed satisfied by the protective layers obtained on gear teeth through such surface treatments as carburizing. It is admitted that thin, and even soft, layers are sometimes used on counterformal surfaces, such as copper deposits on gear teeth; but these are meant only for running-in and not for durability.

Liquids or gases form protective layers which are synonymous with full fluid films. These layers show various interesting aspects from the standpoint of tribodesign, or even from that of machine design in general. In fact, the full fluid film is the most perfect realization of the expedient of the protective layer. In any full fluid film, pressures must be hydrodynamically generated, to the extent where their resultant balances the load to be transmitted through the film from one of the boundary rubbing surfaces to the other.

These two surfaces are thus kept apart, so that contact prevention is indeed complete. Accordingly, any kind of mechanical wear that may be caused by direct contact is eliminated altogether. But, as has already been observed, only with conformal surfaces will the full fluid film, as an interposed force transmitting element, be able to reduce substantially the constriction of the flow of force that would be created in the absence of such a film. In this respect the diffusion of the flow of force, in order to protect both surfaces from the severe surface stressing induced by the constriction of the flow, is best achieved by a fluid film which is far more effective than any solid protective layer. Even with counterformal surfaces where elastohydrodynamic films are exceedingly thin, contact prevention is still perfectly realizable.

It is quite obvious from the discussion presented above that certain general principles, typical for machine design, are also applicable in tribodesign. However, there are certain principles that are specific to tribodesign, but still hardly known amongst machine designers. It is hoped that this book will encourage designers to take advantage of the results, concepts and knowledge offered by tribology.

1.2. Tribological problems in machine design

The view that tribology, in general, and tribodesign, in particular, are intrinsic parts of machine design can be further reinforced by a brief review of tribological problems encountered in the most common machine elements.

1.2.1. Plain sliding bearings

When a journal bearing operates in the hydrodynamic regime of lubrication, a hydrodynamic film develops. Under these conditions conformal surfaces are fully separated and a copious flow of lubricant is provided to prevent overheating. In these circumstances of complete separation, mechanical wear does not take place. However, this ideal situation is not always achieved.

Sometimes misalignment, either inherent in the way the machine is assembled or of a transient nature arising from thermal or elastic distortion, may cause metal–metal contact. Moreover, contact may occur at the instant of starting (before the hydrodynamic film has had the opportunity to develop fully), the bearing may be overloaded from time to time and foreign particles may enter the film space. In some applications, internal combustion engines for example, acids and other corrosive substances may be formed during combustion and transmitted by the lubricant thus inducing a chemical type of wear. The continuous application and removal of hydrodynamic pressure on the shaft may dislodge loosely held particles. In many cases, however, it is the particles of foreign matter which are responsible for most of the wear in practical situations. Most commonly, the hard particles are trapped between the journal and the bearing. Sometimes the particles are embedded in the surface of the softer material, as in the case of white metal, thereby relieving the situation. However, it is commonplace for the hard particles to be embedded in the bearing surface thus constituting a lapping system, giving rise to rapid wear on the hard shaft surface. Generally, however, the wear on hydrodynamically lubricated bearings can be regarded as mild and caused by occasional abrasive action. Chromium plating of crankshaft bearings is sometimes successful in combating abrasive and corrosive wear.

1.2.2. Rolling contact bearings

Rolling contact bearings make up the widest class of machine elements which embody Hertzian contact problems. From a practical point of view, they are usually divided into two broad classes; ball bearings and roller-bearings, although the nature of contact and the laws governing friction and wear behaviour are common to both classes. Although contact is basically a rolling one, in most cases an element of sliding is involved and this is particularly the case with certain types of roller bearings, notably the taper rolling bearings.

Any rolling contact bearing is characterized by two numbers, i.e. the static load rating and L life. The static load-carrying capacity is the load that can be applied to a bearing, which is either stationary or subject to a slight swivelling motion, without impairing its running qualities for subsequent rotation. In practice, this is taken as the maximum load for which the combined deformation of the rolling element and raceways at any point does not exceed 0.001 of the diameter of the rolling element. L_{10} life represents the basic dynamic capacity of the bearing, that is, the load at which the life of a bearing is 1 000 000 revolutions and the failure rate is 10 per cent.

The practising designer will find the overwhelming number of specialized research papers devoted to rolling contact problems somewhat bewildering. He typically wishes to decide his stand regarding the relative importance of elastohydrodynamic (i.e. physical) and boundary (i.e. physico–chemical) phenomena. He requires a frame of reference for the evaluation of the broad array of available contact materials and lubricants, and he will certainly appreciate information indicating what type of application is feasible for rolling contact mechanisms, at what cost, and what is beyond the current state of the art. As in most engineering applications, lubrication of a rolling Hertz contact is undertaken for two reasons: to control the friction forces and to minimize the probability of the contact's failure. With sliding elements, these two purposes are at least co-equal and friction control is often the predominant interest, but failure control is by far the most important purpose of rolling contact lubrication. It is almost universally true that lubrication, capable of providing failure-free operation of a rolling contact, will also confine the friction forces within tolerable limits.

Considering failure control as the primary goal of rolling contact lubrication, a review of contact lubrication technology can be based on the interrelationship between the lubrication and the failure which renders the contact inoperative. Fortunately for the interpretive value of this treatment, considerable advances have recently been made in the analysis and understanding of several of the most important rolling contact failure mechanisms. The time is approaching when, at least for failures detected in their early stages, it will be possible to analyse a failed rolling contact and describe, in retrospect, the lubrication and contact material behaviour which led to or aggravated the failure. These methods of failure analysis permit the engineer to introduce remedial design modifications to this machinery and, specifically, to improve lubrication so as to control premature or avoidable rolling contact failures.

From this point of view, close correlation between lubrication theory and the failure mechanism is also an attractive goal because it can serve to verify lubrication concepts at the level where they matter in practical terms.

1.2.3. Piston, piston rings and cylinder liners

One of the most common machine elements is the piston within a cylinder which normally forms part of an engine, although similar arrangements are also found in pumps, hydraulic motors, gas compressors and vacuum exhausters. The prime function of a piston assembly is to act as a seal and to counterbalance the action of fluid forces acting on the head of the piston. In the majority of cases the sealing action is achieved by the use of piston rings, although these are sometimes omitted in fast running hydraulic machinery finished to a high degree of precision.

Pistons are normally lubricated although in some cases, notably in the chemical industry, specially formulated piston rings are provided to function without lubrication. Materials based on polymers, having intrinsic self-lubricating properties, are frequently used. In the case of fluid lubrication, it is known that the lubrication is of a hydrodynamic nature and, therefore, the viscosity of the lubricant is critical from the point of view of developing the lubricating film and of carrying out its main function, which is to act as a sealing element. Failure of the piston system to function properly is manifested by the occurrence of blow-by and eventual loss of compression. In many cases design must be a compromise, because a very effective lubrication of the piston assembly (i.e. thick oil film, low friction and no blow-by) could lead to high oil consumption in an internal combustion engine. On the other hand, most of the wear takes place in the vicinity of the top-dead-centre where the combination of pressure, velocity and temperature are least favourable to the operation of a hydrodynamic film. Conditions in the cylinder of an internal combustion engine can be very corrosive due to the presence of sulphur and other harmful elements present in the fuel and oil. Corrosion can be particularly harmful before an engine has warmed up and the cylinder walls are below the 'dew-point' of the acid solution.

The normal running-in process can be completed during the period of the works trial, after which the wear rate tends to fall as time goes on. High alkaline oil is more apt to cause abnormal wear and this is attributed to a lack of spreadability at high temperatures. Machined finishes are regarded as having more resistance to scuffing than ground finishes because of the oil-retaining characteristics of the roughened surfaces. The use of taper face rings is effective in preventing scuffing by relieving the edge load in the earliest stages of the process. A high phosphorous lining is better than a vanadium lining in preventing scuffing. The idea of using a rotating piston mechanism to enhance resistance to scuffing is an attractive option.

1.2.4.   Cam and cam followers

Although elastohydrodynamic lubrication theory can now help us to understand how cam-follower contact behaves, from the point of view of its lubrication, it has not yet provided an effective design criterion.

Cam-follower systems are extensively employed in engineering but do not have an extensive literature of their own. One important exception to this is the automotive valve train, a system that contains all the complications possible in a cam-follower contact. The automotive cam and tappet can, therefore, be regarded as a model representing this class of contacts. In automotive cams and tappets the maximum Hertz stress usually lies between 650 and 1300 MPa and the maximum sliding speed may exceed $10 \, \text{m s}^{-1}$. The values of oil film thickness to be expected are comparable with the best surface finish that can be produced by normal engineering processes and, consequently, surface roughness has an important effect on performance.

In a cam and tappet contact, friction is a relatively unimportant factor influencing the performance and its main effect is to generate unwanted heat. Therefore, the minimum attainable value is desired. The important design requirement as far as the contact is concerned is, however, that the working surfaces should support the imposed loads without serious wear or other form of surface failure. Thus it can be said that the development of cams and tappets is dominated by the need to avoid surface failure.

The main design problem is to secure a film of appropriate thickness. It is known that a reduction in nose radius of a cam, which in turn increases Hertzian stress, also increases the relative velocity and thus the oil film thickness. The cam with the thicker film operates satisfactorily in service whereas the cam with the thinner film fails prematurely. Temperature limitations are likely to be important in the case of cams required to operate under intense conditions and scuffing is the most probable mode of failure. The loading conditions of cams are never steady and this fact should also be considered at the design stage.

1.2.5. Friction drives

Friction drives, which are being used increasingly in infinitely variable gears, are the converse of hypoid gears in so far as it is the intention that two smooth machine elements should roll together without sliding, whilst being able to transmit a peripheral force from one to the other. Friction drives normally work in the elastohydrodynamic lubrication regime. If frictional traction is plotted against sliding speed, three principal modes may be identified. First, there is the linear mode in which traction is proportional to the relative velocity of sliding. Then, there is the transition mode during which a maximum is reached and, finally, a third zone with a falling characteristic. The initial region can be shown to relate to the rheological properties of the oil and viscosity is the predominant parameter. However, the fact that a maximum value is observed in the second zone is somewhat surprising. It is now believed that under appropriate circumstances a lubricant within a film, under the high pressure of the Hertzian contact, becomes a glass-like solid which, in common with other solids, has a limiting strength corresponding to the maximum value of traction. Regarding the third zone, the falling-off in traction is usually attributed to the fall in its viscosity associated with an increase in temperature of the lubricant.

Friction drives have received comparatively little attention and the papers available are mainly concerned with operating principles and kinematics. In rolling contact friction drives, the maximum Hertz stress may be in excess of 2600 MPa, but under normal conditions of operation the sliding speed will be of the order of $1 \, \text{m s}^{-1}$ and will be only a small proportion of the rolling speed. The friction drive depends for its effectiveness on the frictional traction transmitted through the lubricated contact and the maximum effective coefficient of friction is required. Because the sliding velocities are relatively low, it is possible to select materials for the working surfaces that are highly resistant to pitting failure and optimization of the frictional behaviour becomes of over-riding importance.

1.2.6. Involute gears

At the instant where the line of contact crosses the common tangent to the pitch circle, involute gear teeth roll one over the other without sliding. During the remaining period of interaction, i.e. when the contact zone lies in the addendum and dedendum, a certain amount of relative sliding occurs. Therefore the surface failure called pitting is most likely to be found on the pitch line, whereas scuffing is found in the addendum and dedendum regions.

There is evidence that with good quality hardened gears, scuffing occurs at the point where deceleration and overload combine to produce the greatest disturbance. However, before reaching the scuffing stage, another type of damage is obtained which is located in the vicinity of the tip of both pinion and gear teeth. This type of damage is believed to be due to abrasion by hard debris detached from the tip wedge. There are indications of subsurface fatigue due to cyclic Hertzian stress. The growth of fatigue cracks can be related to the effect of lubricant trapped in an incipient crack during successive cycles. Because of conservative design factors, the great majority of gear systems now in use is not seriously affected by lubrication deficiency. However, in really compact designs, which require a high degree of reliability at high operating stresses, speeds or temperatures, the lubricant truly becomes an engineering material.

Over the years, a number of methods have been suggested to predict the adequate lubrication of gears. In general, they have served a design purpose but with strong limits to the gear size and operating conditions. The search has continued and, gradually, as the range of speeds and loads continues to expand, designers are moving away from the strictly empirical approach. Two concepts of defining adequate lubrication have received some popularity in recent years. One is the minimum film thickness concept; the other is the critical temperature criteria. They both have a theoretical background but their application to a mode of failure remains hypothetical.

Not long ago, the common opinion was that only a small proportion of the load of counterformal surfaces was carried by hydrodynamic pressure. It was felt that monomolecular or equivalent films, even with non-reactive lubricants, were responsible for the amazing performance of gears. Breakthroughs in the theory of elastohydrodynamic lubrication have shown that this is not likely to be the case. Low-speed gears operating at over 2000 MPa, with a film thickness of several micrometers, show no distress or wear after thousands of hours of operation. High-speed gears operating at computed film thicknesses over $150\,\mu$m frequently fail by scuffing in drives from gas turbines. This, however, casts a shadow over the importance of elastohydrodynamics. The second concept – one gaining acceptance as a design criterion for lubricant failure – is the critical temperature hypothesis. The criterion is very simple. Scuffing will occur when a critical temperature is reached, which is characteristic of the particular combination of the lubricant and the materials of tooth faces.

1.2.7. Hypoid gears

Hypoid gears are normally used in right-angle drives associated with the axles of automobiles. Tooth actions combine the rolling action characteristic of spiral-bevel gears with a degree of sliding which makes this type of gear critical from the point of view of surface loading. Successful operation of a hypoid gear is dependent on the provision of the so-called extreme pressure oils, that is, oils containing additives which form surface protective layers at elevated temperatures. There are several types of additives for compounding hypoid lubricants. Lead-soap, active sulphur additives may prevent scuffing in drives which have not yet been run-in, particularly when the gears have not been phosphated. They are usually not satisfactory under high torque but are effective at high speed. Lead–sulphur chlorine additives are generally satisfactory under high-torque low-speed conditions but are sometimes less so at high speeds. The prevailing modes of failure are pitting and scuffing.

1.2.8.   Worm gears

Worm gears are somewhat special because of the degree of conformity which is greater than in any other type of gear. It can be classified as a screw pair within the family of lower pairs. However, it represents a fairly critical situation in view of the very high degree of relative sliding. From the wear point of view, the only suitable combination of materials is phosphor–bronze with hardened steel. Also essential is a good surface finish and accurate, rigid positioning. Lubricants used to lubricate a worm gear usually contain surface active additives and the prevailing mode of lubrication is mixed or boundary lubrication. Therefore, the wear is mild and probably corrosive as a result of the action of boundary lubricants.

It clearly follows from the discussion presented above that the engineer responsible for the tribological aspect of design, be it bearings or other systems involving moving parts, must be expected to be able to analyse the situation with which he is confronted and bring to bear the appropriate knowledge for its solution. He must reasonably expect the information to be presented to him in such a form that he is able to see it in relation to other aspects of the subject and to assess its relevant to his own system. Furthermore, it is obvious that a correct appreciation of a tribological situation requires a high degree of scientific sophistication, but the same can also be said of many other aspects of modern engineering.

The inclusion of the basic principles of tribology, as well as tribodesign, within an engineering design course generally does not place too great an additional burden on students, because it should call for the basic principles of the material which is required in any engineering course. For example, a study of the dynamics of fluids will allow an easy transition to the theory of hydrodynamic lubrication. Knowledge of thermodynamics and heat transfer can also be put to good use, and indeed a basic knowledge of engineering materials must be drawn upon.

2 Basic principles of tribology

Years of research in tribology justifies the statement that friction and wear properties of a given material are not its intrinsic properties, but depend on many factors related to a specific application. Quantitative values for friction and wear in the forms of friction coefficient and wear rate, quoted in many engineering textbooks, depend on the following basic groups of parameters:

(i) the structure of the system, i.e. its components and their relevant properties;

(ii) the operating variables, i.e. load (stress), kinematics, temperature and time;

(iii) mutual interaction of the system's components.

The main aim of this chapter is a brief review of the basic principles of tribology. Wherever it is possible, these principles are presented in forms of analytical models, equations or formulae rather than in a descriptive, qualitative way. It is felt that this approach is very important for a designer who, by the nature of the design process, is interested in the prediction of performance rather than in testing the performance of an artefact.

2.1. Origins of sliding friction

Whenever there is contact between two bodies under a normal load, W, a force is required to initiate and maintain relative motion. This force is called frictional force, F. Three basic facts have been experimentally established:

(i) the frictional force, F, always acts in a direction opposite to that of the relative displacement between the two contacting bodies;

(ii) the frictional force, F, is a function of the normal load on the contact, W,

$$F = fW \tag{2.1}$$

where f is the coefficient of friction;

(iii) the frictional force is independent of a nominal area of contact.

These three statements constitute what is known as the laws of sliding friction under dry conditions.

Studies of sliding friction have a long history, going back to the time of Leonardo da Vinci. Luminaries of science such as Amontons, Coulomb and Euler were involved in friction studies, but there is still no simple model which could be used by a designer to calculate the frictional force for a given pair of materials in contact. It is now widely accepted that friction results from complex interactions between contacting bodies which include the effects of surface asperity deformation, plastic gross deformation of a weaker material by hard surface asperities or wear particles and molecular interaction leading to adhesion at the points of intimate contact. A number of factors, such as the mechanical and physico–chemical properties of the materials in contact, surface topography and environment determine the relative importance of each of the friction process components.

At a fundamental level there are three major phenomena which control the friction of unlubricated solids:

(i) the real area of contact;
(ii) shear strength of the adhesive junctions formed at the points of real contact;
(iii) the way in which these junctions are ruptured during relative motion.

Friction is always associated with energy dissipation, and a number of stages can be identified in the process leading to energy losses.

Stage I. Mechanical energy is introduced into the contact zone, resulting in the formation of a real area of contact.

Stage II. Mechanical energy is transformed within the real area of contact, mainly through elastic deformation and hysteresis, plastic deformation, ploughing and adhesion.

Stage III. Dissipation of mechanical energy which takes place mainly through: thermal dissipation (heat), storage within the bulk of the body (generation of defects, cracks, strain energy storage, plastic transformations) and emission (acoustic, thermal, exo-electron generation).

2.2. Contact between bodies in relative motion

Nowadays it is a standard requirement to take into account, when analysing the contact between two engineering surfaces, the fact that they are covered with asperities having random height distribution and deforming elastically or plastically under normal load. The sum of all micro-contacts created by individual asperities constitutes the real area of contact which is usually only a tiny fraction of the apparent geometrical area of contact (Fig. 2.1). There are two groups of properties, namely, deformation properties of the materials in contact and surface topography characteristics, which define the magnitude of the real contact area under a given normal load W. Deformation properties include: elastic modulus, E, yield pressure, P_y and hardness, H. Important surface topography parameters are: asperity distribution, tip radius, β, standard deviation of asperity heights, σ, and slope of asperity Θ.

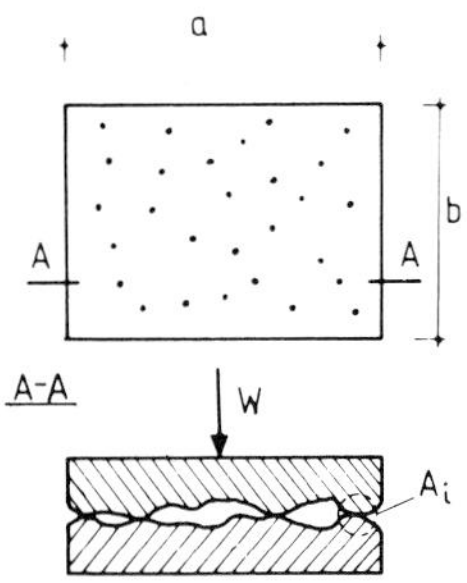

Generally speaking, the behaviour of metals in contact is determined by:

the so-called plasticity index

$$\psi = \frac{E}{P_y}\left(\frac{\sigma}{\beta}\right)^{\frac{1}{2}}. \tag{2.2}$$

If the plasticity index $\psi < 0.6$, then the contact is classified as elastic. In the case when $\psi > 1.0$, the predominant deformation mode within the contact

$A_n = a \times b$ (nominal contact area)

$A_r = \Sigma A_i$ (real contact area)

Figure 2.1

zone is called *plastic deformation*. Depending on the deformation mode within the contact, its real area can be estimated from:

the elastic contact

$$A_e = C \left(\frac{W}{E} \right)^n,$$
(2.3)

where $\frac{2}{3} < n < 1$;

the plastic contact

$$A_p = C \frac{W}{P_y} \approx \frac{W}{H},$$
(2.4)

where C is the proportionality constant.

The introduction of an additional tangential load produces a phenomenon called *junction growth* which is responsible for a significant increase in the asperity contact areas. The magnitude of the junction growth of metallic contact can be estimated from the expression

$$A = A_p \left[1 + \alpha \left(\frac{F}{W} \right)^2 \right]^{\frac{1}{2}},$$
(2.5)

where $\alpha \approx 9$ for metals.

In the case of organic polymers, additional factors, such as viscoelastic and viscoplastic effects and relaxation phenomena, must be taken into account when analysing contact problems.

2.3. Friction due to adhesion

One of the most important components of friction originates from the formation and rupture of interfacial adhesive bonds. Extensive theoretical and experimental studies have been undertaken to explain the nature of adhesive interaction, especially in the case of clean metallic surfaces. The main emphasis was on the electronic structure of the bodies in frictional contact. From a theoretical point of view, attractive forces within the contact zone include all those forces which contribute to the cohesive strength of a solid, such as the metallic, covalent and ionic short-range forces as well as the secondary van der Waals bonds which are classified as long-range forces. An illustration of a short-range force in action provides two pieces of clean gold in contact and forming metallic bonds over the regions of intimate contact. The interface will have the strength of a bulk gold. In contacts formed by organic polymers and elastomers, long-range van der Waals forces operate. It is justifiable to say that interfacial adhesion is as natural as the cohesion which determines the bulk strength of materials.

The adhesion component of friction is usually given as: the ratio of the interfacial shear strength of the adhesive junctions to the yield strength of the asperity material

$$f_a = \frac{F_a}{W} \approx \frac{\tau_{12}}{P_y}.$$
(2.6)

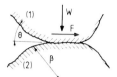

Figure 2.2

For most engineering materials this ratio is of the order of 0.2 and means that the friction coefficient may be of the same order of magnitude. In the case of clean metals, where the junction growth is most likely to take place, the adhesion component of friction may increase to about 10–100. The presence of any type of lubricant disrupting the formation of the adhesive junction can dramatically reduce the magnitude of the adhesion component of friction. This simple model can be supplemented by the surface energy of the contacting bodies. Then, the friction coefficient is given by (see Fig. 2.2)

$$f_a = (\tau_{12}/P_y)\left[1 - 2\frac{W_{12}\tan\Theta}{P_y}\right]^{-1}, \tag{2.7}$$

where $W_{12} = \gamma_1 + \gamma_2 - \gamma_{12}$ is the surface energy.

Recent progress in fracture mechanics allows us to consider the fracture of an adhesive junction as a mode of failure due to crack propagation

$$f_a = C\frac{\sigma_{12}\delta_c}{n^2(WH)^{\frac{1}{2}}}, \tag{2.8}$$

where σ_{12} is the interfacial tensile strength, δ_c is the critical crack opening displacement, n is the work-hardening factor and H is the hardness.

It is important to remember that such parameters as the interfacial shear strength or the surface energy characterize a given pair of materials in contact rather than the single components involved.

2.4. Friction due to ploughing

Ploughing occurs when two bodies in contact have different hardness. The asperities on the harder surface may penetrate into the softer surface and produce grooves on it, if there is relative motion. Because of ploughing a certain force is required to maintain motion. In certain circumstances this force may constitute a major component of the overall frictional force observed. There are two basic reasons for ploughing, namely, ploughing by surface asperities and ploughing by hard wear particles present in the contact zone (Fig. 2.3). The case of ploughing by the hard conical asperity is shown in Fig. 2.3(a), and the formula for estimating the coefficient of friction is as follows:

$$f_p = \frac{2}{\pi}\tan\Theta. \tag{2.9}$$

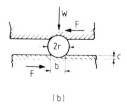

Figure 2.3

Asperities on engineering surfaces seldom have an effective slope, given by Θ, exceeding 5 to 6; it follows, therefore, that the friction coefficient, according to eqn (2.9), should be of the order of 0.04. This is, of course, too low a value, mainly because the piling up of the material ahead of the moving asperity is neglected. Ploughing of a brittle material is inevitably associated with micro-cracking and, therefore, a model of the ploughing process based on fracture mechanics is in place. Material properties such as fracture toughness, elastic modulus and hardness are used to estimate the coefficient of friction, which is given by

$$f_p = \frac{F_p}{W} \approx C \frac{K_{Ic}^2}{E(HW)^{\frac{1}{2}}}, \tag{2.10}$$

where K_{Ic} is the fracture toughness, E is the elastic modulus and H is the hardness.

The ploughing due to the presence of hard wear particles in the contact zone has received quite a lot of attention because of its practical importance. It was found that the frictional force produced by ploughing is very sensitive to the ratio of the radius of curvature of the particle to the depth of penetration. The formula for estimating the coefficient of friction in this case has the following form:

$$f_p = \frac{2}{\pi}\left[\left(\frac{2r}{b}\right)^2 \sin^{-1}\frac{b}{2r} - \left\{\left(\frac{2r}{b}\right)^2 - 1\right\}^{\frac{1}{2}}\right]. \tag{2.11}$$

2.5 Friction due to deformation

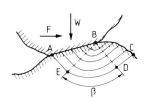

Figure 2.4

Mechanical energy is dissipated through the deformations of contacting bodies produced during sliding. The usual technique in analysing the deformation of the single surface asperity is the slip-line field theory for a rigid, perfectly plastic material. A slip-line deformation model of friction, shown in Fig. 2.4, is based on a two-dimensional stress analysis of Prandtl. Three distinct regions of plastically deformed material may develop and, in Fig. 2.4, they are denoted ABE, BED and BDC. The flow shear stress of the material defines the maximum shear stress which can be developed in these regions. The coefficient of friction is given by the expression

$$f_d = \frac{F}{W} = \lambda \tan\left\{\arcsin\left[\frac{\sqrt{2}}{4}\frac{(2+\beta)}{(1+\beta)}\right]\right\}, \tag{2.12}$$

where $\lambda = \lambda(E; H)$ is the portion of plastically supported load, E is the elastic modulus and H is the hardness.

The proportion of load supported by the plastically deformed regions and related, in a complicated way, to the ratio of the hardness to the elastic modulus is an important parameter in this model. For completely plastic asperity contact and an asperity slope of 45°, the coefficient of friction is 1.0. It decreases to 0.55 for an asperity slope approaching zero.

Another approach to this problem is to assume that the frictional work performed is equal to the work of the plastic deformation during steady-state sliding. This energy-based plastic deformation model of friction gives the following expression for the coefficient of friction:

$$f_d = \frac{A_r}{W}\tau_{max} F\left(\frac{\tau_s}{\tau_{max}}\right), \tag{2.13}$$

$$F\left(\frac{\tau_s}{\tau_{max}}\right) = 1 - 2\frac{\ln\left(1 + \frac{\tau_s}{\tau_{max}}\right) - \left(\frac{\tau_s}{\tau_{max}}\right)}{\ln[1 - (\tau_s/\tau_{max})^2]},$$

where A_r is the real area of contact, τ_{max} denotes the ultimate shear strength of a material and τ_s is the average interfacial shear strength.

2.6. Energy dissipation during friction

In a practical engineering situation all the friction mechanisms, discussed so far on an individual basis, interact with each other in a complicated way. Figure 2.5 is an attempt to visualize all the possible steps of friction-induced energy dissipations. In general, frictional work is dissipated at two different locations within the contact zone. The first location is the interfacial region characterized by high rates of energy dissipation and usually associated with an adhesion model of friction. The other one involves the bulk of the body and the larger volume of the material subjected to deformations. Because of that, the rates of energy dissipation are much lower. Energy dissipation during ploughing and asperity deformations takes place in this second location. It should be pointed out, however, that the distinction of two locations being completely independent of one another is artificial and serves the purpose of simplification of a very complex problem. The various processes depicted in Fig. 2.5 can be briefly characterized as follows:

(i) plastic deformations and micro-cutting;
(ii) viscoelastic deformations leading to fatigue cracking and tearing, and subsequently to subsurface excessive heating and damage;
(iii) true sliding at the interface leading to excessive heating and thus creating the conditions favourable for chemical degradation (polymers);
(iv) interfacial shear creating transferred films;
(v) true sliding at the interface due to the propagation of Schallamach waves (elastomers).

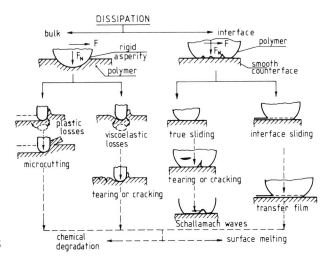

Figure 2.5

2.7. Friction under complex motion conditions

Complex motion conditions arise when, for instance, linear sliding is combined with the rotation of the contact area about its centre (Fig. 2.6). Under such conditions, the frictional force in the direction of linear motion

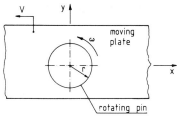

Figure 2.6

is not only a function of the usual variables, such as load, contact area diameter and sliding velocity, but also of the angular velocity. Furthermore, there is an additional force orthogonal to the direction of linear motion. In Fig. 2.6, a spherically ended pin rotates about an axis normal to the plate with angular velocity ω and the plate translates with linear velocity V. Assuming that the slip at the point within the circular area of contact is opposed by simple Coulomb friction, the plate will exert a force $\tau \, dA$ in the direction of the velocity of the plate relative to the pin at the point under consideration. To find the components of the total frictional force in the x and y directions it is necessary to sum the frictional force vectors, $\tau \, dA$, over the entire contact area A. Here, τ denotes the interfacial shear strength. The integrals for the components of the total frictional force are elliptical and must be evaluated numerically or converted into tabulated form.

2.8. Types of wear and their mechanisms

Friction and wear share one common feature, that is, complexity. It is customary to divide wear occurring in engineering practice into four broad general classes, namely: adhesive wear, surface fatigue wear, abrasive wear and chemical wear. Wear is usually associated with the loss of material from contracting bodies in relative motion. It is controlled by the properties of the material, the environmental and operating conditions and the geometry of the contacting bodies. As an additional factor influencing the wear of some materials, especially certain organic polymers, the kinematic of relative motion within the contact zone should also be mentioned. Two groups of wear mechanism can be identified; the first comprising those dominated by the mechanical behaviour of materials, and the second comprising those defined by the chemical nature of the materials. In almost every situation it is possible to identify the leading wear mechanism, which is usually determined by the mechanical properties and chemical stability of the material, temperature within the contact zone, and operating conditions.

2.8.1. Adhesive wear

Adhesive wear is invariably associated with the formation of adhesive junctions at the interface. For an adhesive junction to be formed, the interacting surfaces must be in intimate contact. The strength of these junctions depends to a great extent on the physico–chemical nature of the contacting surfaces. A number of well-defined steps leading to the formation of adhesive-wear particles can be identified:

  (i) deformation of the contacting asperities;
  (ii) removal of the surface films;
  (iii) formation of the adhesive junction (Fig. 2.7);
  (iv) failure of the junctions and transfer of material;
  (v) modification of transferred fragments;
  (vi) removal of transferred fragments and creation of loose wear particles.

The volume of material removed by the adhesive-wear process can be

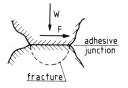

Figure 2.7

estimated from the expression proposed by Archard

$$V_a = k \frac{W}{H} L, \tag{2.14}$$

where k is the wear coefficient, L is the sliding distance and H is the hardness of the softer material in contact.

The wear coefficient is a function of various properties of the materials in contact. Its numerical value can be found in textbooks devoted entirely to tribology fundamentals. Equation (2.14) is valid for dry contacts only. In the case of lubricated contacts, where wear is a real possibility, certain modifications to Archard's equation are necessary. The wear of lubricated contacts is discussed elsewhere in this chapter.

While the formation of the adhesive junction is the result of interfacial adhesion taking place at the points of intimate contact between surface asperities, the failure mechanism of these junctions is not well defined. There are reasons for thinking that fracture mechanics plays an important role in the adhesive junction failure mechanism. It is known that both adhesion and fracture are very sensitive to surface contamination and the environment, therefore, it is extremely difficult to find a relationship between the adhesive wear and bulk properties of a material. It is known, however, that the adhesive wear is influenced by the following parameters characterizing the bodies in contact:

 (i) electronic structure;
 (ii) crystal structure;
 (iii) crystal orientation;
 (iv) cohesive strength.

For example, hexagonal metals, in general, are more resistant to adhesive wear than either body-centred cubic or face-centred cubic metals.

2.8.2. Abrasive wear

Abrasive wear is a very common and, at the same time, very serious type of wear. It arises when two interacting surfaces are in direct physical contact, and one of them is significantly harder than the other. Under the action of a normal load, the asperities on the harder surface penetrate the softer surface thus producing plastic deformations. When a tangential motion is introduced, the material is removed from the softer surface by the combined action of micro-ploughing and micro-cutting. Figure 2.8 shows the essence of the abrasive-wear model. In the situation depicted in Fig. 2.8, a hard conical asperity with slope, Θ, under the action of a normal load, W, is traversing a softer surface. The amount of material removed in this process can be estimated from the expression

$$\text{simplified} \quad V_{\text{abr}} = \frac{2}{\pi} \frac{\tan \Theta}{H} WL, \tag{2.15}$$

$$\text{refined} \quad V_{\text{abr}} = n^2 \frac{P_y E W^{3/2}}{K_{\text{Ic}}^2 H^{3/2}} L, \tag{2.16}$$

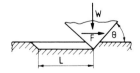

Figure 2.8

where E is the elastic modulus, H is the hardness of the softer material, K_{Ic} is the fracture toughness, n is the work-hardening factor and P_y is the yield strength.

The simplified model takes only hardness into account as a material property. Its more advanced version includes toughness as recognition of the fact that fracture mechanics principles play an important role in the abrasion process. The rationale behind the refined model is to compare the strain that occurs during the asperity interaction with the critical strain at which crack propagation begins.

In the case of abrasive wear there is a close relationship between the material properties and the wear resistance, and in particular:

(i) there is a direct proportionality between the relative wear resistance and the Vickers hardness, in the case of technically pure metals in an annealed state;

(ii) the relative wear resistance of metallic materials does not depend on the hardness they acquire from cold work-hardening by plastic deformation;

(iii) heat treatment of steels usually improves their resistance to abrasive wear;

(iv) there is a linear relationship between wear resistance and hardness for non-metallic hard materials.

The ability of the material to resist abrasive wear is influenced by the extent of work-hardening it can undergo, its ductility, strain distribution, crystal anisotropy and mechanical stability.

2.8.3 Wear due to surface fatigue

Load carrying nonconforming contacts, known as Hertzian contacts, are sites of relative motion in numerous machine elements such as rolling bearings, gears, friction drives, cams and tappets. The relative motion of the surfaces in contact is composed of varying degrees of pure rolling and sliding. When the loads are not negligible, continued load cycling eventually leads to failure of the material at the contacting surfaces. The failure is attributed to multiple reversals of the contact stress field, and is therefore classified as a fatigue failure. Fatigue wear is especially associated with rolling contacts because of the cycling nature of the load. In sliding contacts, however, the asperities are also subjected to cyclic stressing, which leads to stress concentration effects and the generation and propagation of cracks. This is schematically shown in Fig. 2.9. A number of steps leading to the generation of wear particles can be identified. They are:

(i) transmission of stresses at contact points;

(ii) growth of plastic deformation per cycle;

(iii) subsurface void and crack nucleation;

(iv) crack formation and propagation;

(v) creation of wear particles.

A number of possible mechanisms describing crack initiation and propagation can be proposed using postulates of the dislocation theory. Analytical

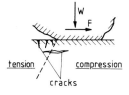

Figure 2.9

models of fatigue wear usually include the concept of fatigue failure and also of simple plastic deformation failure, which could be regarded as low-cycle fatigue or fatigue in one loading cycle. Theories for the fatigue-life prediction of rolling metallic contacts are of long standing. In their classical form, they attribute fatigue failure to subsurface imperfections in the material and they predict life as a function of the Hertz stress field, disregarding traction. In order to interpret the effects of metal variables in contact and to include surface topography and appreciable sliding effects, the classical rolling contact fatigue models have been expanded and modified. For sliding contacts, the amount of material removed due to fatigue can be estimated from the expression

$$V_f = C \frac{\eta \gamma}{\bar{\varepsilon}_1^2 H} WL, \tag{2.17}$$

where η is the distribution of asperity heights, γ is the particle size constant, $\bar{\varepsilon}_1$ is the strain to failure in one loading cycle and H is the hardness.

It should be mentioned that, taking into account the plastic–elastic stress fields in the subsurface regions of the sliding asperity contacts and the possibility of dislocation interactions, wear by delamination could be envisaged.

2.8.4. Wear due to chemical reactions

It is now accepted that the friction process itself can initiate a chemical reaction within the contact zone. Unlike surface fatigue and abrasion, which are mainly controlled by stress interactions and deformation properties, wear resulting from chemical reactions induced by friction is influenced mainly by the environment and its active interaction with the materials in contact. There is a well-defined sequence of events leading to the creation of wear particles (Fig. 2.10). At the beginning, the surfaces in contact react with the environment, creating reaction products which are deposited on the surfaces. The second step involves the removal of the reaction products due to crack formation and abrasion. In this way, a parent material is again exposed to environmental attack. The friction process itself can lead to thermal and mechanical activation of the surface layers inducing the following changes:

 (i) increased reactivity due to increased temperature. As a result of that the formation of the reaction product is substantially accelerated;
(ii) increased brittleness resulting from heavy work-hardening.

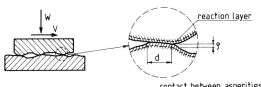

Figure 2.10

A simple model of chemical wear can be used to estimate the amount of material loss

$$V_t = \frac{k}{\xi^2 \rho^2} \frac{d}{H} \frac{W}{V} \mathrm{L},$$ (2.18)

where k is the velocity factor of oxidation, d is the diameter of asperity contact, ρ is the thickness of the reaction layer (Fig. 2.10), ξ is the critical thickness of the reaction layer and H is the hardness.

The model, given by eqn (2.18), is based on the assumption that surface layers formed by a chemical reaction initiated by the friction process are removed from the contact zone when they attain certain critical thicknesses.

2.9. Sliding contact between surface asperities

The problem of relating friction to surface topography in most cases reduces to the determination of the real area of contact and studying the mechanism of mating micro-contacts. The relationship of the frictional force to the normal load and the contact area is a classical problem in tribology. The adhesion theory of friction explains friction in terms of the formation of adhesive junctions by interacting asperities and their subsequent shearing. This argument leads to the conclusion that the friction coefficient, given by the ratio of the shear strength of the interface to the normal pressure, is a constant of an approximate value of 0.17 in the case of metals. This is because, for perfect adhesion, the mean pressure is approximately equal to the hardness and the shear strength is usually taken as 1/6 of the hardness. This value is rather low compared with those observed in practical situations. The controlling factor of this apparent discrepancy seems to be the type or class of an adhesive junction formed by the contacting surface asperities. Any attempt to estimate the normal and frictional forces, carried by a pair of rough surfaces in sliding contact, is primarily dependent on the behaviour of the individual junctions. Knowing the statistical properties of a rough surface and the failure mechanism operating at any junction, an estimate of the forces in question may be made.

The case of sliding asperity contact is a rather different one. The practical way of approaching the required solution is to consider the contact to be of a quasi-static nature. In the case of exceptionally smooth surfaces the deformation of contacting asperities may be purely elastic, but for most engineering surfaces the contacts are plastically deformed. Depending on whether there is some adhesion in the contact or not, it is possible to introduce the concept of two further types of junctions, namely, welded junctions and non-welded junctions. These two types of junctions can be defined in terms of a stress ratio, β, which is given by the ratio of, s, the shear strength of the junction to, k, the shear strength of the weaker material in contact

$$\beta = s/k.$$

For welded junctions, the stress ratio is

$$\beta = s/k = 1,$$

i.e., the ultimate shear strength of the junction is equal to that of the weaker material in contact.

For non-welded junctions, the stress ratio is

$$\beta = s/k < 1.$$

A welded junction will have adhesion, i.e. the pair of asperities will be welded together on contact. On the other hand, in the case of a non-welded junction, adhesive forces will be less important.

For any case, if the actual contact area is A, then the total shear force is

$$S = sA = \beta kA, \tag{2.19}$$

where $0 \leqslant \beta \leqslant 1$, depending on whether we have a welded junction or a non-welded one. There are no direct data on the strength of adhesive bonds between individual microscopic asperities. Experiments with field-ion tips provide a method for simulating such interactions, but even this is limited to the materials and environments which can be examined and which are often remote from practical conditions. Therefore, information on the strength of asperity junctions must be sought in macroscopic experiments. The most suitable source of data is to be found in the literature concerning pressure welding. Thus the assumption of elastic contacts and strong adhesive bonds seems to be incompatible. Accordingly, the elastic contacts lead to non-welded junctions only and for them $\beta < 1$. Plastic contacts, however, can lead to both welded and non-welded junctions. When modelling a single asperity as a hemisphere of radius equal to the radius of the asperity curvature at its peak, the Hertz solution for elastic contact can be employed.

The normal load, supported by the two hemispherical asperities in contact, with radii R_1 and R_2, is given by

$$P = \tfrac{4}{3} E' w^{3/2} \sqrt{R_1 R_2/(R_1 + R_2)}, \tag{2.20}$$

and the area of contact is given by

$$A = \pi w [R_1 R_2/(R_1 + R_2)]. \tag{2.21}$$

Here w is the geometrical interference between the two spheres, and E' is given by the relation

$$\frac{1}{E'} = \frac{1 - v_1^2}{E_1} + \frac{1 - v_2^2}{E_2},$$

where E_1, E_2 and v_1, v_2 are the Young moduli and the Poisson ratios for the two materials. The geometrical interference, w, which equals the normal compression of the contacting hemispheres is given by

$$w = (R_1 + R_2) - \sqrt{d^2 + x^2}, \tag{2.22}$$

where d is the distance between the centres of the two hemispheres in contact and x denotes the position of the moving hemisphere. By substitution of eqn (2.22) into eqns (2.20) and (2.21), the load, P, and the area of contact, A, may be estimated at any time.

Denoting by α the angle of inclination of the load P on the contact with the horizontal, it is easy to find that

$$\sin \alpha = \frac{d}{(d^2 + x^2)^{\frac{1}{2}}},$$

$$\cos \alpha = \frac{x}{(d^2 + x^2)^{\frac{1}{2}}}. \qquad (2.23)$$

The total horizontal and vertical forces, H and V, at any position defined by x of the sliding asperity (moving linearly past the stationary one), are given by

$$V = P \sin \alpha - S \cos \alpha,$$
$$H = P \cos \alpha + S \sin \alpha. \qquad (2.24)$$

Equation (2.24) can be solved for different values of d and β.

A limiting value of the geometrical interference w can be estimated for the initiation of plastic flow. According to the Hertz theory, the maximum contact pressure occurs at the centre of the contact spot and is given by

$$q_0 = \frac{3P}{2A}.$$

The maximum shear stress occurs inside the material at a depth of approximately half the radius of the contact area and is equal to about $0.31q_0$. From the Tresca yield criterion, the maximum shear stress for the initiation of plastic deformation is $Y/2$, where Y is the tensile yield stress of the material under consideration. Thus

$$Y/2 = 0.31 \frac{3P}{2A}.$$

Substituting P and A from eqns (2.20) and (2.21) gives

$$w = 6.4 \left(\frac{Y}{E'} \right)^2 R_1 R_2 / (R_1 + R_2).$$

Since Y is approximately equal to one third of the hardness for most materials, we have

$$w \approx 0.7 \left(\frac{H_b}{E'} \right)^2 \phi,$$

where $\phi = R_1 R_2 / (R_1 + R_2)$ and H_b denotes Brinell hardness.

The foregoing equation gives the value of geometrical interference, w, for the initiation of plastic flow. For a fully plastic junction or a noticeable plastic flow, w will be rather greater than the value given by the previous relation. Thus the criterion for a fully plastic junction can be given in terms of the maximum geometric interference

$$w_{max} > w_p$$

and

$$w_p = \left(\frac{H_b}{E'}\right)^2 \phi. \tag{2.25}$$

Hence, for the junction to be completely plastic, w_{max} must be greater than w_p. An approximate solution for normal and shear stresses for the plastic contacts can be determined through slip-line theory, where the material is assumed to be rigid-plastic and nonstrain hardening. For hemispherical asperities, the plane-strain assumption is not, strictly speaking, valid. However, in order to make the analysis feasible, the Green's plane-strain solution for two wedge-shaped asperities in contact is usually used. Plastic deformation is allowed in the softer material, and the equivalent junction angle α is determined by geometry. Quasi-static sliding is assumed and the solution proposed by Green is used at any time of the junction life. The stresses, normal and tangential to the interface, are

$$\begin{aligned} p &= k(1 + \sin 2\gamma + \tfrac{1}{2}\pi + 2\gamma - 2\alpha), \\ s &= k\cos 2\gamma, \end{aligned} \tag{2.26}$$

where α is the equivalent junction angle and γ is the slip-line angle. Assuming that the contact spot is circular with radius a, even though the Green's solution is strictly valid for the plane strain, we get

$$\begin{aligned} P &= p\pi a^2, \\ S &= s\pi a^2, \end{aligned} \tag{2.27}$$

where $a = \sqrt{2\phi w}$ and $\phi = R_1 R_2/(R_1 + R_2)$. Resolution of forces in two fixed directions gives

$$\begin{aligned} \text{(vertical direction)} \quad & V = P\cos\delta - S\sin\delta, \\ \text{(horizontal direction)} \quad & H = P\sin\delta + S\cos\delta, \end{aligned} \tag{2.28}$$

where δ is the inclination of the interface to the sliding velocity direction. Thus V and H may be determined as a function of the position of the moving asperity if all the necessary angles are determined by geometry.

2.10 The probability of surface asperity contact

As stated earlier, the degree of separation of the contacting surfaces can be measured by the ratio h/σ, frequently called the lambda ratio, λ. In this section the probability of asperity contact for a given lubricant film of thickness h is examined. The starting point is the knowledge of asperity height distributions. It has been shown that most machined surfaces have nearly Gaussian distribution, which is quite important because it makes the mathematical characterization of the surfaces much more tenable.

Thus if x is the variable of the height distribution of the surface contour, shown in Fig. 2.11, then it may be assumed that the function $F(x)$, for the cumulative probability that the random variable x will not exceed the

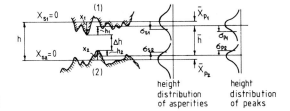

height
distribution
of asperities

height
distribution
of peaks

Figure 2.11

specific value X, exists and will be called the distribution function. Therefore, the probability density function $f(x)$ may be expressed as

$$f(x) = \frac{\mathrm{d}F(x)}{\mathrm{d}x}.$$

The probability that the variable x_i will not exceed a specific value X can be expressed as

$$F(x) = P(x \leqslant X) = \int_{-\infty}^{X} f(x)\,\mathrm{d}x. \tag{2.29}$$

The mean or expected value $\bar{X}$ of a continuous surface variable x_i may be expressed as

$$\bar{X}_s = \int_{-\infty}^{\infty} xf(x)\,\mathrm{d}x. \tag{2.30}$$

The variance can be defined as

$$\sigma^2 = \int_{-\infty}^{\infty} (x - \bar{X})^2 f(x)\,\mathrm{d}x,$$

where σ is equal to the square root of the variance and can be defined as the standard deviation of x.

From Fig. 2.11, x_1 and x_2 are the random variables for the contacting surfaces. It is possible to establish the statistical relationship between the surface height contours and the peak heights for various surface finishes by comparison with the comulative Gaussian probability distributions for surfaces and for peaks. Thus, the mean of the peak distribution can be expressed approximately as

$$X_p = B_m \sigma_s \tag{2.31}$$

and the standard deviation of peak heights can be represented as

$$\sigma_p = \left[\frac{\sum_{i=1}^{n} (h_i - \bar{X}_{pi})^2}{n} \right]^{\frac{1}{2}}, \tag{2.32}$$

when such measurements are available, or it can be approximated by

$$\sigma_p = B_d \sigma_s. \tag{2.33}$$

When surface contours are Gaussian, their standard deviations can be represented as

$$\sigma_{s1} = \left[\frac{\int_{-\alpha}^{\alpha} (h_1 - \bar{X}_{s1})^2 \phi(h_1) \, dh_1}{\int_{-\alpha}^{\alpha} \phi(h_1) \, dh_1} \right]^{\frac{1}{2}}, \tag{2.34}$$

$$\sigma_{s2} = \left[\frac{\int_{-\alpha}^{\alpha} (h_2 - \bar{X}_{s2})^2 \phi(h_2) \, dh_2}{\int_{-\alpha}^{\alpha} \phi(h_2) \, dh_2} \right]^{\frac{1}{2}}, \tag{2.35}$$

or approximated by

$$\sigma_{s1} = (\text{r.m.s.})_1 \approx 1.11 (\text{c.l.a.})_1,$$
$$\sigma_{s2} = (\text{r.m.s.})_2 \approx 1.11 (\text{c.l.a.})_2,$$

where r.m.s. indicates the root mean square, c.l.a. denotes the centre-line average, B_m is the surface-to-peak mean proportionality factor, and B_d is the surface-to-peak standard deviation factor. To determine the statistical parameters, B_m and B_d, cumulative frequency distributions of both asperities and peaks are required or, alternatively, the values of X_p, σ_s and σ_p. This information is readily available from the standard surface topography measurements.

Referring to Fig. 2.11, if the distance between the mean lines of asperital peaks is $\bar{h}$, then

$$\bar{h} = h - \bar{X}_{p1} - \bar{X}_{p2} \tag{2.36}$$

and the clearance may be expressed as

$$\Delta h = \bar{h} - h_1 - h_2, \tag{2.37}$$

where h_1 and h_2 are random variables and h is the thickness of the lubricant film. If it is assumed that the probability density function is equal to $\phi(h_1 + h_2)$, then the probability that a particular pair of asperities has a sum height, between $h_1 + h_2$ and $(h_1 + h_2) + d(h_1 + h_2)$, will be $\phi(h_1 + h_2) d(h_1 + h_2)$. Thus, the probability of interference between any two asperities is

$$P(h_1 + h_2 \geqslant \bar{h}) = \int_{\bar{h}}^{\infty} \phi(h_1 + h_2) d(h_1 + h_2)$$

or

$$P(h_1 + h_2 \geqslant \bar{h}) = P(\Delta h \leqslant 0) = \int_{0}^{\infty} \phi(-\Delta h) d(-\Delta h). \tag{2.38}$$

Thus $(-\Delta h)$ is a new random variable that has a Gaussian distribution with a probability density function

$$\phi(-\Delta h) = \frac{1}{(2\pi)^{\frac{1}{2}} \sigma^*} \exp\left[\frac{-(\bar{h} - \Delta h)^2}{2\sigma^{*2}} \right],$$

so that

$$P(\Delta h \leqslant 0) = \frac{1}{(2\pi)^{\frac{1}{2}}\sigma^*} \int_{0}^{\alpha} \exp\left[\frac{-(\bar{h}-\Delta h)^2}{2\sigma^{*2}}\right] \mathrm{d}(-\Delta h) \tag{2.39}$$

is the probability that Δh is negative, i.e. the probability of asperity contact. In the foregoing, $\bar{h}$ is the mean value of the separation (see Fig. 2.11) and $\sigma^* = (\sigma_{p_1}^2 + \sigma_{p_2}^2)^{\frac{1}{2}}$ is the standard deviation.

The probability $P(\Delta h \leqslant 0)$ of asperital contact can be found from the normalized contact parameter $\bar{h}$, where $\Delta\bar{h} = \bar{h}/\sigma^*$ is the number of standard deviations from mean $\bar{h}$. For this purpose, standard tables of normal probability functions are used. The values of $\Delta\bar{h}$ represent the number of standard deviations for specific probabilities of asperity contact, $P(\Delta h \leqslant 0)$. They can be described mathematically in terms of the specific film thickness or the lambda ratio, λ, and the r.m.s. surface roughness R. Thus, from the definition of the lambda ratio

$$\lambda = \frac{2h}{R_1 + R_2}, \tag{2.40}$$

where R_1 and R_2 are the r.m.s. roughnesses of surfaces 1 and 2, respectively. If it is assumed that $\sigma_{s_1} \approx R_1$ and $\sigma_{s_2} \approx R_2$, and that $\bar{h}$, B_d and B_m are defined as shown above, then

$$\bar{h} = f(R_1, R_2)\lambda$$

or

$$f(R_1, R_2) = \frac{\Delta\bar{h}}{\lambda} = \frac{\bar{h}}{\sigma^*} \frac{R_1 + R_2}{2h},$$

and finally

$$f(R_1, R_2) = \frac{(h - B_{m1}\sigma_{s1} - B_{m2}\sigma_{s2})(\sigma_{s1} + \sigma_{s2})}{2(B_{d1}^2\sigma_{s1}^2 + B_{d2}^2\sigma_{s2}^2)^{\frac{1}{2}}h}. \tag{2.41}$$

The general expression for the lambda ratio has the following form

$$\lambda = \frac{2\Delta\bar{h}h(B_{d_1}^2\sigma_{s_1}^2 + B_{d_2}^2\sigma_{s_2}^2)^{\frac{1}{2}}}{(\sigma_{s_1} + \sigma_{s_2})(h - B_{m_1}\sigma_{s_1} - B_{m_2}\sigma_{s_2})}. \tag{2.42}$$

If the contacting surfaces have the same surface roughness, then

$$B_{m_1} = B_{m_2} = B_m \quad\text{and}\quad B_{d_1} = B_{d_2} = B_d.$$

Taking into account the above assumptions

$$\lambda = \frac{2\Delta\bar{h}hB_d(\sigma_{s_1}^2 + \sigma_{s_2}^2)^{\frac{1}{2}}}{(\sigma_{s_1} + \sigma_{s_2})(h - B_m(\sigma_{s_1} + \sigma_{s_2}))}.$$

If it is further assumed that $R_1 = R_2 = R$ and therefore $p\sigma_{s1} = \sigma_{s2} = \sigma_s$, then

$$\lambda = \frac{2\Delta\bar{h}hB_d(2\sigma_s^2)^{\frac{1}{2}}}{(h - B_m2\sigma_s)2\sigma_s} = \frac{\Delta\bar{h}hB_d\sqrt{2}}{(h - 2B_mR)}$$

or

$$\lambda = \frac{\Delta \bar{h} B_{\rm d} h \sqrt{2}}{\bar{h}}, \tag{2.43}$$

where $\bar{h} = h - 2B_{\rm m}R$. If λ is known, then

$$\Delta \bar{h} = \frac{\lambda(h - 2B_{\rm m}R)}{B_{\rm d} h \sqrt{2}} = \frac{\lambda \bar{h}}{B_{\rm d} h \sqrt{2}}. \tag{2.44}$$

In the case of heavily loaded contacts, plastic deformation of interacting asperities is very likely. Therefore, it is desirable to determine the probability of plastic asperity contact.

The probability of plastic contact may be expressed as

$$P(h_1 + h_2 > \bar{h} + \delta_{\rm p}) = \int\limits_{\bar{h} + \delta_{\rm p}}^{\infty} \phi(h_1 + h_2) {\rm d}(h_1 + h_2), \tag{2.45}$$

where plastic asperity deformation, $\delta_{\rm p}$, is calculated from

$$\delta_{\rm p} = \left(\frac{\pi}{2}\right)^2 r \left(\frac{0.6 p_{\rm m}}{E'}\right)^2$$

or

$$\delta_{\rm p} = 0.89 r \left(\frac{p_{\rm m}}{E'}\right)^2, \tag{2.46}$$

where r is the average radius of the asperity peaks, $p_{\rm m}$ is the flow hardness of the softer material and $E' = [(1 - v_1^2)/E_1 + (1 - v_2^2)/E_2]^{-1}$. By normalizing the expression for $\delta_{\rm p}$

$$\frac{\delta_{\rm p}}{\sigma^*} = 0.89 \frac{r}{\sigma^*} \left(\frac{p_{\rm m}}{E'}\right)^2,$$

and introducing the plasticity index, defined as

$$\psi = \frac{E'}{p_{\rm m}} \left(\frac{\sigma^*}{r}\right)^{\frac{1}{2}},$$

the normalized plastic asperity deformation, δ_{ψ}, can be written as

$$\delta_{\psi} = \frac{\delta_{\rm p}}{\sigma^*} = \frac{0.89}{\psi^2}$$

and finally

$$\delta_{\rm p} = \frac{0.89 \sigma^*}{\psi^2}. \tag{2.47}$$

Thus the probability of plastic contact is

$$P(\Delta h \leqslant \delta_{\rm p}) = \frac{1}{(2\pi)^{\frac{1}{2}} \sigma^*} \int\limits_{\delta_{\rm p}}^{\infty} \exp\left[\frac{-(\bar{h} - \Delta h)^2}{2\sigma^{*2}}\right] {\rm d}(-\Delta h). \tag{2.48}$$

If $\Delta h' = \Delta h + \delta_p$, then the probability density function is

$$\phi(-\Delta h') = \frac{1}{(2\pi)^{\frac{1}{2}}\sigma*} \exp\left[\frac{-(\bar{h} + \delta_p - \Delta h')^2}{2\sigma*^2}\right].$$

The probability that $\Delta h'$ is negative, i.e. the probability of asperity contact, is given by

$$P(\Delta h' \leqslant 0) = \frac{1}{(2\pi)^{\frac{1}{2}}\sigma*} \int_0^\infty \exp\left[\frac{-(h + \delta_p - \Delta h')^2}{2\delta*^2}\right] d(-\Delta h'). \quad (2.49)$$

2.11 The wear in lubricated contacts

Wear occurs as a result of interaction between two contacting surfaces. Although understanding of the various mechanisms of wear, as discussed earlier, is improving, no reliable and simple quantitative law comparable with that for friction has been evolved. An innovative and rational design of sliding contacts for wear prevention can, therefore, only be achieved if a basic theoretical description of the wear phenomenon exists.

In lubricated contacts, wear can only take place when the lambda ratio is less than 1. The predominant wear mechanism depends strongly on the environmental and operating conditions. Usually, more than one mechanism may be operating simultaneously in a given situation, but often the wear rate is controlled by a single dominating process. It is reasonable to assume, therefore, that any analytical model of wear for partially lubricated contacts should contain adequate expressions for calculating the volume of worn material resulting from the various modes of wear. Furthermore, it is essential, in the case of lubricated contacts, to realize that both the contacting asperities and the lubricating film contribute to supporting the load. Thus, only the component of the total load, on the contact supported directly by the contacting asperities, contributes to the wear on the interacting surfaces.

First, let us consider the wear of partially lubricated contacts as a complex process consisting of various wear mechanisms. This involves setting up a compound equation of the type

$$V = V_f + V_a + V_c + V_d + V_{fa} + V_{fc} + V_{ac} + V_{fac}, \quad (2.50)$$

where V denotes the volume of worn material and the subscripts f, a, c and d refer to fatigue, adhesion, corrosion and abrasion, respectively. This not only recognizes the prevalence of mixed modes but also permits compensation for their interactions. In eqn (2.50), abrasion has a unique role. Because all the available mathematical models for primary wear assume clean components and a clean lubricating medium, there will therefore be no abrasion until wear particles have accumulated in the contact zone. Thus V_d becomes a function of the total wear V of uncertain form, but is probably a step function. It appears that if V_d is dominant in the wear process, it must overshadow all other terms in eqn (2.50).

When V_d does not dominate eqn (2.50) it is possible to make some predictions about the interaction terms. Thus it is known that corrosion greatly accelerates fatigue, for example, by hydrogen embrittlement of iron, so that V_{fc} will tend to be large and positive. On the other hand, adhesion and fatigue rarely, if ever, coexist, and this is presumably because adhesive wear destroys the microcracks from which fatigue propagates. Hence, the wear volume V_{fa} due to the interaction between fatigue and adhesion will always be zero. Since adhesion and corrosion are dimensionally similar, it may be hoped that V_{ac} and V_{fac} will prove to be negligible. If this is so, only V_{fc} needs to be evaluated. By assuming that the lubricant is not corrosive and that the environment is not excessively humid, it is possible to simplify eqn (2.50) further, and to reduce it to the form

$$V = V_a + V_f. \qquad (2.51)$$

According to the model presented here adhesive wear takes place on the metal–metal contact area, A_m, whereas fatigue wear should take place on the remaining real area of contact, that is, $A_r - A_m$. Repeated stressing through the thin adsorbed lubricant film existing on these micro-areas of contact would be expected to produce fatigue wear.

The block diagram of the model for evaluating the wear in lubricated contacts is shown in Fig. 2.12. It is provided in order to give a graphical decision tree as to the steps that must be taken to establish the functional lubrication regimes within which the sliding contact is operating. This block diagram can be used as a basis for developing a computer program facilitating the evaluation of the wear.

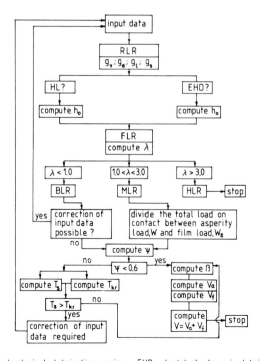

RLR– rheological lubrication regime; EHD– elastohydrodynamic lubrication
HL – hydrodynamic lubrication; FLR– functional lubrication regime
BLR– boundary lubrication regime; MLR– mixed lubrication regime
HLR– hydrodynamic lubrication regime

Figure 2.12

2.11.1. Rheological lubrication regime

As a first step in a calculating procedure the operating rheological lubrication regime must be determined. It can be examined by evaluating the viscosity parameter g_v and the elasticity parameter g_e

$$g_v = \frac{\alpha w}{R} \left(\frac{w}{\mu_0 V} \right)^{\frac{1}{2}}, \tag{2.52}$$

$$g_e = \left(\frac{w}{E'R} \right)^{\frac{1}{2}} \left(\frac{w}{\mu_0 V} \right)^{\frac{1}{2}}, \tag{2.53}$$

where w is the normal load per unit width of the contact, R is the relative radius of curvature of the contacting surfaces, E' is the effective elastic modulus, μ_0 is the lubricant viscosity at inlet conditions and V is the relative surface velocity.

The range of hydrodynamic lubrication is expressed by eqns (2.52) and (2.53) for the g_v and g_e inequalities as follows:

$$g_v < 1.5 \quad \text{and} \quad g_e < 0.6.$$

Operating conditions outside the limitations for g_v and g_e are defined as elastohydrodynamic lubrication. The range of the speed parameter g_s and the load parameter g_1 for practical elastohydrodynamic lubrication must be limited to within the following range of inequalities:

$$1.8 < g_s < 100,$$

where

$$g_s = \alpha \left(\frac{E'^3 \mu_0 V}{R} \right)^{\frac{1}{4}} \tag{2.54}$$

and

$$1.0 < g_1 < 100,$$

where

$$g_1 = \alpha \left(\frac{wE'}{2\pi R} \right)^{\frac{1}{2}}, \tag{2.55}$$

where α is the pressure–viscosity coefficient. Equations (2.52), (2.53), (2.54) and (2.55) help to establish whether or not the lubricated contact is in the hydrodynamic or elastohydrodynamic lubrication regime.

2.11.2. Functional lubrication regime

In the hydrodynamic lubrication regime, the minimum film thickness for smooth surfaces can be calculated from the following formula:

$$h_0 = 4.9 \frac{\mu_0 VR}{w}, \tag{2.56}$$

where 4.9 is a constant referring to a rigid solid with an isoviscous lubricant.

Under elastohydrodynamic conditions, the minimum film thickness for cylindrical contacts of smooth surfaces can be calculated from

$$h_0 = 2.65 \frac{\alpha^{0.54}(\mu_0 V)^{0.7} R^{0.43}}{w^{0.13} E'^{0.03}}. \tag{2.57}$$

In the case of point contacts on smooth surfaces the minimum film thickness can be calculated from the expression

$$h_0 = 0.84(\mu_0 \alpha V)^{0.74} R^{0.41} (E'/w)^{0.074}. \tag{2.58}$$

When operating sliding contacts with thin films, it is necessary to ascertain that they are not in the boundary lubrication regime. This can be done by calculating the specific film thickness or the lambda ratio

$$\lambda = h_0/S. \tag{2.59}$$

It is usual that $S = (R_1 + R_2)/2 = R_{sk}$, where $R_{sk} = 1.11R_a$ is the r.m.s. height of surface roughness.

If the lambda ratio is larger than 3 it is usual to assume that the probability of the metal–metal asperity contact is insignificant and therefore no adhesive wear is possible. Similarly, the lubricating film is thick enough to prevent fatigue failure of the rubbing surfaces. However, if λ is less than 1.0, the operating regime is boundary lubrication and some adhesive and fatigue wear would be likely. Thus, the change in the operating conditions of the contact should be seriously considered. If this is not possible for practical reasons, the mode of asperity contact should be determined by examining the plasticity index, ψ.

However, in the mixed lubrication regime in which λ is in the range 1.0–3.0, where most machine sliding contacts or sliding/rolling contacts operate, the total load is shared between the asperity load W and the film load W_s, and only the load supported by the contacting asperities should contribute to wear. When ψ is less than 0.6 the contact between asperities will be considered to be elastic under all practical loads, and when it is greater than 1.0 the contact will be regarded as being partially plastic even under the lightest load. When the range is between 0.6 and 1.0, the mode of contact is mixed and an increase in load can change the contact of some asperities from elastic to plastic. When $\psi < 0.6$, seizure is rather unlikely but metal–metal asperity contact is probable because of the fluctuation of the adsorbed lubricant molecules, and therefore the idea of fractional film defects should be introduced and examined.

2.11.3. Fractional film defects

(i) Simple lubricant

A property of some measurable influence, which has a critical effect on wear in the lubricated contact, is the heat of adsorption of the lubricant. This is particularly true in the case of the adhesive wear resulting from direct metal–metal asperity contacts. If lubricant molecules remain attached to

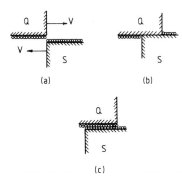

Figure 2.13

the load-bearing surfaces, then the probability of forming an adhesive wear particle is reduced. Figure 2.13 is an idealized representation of two opposing surface asperities and their adsorbed species coming into contact. At slow rate of approach the adsorbed molecules will have ample time to desorb, thus permitting direct metal–metal contact (case (b) in Fig. 2.13). At high rates of approach the time will be insufficient for desorption and metal–metal contact will be prevented (case (c) in Fig. 2.13).

In physical terms, the fractional film defect, β, can be defined as a ratio of the number of sites on the friction surface unoccupied by lubricant molecules to the total number of sites on the friction surface, i.e.

$$\beta = A_m/A_r, \tag{2.60}$$

where A_m is the metal–metal area of contact and A_r is the real area of contact. The relationship between the fractional film defect and the ratio of the time for the asperity to travel a distance equivalent to the diameter of the adsorbed molecule, t_z, and the average time that a molecule remains at a given surface site, t_r, has the form

$$(1 - \beta) = \exp(-t_z/t_r). \tag{2.61}$$

Time t_z is given by

$$t_z = Z/V. \tag{2.62}$$

Values of Z – the diameter of a molecule in an adsorbed state – are not generally available, but some rough estimate of Z can be given using the following expression:

$$Z = \left[\frac{6V_m}{\pi N_a}\right]^{\frac{1}{3}}. \tag{2.63}$$

Taking the Avogadro number as $N_a = 6.02 \times 10^{23}$

$$Z = 1.47 \times 10^{-8} V_m, \tag{2.64}$$

where V_m is the molecular volume of the lubricant. It is clear that $\beta \to 1.0$ if $t_z \gg t_r$. Also, $\beta \to 0$ if $t_z \ll t_r$. The average time, t_r, spent by one molecule in the same site, is given by the following expression:

$$t_r = t_0 \exp(E_c/RT_s), \tag{2.65}$$

where E_c is the heat of adsorption of the lubricant, R is the gas constant and T_s is the absolute temperature at the contact zone. Here, t_0 can be considered to a first approximation as the period of vibration of the adsorbed molecule. Again, t_0 can be estimated using the following formula:

$$t_0 = 4.75 \times 10^{-13} \left[\frac{MV_m^{\frac{2}{3}}}{T_m}\right]^{\frac{1}{2}}, \tag{2.66}$$

where M is the molecular weight of the lubricant and T_m is its melting point. Values of T_m are readily available for pure compounds but for mixtures such as commercial oils they simply do not exist. In such cases, a generalized melting point based on the liquid/vapour critical point will be used

$$T_m = 0.4T_c,$$

where T_c is the critical temperature. Taking into account the expressions discussed above, the final formula for the fractional film defect, β, has the form

$$\beta = 1 - \exp\left\{-\left[\frac{30.9 \times 10^5 T_m^{\frac{1}{2}}}{V M^{\frac{1}{2}}}\right]\exp(-E_c/RT_s)\right\}. \tag{2.67}$$

Equation (2.67) is only valid for a simple lubricant without any additives.

(ii) *Compounded lubricant*

To remove the limitation imposed by eqn (2.67) and extending the concept of the fractional film defect on compounded lubricants, it is necessary to introduce the idea of temporary residence for both additive and base fluid molecules on the lubricated metal surface in a dynamic equilibrium. For a lubricant containing two components, additive (a) and base fluid (b), the area A_m arises from the spots originally occupied by both (a) and (b). Thus,

$$A_m = A_{m,a} + A_{m,b}. \tag{2.68}$$

The fractional film defect for both (a) and (b) can be defined as

$$\beta_a = A_{m,a}/A_a, \tag{2.69}$$

$$\beta_b = A_{m,b}/A_b, \tag{2.70}$$

where A_a and A_b represent the original areas covered by (a) and (b), respectively. The fraction of surface covered originally by the additive, before contact, is

$$\delta = A_a/A_r, \tag{2.71}$$

where $A_r = A_a + A_b$ is the real area of contact.

According to eqn (2.60), the fractional film defect of the compounded lubricant can be expressed as

$$\Theta = A_m/A_r.$$

From eqn (2.69)

$$A_{m,a} = \beta_a A_a.$$

From eqn (2.70)

$$A_{m,b} = \beta_b A_b.$$

Taking the above into account, eqn (2.68) becomes

$$A_m = \beta_a A_a + \beta_b A_b. \tag{2.72}$$

Reorganized, eqn (2.71) becomes

$$A_b = A_r(1 - \Theta).$$

Thus, eqn (2.72) becomes

$$A_m = \beta_a A_a + \beta_b A_r (1 - \Theta)$$

and finally

$$A_m = [\beta_b + (\beta_a - \beta_b)\Theta] A_r.$$

Thus, the fractional film defect of the compounded lubricant is given by

$$\delta = A_m / A_r = \beta_b + (\beta_a - \beta_b)\Theta. \tag{2.73}$$

Following the same argument as in the case of the simple lubricant, it is possible to relate the fractional film defect for both (a) and (b) to the heat of adsorption, E_c,

for additive (a)

$$\beta_a = 1 - \exp\left\{-\left[\frac{30.9 \times 10^5 T_{m,a}^{\frac{1}{2}}}{V M_a^{\frac{1}{2}}}\right]\exp\left(-\frac{E_{c,a}}{RT_s}\right)\right\}, \tag{2.74}$$

for the base fluid (b)

$$\beta_b = 1 - \exp\left\{-\left[\frac{30.9 \times 10^5 T_{m,b}^{\frac{1}{2}}}{V M_b^{\frac{1}{2}}}\right]\exp\left(-\frac{E_{c,b}}{RT_s}\right)\right\}. \tag{2.75}$$

2.11.4. Load sharing in lubricated contacts

The adhesive wear of lubricated contacts, and in particular lubricated concentrated contacts, is now considered. The solution of the problem is based on partial elastohydrodynamic lubrication theory. In this theory, both the contacting asperities and the lubricating film contribute to supporting the load. Thus

$$W_c = W_s + W, \tag{2.76}$$

where W_c is the total load, W_s is the load supported by the lubricating film and W is the load supported by the contacting asperities. Only part of the total load, namely W, can contribute to the adhesive wear. In view of the experimental results this assumption seems to be justified. Load W supported by the contacting asperities results in the asperity pressure p, given by

$$p = 2/3 E(Nr\sigma_*)(\sigma_*/r)^{\frac{1}{2}} F_{\frac{3}{2}}(d_c/\sigma_*). \tag{2.77}$$

The total pressure resulting from the load W_c is given by

$$p_c = 0.39 E^{\frac{2}{3}} W_c^{\frac{1}{3}} R_e^{-\frac{2}{3}}. \tag{2.78}$$

Thus the ratio p/p_c is given by

$$p/p_c = 1.7 R_e^{\frac{2}{3}} W_c^{-\frac{1}{3}} E^{\frac{1}{3}} (Nr\sigma_*)(\sigma_*/r)^{\frac{1}{2}} F_{\frac{3}{2}}(d_c/\sigma_*), \tag{2.79}$$

where $F_{\frac{3}{2}}(d_c\sigma_*)$ is a statistical function in the Greenwood–Williamson model of contact between two real surfaces, R_e is the relative radius of curvature of the contacting surfaces, E is the effective elastic modulus, N is the asperity density, r is the average radius of curvature at the peak of asperities, σ_* is the standard deviation of the peaks and d_e is the equivalent separation between the mean height of the peaks and the flat smooth surface. The ratio of lubricant pressure to total pressure is given by

$$p_s/p_c = \frac{1}{\lambda}\left(\frac{h_0}{\bar{h}}\right)^{6.3}, \qquad (2.80)$$

where λ is the specific film thickness defined previously, $\bar{h}$ is the mean thickness of the film between two actual rough surfaces and h_0 is the film thickness with smooth surfaces.

It should be remembered however that eqn (2.80) is only applicable for values of the lambda ratio very near to unity. For rougher surfaces, a more advanced theory is clearly required. The fraction of the total pressure, p_c, carried by the asperities is a function of d_e/σ_* and the fraction carried hydrodynamically by the lubricant film is a function of $h_0/\bar{h}$. To combine these two results the relationship between d_e and $\bar{h}$ is required. The separation d_e in the single rough surface model is related to the actual separation of the two rough surfaces by

$$d_e \approx d + 0.5\sigma_s,$$

where σ_s is the standard deviation of the surface height. The separation of the surface is related to the separation of the peaks by

$$(s - d) \approx 0.8(\sigma_{s_1} + \sigma_{s_2}) \approx 1.1\sigma_s$$

for surfaces of comparable roughness, and for $\sigma_* \approx 0.7\sigma_s$. Combining these relationships, we find that

$$d_e/\sigma_* = 1.4(s/\sigma_s) - 0.9. \qquad (2.81)$$

Because the space between the two contacting surfaces should accommodate the quantity of lubricant delivered by the entry region to the contacting surfaces it is thus possible to relate the mean film thickness, $\bar{h}$, to the mean separation between the surfaces, s. Using the condition of continuity the mean height of the gap between two rough surfaces, $\bar{h}$, can be calculated from

$$\bar{h}/\sigma_s = (s/\sigma_s) + F_1(s/\sigma_s),$$

where $F_1(s/\sigma_s)$ is the statistical function in the Greenwood–Williamson model of contact between nominally flat rough surfaces.

It is possible, therefore, to plot both the asperity pressure and the film pressure with a datum of $(\bar{h}/\sigma_s)$. The point of intersection between the appropriate curves of asperity pressure and film pressure determines the division of total load between the contacting asperities and the lubricating film. The analytical solution requires a value of $\bar{h}/\sigma_s$ to be found by iteration, for which

$$(p/p_c) + (p_s/p_c) = 1. \qquad (2.82)$$

2.11.5. Adhesive wear equation

Theoretically, the volume of adhesive wear should strictly be a function of the metal–metal contact area, A_m, and the sliding distance. This hypothesis is central to the model of adhesive wear. Thus, it can be written as

$$V = k_m A_m L, \qquad (2.83)$$

where k_m is a dimensionless constant specific to the rubbing materials and independent of any surface contaminants or lubricants.

Expressing the real area of contact, A_r, in terms of W and P and taking into account the concept of fractional surface film defect, β, eqn (2.83) becomes

$$V/L = k_m \beta (W/P), \qquad (2.84)$$

where W is the load supported by the contacting asperities and P is the flow pressure of the softer material in contact. Equation (2.84) contains a parameter k_m which characterizes the tendency of the contacting surfaces to wear by the adhesive process, and a parameter β indicating the ability of the lubricant to reduce the metal–metal contact area, and which is variable between zero and one.

Although it has been customary to employ the yield pressure, P, which is obtained under static loading, the value under sliding will be less because of the tangential stress. According to the criterion of plastic flow for a two-dimensional body under combined normal and tangential stresses, yielding of the friction junction will follow the expression

$$P^2 + \alpha S^2 = P_m^2, \qquad (2.85)$$

where P is now the flow pressure under combined stresses, S is the shear strength, P_m is the flow pressure under static load and α may be taken as 3. An exact theoretical solution for a three-dimensional friction junction is not known. In these circumstances however, the best approach is to assume the two-dimensional junction.

From friction theory

$$S = (F/A_r) = (fW/A_r) = fP,$$

where F is the total frictional force. Thus

$$P = \frac{P_m}{(1 + 3f^2)^{\frac{1}{2}}}, \qquad (2.86)$$

and eqn (2.84) becomes

$$(V/L) = k_m (1 + 3f^2)^{\frac{1}{2}} \beta (W/P_m). \qquad (2.87)$$

Equation (2.87) now has the form of an expression for the adhesive wear of lubricated contacts which considers the influence of tangential stresses on the real area of contact. The values of W and β can be calculated from the equations presented and discussed earlier.

2.11.6. Fatigue wear equation

It is known that conforming and nonconforming surfaces can be lubricated
hydrodynamically and that if the surfaces are smooth enough they will not
touch. Wear is not then expected unless the loads are large enough to bring
about failure by fatigue. For real surface contact the point of maximum
shear stress lies beneath the surface. The size of the region where flow occurs
increases with load, and reaches the surface at about twice the load at which
flow begins, if yielding does not modify the stresses. Thus, for a friction
coefficient of 0.5 the load required to induce plastic flow is reduced by a
factor of 3 and the point of maximum shear stress rises to the surface. The
existence of tensile stresses is important with respect to the fatigue wear of
metals. The fact, that there is a range of loads under which plastic flow can
occur without extending to the surface, implies that under such conditions,
protective films such as the lubricant boundary layers will remain intact.
Thus, the obvious question is, how can wear occur when asperities are
always separated by intact lubricant layers. The answer to this question
appears to lie in the fact that some wear processes can occur in the presence
of surface films. Surface films protect the substrate materials from damage
in depth but they do not prevent subsurface deformation caused by
repeated asperity contact. Each asperity contact is associated with a wave
of deformation. Each cross-section of the rubbing surfaces is therefore
successively subjected to compressive and tensile stresses. Assuming that
adhesive wear takes place in the metal–metal contact area, A_m, it is logical
to conclude that fatigue wear takes place on the remaining part, that is
$(A_r - A_m)$, of the real contact area. Repeated stresses through the thin
adsorbed lubricant film existing on these micro-areas are expected to cause
fatigue wear. To calculate the amount of fatigue wear in a lubricated
contact, an engineering wear model, developed at IBM, can be adopted.
The basic assumptions of the non-zero wear model are consistent with the
Palmgren function, since the coefficient of friction is assumed to be constant
for any given combination of materials irrespective of load and geometry.
Thus the model has the correct dimensional relationship for fatigue wear.

   Non-zero wear is a change in the contour which is more marked than the
surface finish. The basic measure of wear is the cross-sectional area, Q, of a
scar taken in a plane perpendicular to the direction of motion. The model
for non-zero wear is formulated on the assumption that wear can be related to
a certain portion, U, of the energy expanded in sliding and to the number N of
passes, by means of a differential equation of the type

$$dQ = \left(\frac{\partial Q}{\partial N}\right)_N dU + \left(\frac{\partial Q}{\partial N}\right)_U dN. \tag{2.88}$$

For fatigue wear an equation can be developed from eqn (2.88);

$$dQ = C''(\tau_{max}S)^{\frac{9}{2}} dN + \frac{9}{2}\frac{Q}{\tau_{max}S} d(\tau_{max}S), \tag{2.89}$$

where C'' is a parameter which is independent of N, S is the maximum width of the contact region taken in a plane parallel to the direction of motion and τ_{max} is the maximum shear stress occurring in the vicinity of the contact region.

For non-zero wear it is assumed that a certain portion of the energy expanded in sliding and used to create wear debris is proportional to $\tau_{max}S$. Integration of eqn (2.89) results in an expression which shows how wear progresses as the number of operations of a mechanism increases. The manner in which such an expression is obtained for the pin-on-disc configuration is illustrated by a numerical example.

The procedure for calculating non-zero wear is somewhat complicated because there is no simple algebraic expression available for relating lifetime to design parameters for the general case. The development of the necessary expressions for the determination of suitable combinations of design parameters is a step-like procedure. The first step involves integration of the particular form of the differential equation of which eqn (2.89) is the general form. This step results in a relationship between Q and the allowable total number L of sliding passes and usually involves parameters which depend on load, geometry and material properties. The second step is the determination of the dependence of the parameters on these properties. From these steps, expressions are derived to determine whether a given set of design parameters is satisfactory, and the values that certain parameters must assume so that the wear will be acceptable.

2.11.7. Numerical example

Let us consider a hemispherically-ended pin of radius $R = 5$ mm, sliding against the flat surface of a disc. The system under consideration is shown in Fig. 2.14. The radius, r, of the wear track is 75 mm. The material of the disc is steel, hardened to a Brinell hardness of 75×10^2 N/mm^2. The pin is made of brass of Brinell hardness of 11.5×10^2 N/mm^2. The yield point in shear of the steel is 10.5×10^2 N/mm^2 and of the brass is 1.25×10^2 N/mm^2. The disc is rotated at $n = 12.7$ rev min^{-1} which corresponds to $V = 0.1$ m s^{-1}. The load W on the system is 10 N. The system is lubricated with n-hexadecane. It is assumed, with some justification, that the wear on the disc is zero.

When a lubricant is used it is necessary to develop expressions for Q and $\tau_{max}S$ in terms of a common parameter so that eqn (2.89) may be integrated. This is done by expressing these quantities in terms of the width T of the wear scar (see Fig. 2.14). If the depth, h, of the wear scar is small in comparison with the radius of the pin, the scar shape may be approximated to a triangle and

$$Q \approx (1/2)hT. \tag{2.90}$$

If h is larger, eqn (2.90) will become more complex. From the geometry of the system shown in Fig. 2.14

$$h = T^2/8R, \tag{2.91}$$

$$Q = T^3/16R. \tag{2.92}$$

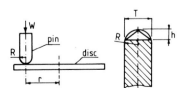

Figure 2.14

Since the contact conforms

$$\tau_{max} = (KW/A)(\tfrac{1}{4} + f^2)^{\frac{1}{2}}. \tag{2.93}$$

Using $A = \pi(T/2)^2$,

$$\tau_{max} = \frac{4KW(\tfrac{1}{4} + f^2)^{\frac{1}{2}}}{\pi T^2}. \tag{2.94}$$

In the case under consideration, $S = T$ and therefore

$$\tau_{max}S = \frac{4KW(\tfrac{1}{4} + f^2)^{\frac{1}{2}}}{\pi T}. \tag{2.95}$$

Equations (2.92) and (2.93) allow eqn (2.89) to be integrated because they express Q and $\tau_{max}S$ respectively in terms of a single variable T. Thus

$$dQ = \sim \frac{3T^2}{16R} dT, \tag{2.96}$$

$$d(\tau_{max}S) = -\frac{4KW(\tfrac{1}{4} + f^2)^{\frac{1}{2}}}{\pi T^2} dT. \tag{2.97}$$

Before eqn (2.89) can be integrated it is necessary to consider the variation in Q with N. Since the size of the contact changes with wear, it is possible to change the number of passes experienced by a pin in one operation

$$n_p = B/S,$$

where $B = 2\pi r$ is the sliding distance during one revolution of the disc. Because $dN = n_p dL$, where L is the total number of disc revolutions during a certain period of time, we obtain

$$dN = \frac{2\pi r}{T} dL.$$

Substituting the above expressions into eqn (2.89) gives:

$$\frac{3T^2}{16R} dT = C'' \left\{ \frac{4KW(\tfrac{1}{4} + f^2)^{\frac{1}{2}}}{\pi T} \right\}^{\frac{9}{2}} \frac{2\pi r}{T} dL +$$
$$+ \frac{9}{2} \frac{T^3}{16R} \left\{ \frac{4KW(\tfrac{1}{4} + f^2)^{\frac{1}{2}}}{\pi T} \right\}^{-1} \left\{ -\frac{4KW(\tfrac{1}{4} + f^2)^{\frac{1}{2}}}{\pi T^2} \right\} dT. \tag{2.98}$$

After rearranging, eqn (2.98) becomes

$$T^{\frac{15}{2}} dT = \frac{32}{15} C'' 2^{10} r K^{\frac{9}{2}} W^{\frac{9}{2}} (\tfrac{1}{4} + f^2)^{\frac{9}{4}} \frac{1}{\pi^{\frac{7}{2}}} R \, dL. \tag{2.99}$$

Integration of eqn (2.99) gives

$$\frac{2T^{\frac{17}{2}}}{17} = \frac{32}{15} C'' 2^{10} r K^{\frac{9}{2}} W^{\frac{9}{2}} (\tfrac{1}{4} + f^2)^{\frac{9}{4}} \frac{1}{\pi^{\frac{7}{2}}} RL + C_2. \tag{2.100}$$

Because $Q = T^3/16R$ and therefore $T = (16QR)^{\frac{1}{3}}$. Substituting the expression for T into eqn (2.100) and rearranging gives

$$Q^{\frac{17}{6}} = 0.1312 \frac{C''rK^{\frac{9}{2}}(\frac{1}{4}+f^2)^{\frac{9}{4}}L}{R^{\frac{11}{6}}} + C_2 \qquad (2.101)$$

and finally,

$$Q^{\frac{17}{6}} = C_1 L + C_2, \qquad (2.102)$$

where

$$C_1 = 0.1312 \frac{C''rK^{\frac{9}{2}}W^{\frac{9}{2}}(\frac{1}{4}+f^2)^{\frac{9}{4}}}{R^{\frac{11}{6}}}$$

and C_2 is a constant of integration. Equation (2.102) gives the dependence of Q on L. The dependence of Q on the other parameters of the system is contained in the quantities C_1 and C_2 of eqn (2.102).

Equation (2.102) implicitly defines the allowed ranges of certain parameters. In using this equation these parameters cannot be allowed to assume values for which the assumptions made in obtaining eqn (2.102) are invalid.

One way of determining C_1 and C_2 in eqn (2.102), is to perform a series of controlled experiments, in which Q is determined for two different numbers of operations for various values and combinations of the parameters of interest. These values of Q for different values of L enable C_1 and C_2 to be determined. In certain cases, however, C_1 and C_2 can be determined on an analytical basis. One analytical approach is for the case in which there is a period of at least 2000 passes of what may be called zero wear before the wear has progressed to beyond the surface finish. This is done by taking C_2 to be zero and determining C_1 from the model for zero wear. C_1 is determined by first finding the maximum number L_1 of operations for which there will be zero wear for the load, geometry etc. of interest. L_1 is then given by:

$$L_1 = \frac{2 \times 10^3}{n_p} \left(\frac{\gamma_R \tau_y}{\tau_{max}} \right)^9, \qquad (2.103)$$

where τ_{max} is the maximum shear stress computed using the unworn geometry, τ_y is the yield point in shear of the weaker material and γ_R is a quantity characteristic of the mode of lubrication. The geometry of the wear scar produced during the number L_1 of passes, is taken to be a scar of the profile assumed in deriving eqn (2.102) and of a depth equal to one-half of the peak-to-peak surface roughness of the material of the pin. In the particular case under consideration it is assumed that

$$\gamma_R = 0.20 \quad \text{(fatigue mode of wear)},$$
$$f = 0.26 \quad \text{(coefficient of friction)}.$$

For the material of the pin

$$E = 1.1 \times 10^5 \, \text{N/mm}^2,$$
$$v = 0.33,$$

and for the material of the disc

$$E = 2.1 \times 10^2 \, \text{N/mm}^2,$$
$$v = 0.33.$$

The maximum shear stress $\tau_{max} = 0.31 q_0$, where $q_0 = 3W/2\pi ab$, and a is the semimajor axis and b is the semiminor axis of the pressure ellipse. For the assumed data $q_0 = 789.5 \, \text{N/mm}^2$ and $\tau_{max} = 245 \, \text{N/mm}^2$.

The number of sliding passes for the pin during one operation is

$$n_{\text{p}} = B/S = \frac{2\pi r}{T} = 3019.2 \quad 1/\text{rev}.$$

The number, L, of operations is given by

$$L_1 = \frac{2 \times 10^3}{n_{\text{p}}} \left(\frac{\gamma_R \tau_y}{\tau_{max}} \right)^9 = 0.662(0.102)^9.$$

For zero wear, Q is given by

$$Q = \tfrac{1}{2} h 2 (2Rh)^{\frac{1}{2}} = 0.018 \times 10^{-3} \, \text{mm}^2.$$

Therefore, the constant of integration C_1, is given by

$$C_1 = \frac{Q^{\frac{17}{6}}}{L_1} = 0.87 \times 10^{-5}.$$

Having determined C_1, Q can be calculated for $L = 10^6$ revolutions

$$Q^{\frac{17}{6}} = 0.87 \times 10^{-5} \times 10^6 = 8.7,$$
$$Q = (8.7)^{\frac{6}{17}} = 2.146 \, \text{mm}^2.$$

The volume of the wear debris is given by

$$V_{\text{f}} = \pi h^2 (R - \tfrac{1}{3} h)$$

and using $h = T^2/8R$ and $T = (16QR)^{\frac{1}{3}}$ gives

$$V_{\text{f}} = 3.14(0.77)^2 (5 - \tfrac{0.77}{3}) = 8.82 \, \text{mm}^3.$$

As mentioned earlier, in the case of a lubricated system it is reasonable to expect additional wear resulting from the adhesive process. The volume, V_{a}, of the wear debris resulting from adhesive wear must be determined using relationships discussed earlier. For n-hexadecane as a lubricant we have: $t_0 = 2.8 \times 10^{-12} \, \text{s}$, $Z = 1.13 \times 10^{-7} \, \text{cm}$ and $E_{\text{c}} = 11\,700 \, \text{cal/mol}$. Furthermore, $f = 0.26$, $k_m = 0.23$, $R = 1.9872 \, \text{cal/mol K}$ and $T_{\text{s}} = 295.7 \, \text{K}$. For these parameters characterizing the system under consideration, the fractional film defect β is

$$\beta = 1 - \exp\left\{-\frac{1.13 \times 10^{-7}}{10 \times 2.8 \times 10^{-12}} \exp\left(-\frac{11\,700}{1.9872 \times 295.7}\right)\right\} = 0.000\,01.$$

Finally, V_a is

$$V_a = 0.23[1 + 3(0.26)^2]^{\frac{1}{2}} \times 0.000\,01 \frac{1000}{11.5 \times 10^6} 47 \times 10^6 = 10.3\,\text{mm}^3.$$

The total volume of wear debris is

$$V = V_a + V_f = 10.3 + 8.82 = 19.12\,\text{mm}^3.$$

2.12. Relation between fracture mechanics and wear

An analytical description of the fracture aspects of wear is quite difficult. The problems given here are particularly troublesome:

(i) debris is generated by crack formation in material which is highly deformed and whose mechanical properties are poorly understood;

(ii) the cracks are close to the surface and local stresses cannot be accurately specified;

(iii) the crack size can be of the same order of magnitude as microstructural features which invalidates the continuum assumption on which fracture mechanics is based.

The first attempt to introduce fracture mechanics concepts to wear problems was made by Fleming and Suh some 10 years ago. They analysed a model of a line contact force at an angle to the free surface as shown in Fig. 2.15. The line force represents an asperity contact under a normal load, W, with a friction component $W \tan \alpha$. Then the stress intensity associated with a subsurface crack is calculated by assuming that it forms in a perfectly elastic material. While the assumption appears to be somewhat unrealistic, it has, however, some merit in that near-surface material is strongly work-hardened and the stress-strain response associated with the line force passing over it is probably close to linear.

The Fleming–Suh model envisages crack formation behind the line load where small tensile stresses occur. However, it is reasonable to assume that the more important stresses are the shear-compression combination which is associated with crack formation ahead of the line force as illustrated in Fig. 2.15. For the geometry of Fig. 2.15, the crack is envisioned to form as a result of shear stresses and its growth is inhibited by friction between the opposing faces of the crack. In this way the coefficient of friction of the material subjected to the wear process and sliding on itself enters the analysis. The elastic normal stress at any point below the surface in the absence of a crack is given by

$$\sigma_{yy} \approx -\frac{2W}{\pi y} \frac{\cos(\alpha - \Theta)\cos^3\Theta}{\cos\alpha}. \tag{2.104}$$

The terms in eqn (2.104) are defined in Fig. 2.15. In particular, the friction coefficient between the contact and the surface is given by $\tan\alpha$.

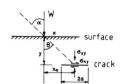

Figure 2.15

A non-dimensional normal stress T can be defined as

$$T = \frac{\pi y \sigma_{yy}}{2W}.$$ (2.105)

The shear stress acting at the same point is

$$\sigma_{xy} = \sigma_{yy} \tan \Theta$$ (2.106)

and the corresponding nondimensional stress is

$$S = \frac{\pi y \sigma_{xy}}{2W}.$$ (2.107)

Figure 2.16 shows the distribution of shear stress along a plane parallel to the surface (y is constant). It is seen that that shear stress distribution is asymmetrical, with larger stresses being developed ahead of the contact line than behind it, and with the sense of the stress changing sign directly below the contact line. Thus any point below the surface will experience a cyclic stress history from negative to positive shear as the contact moves along the surface. The shear asymmetry becomes more pronounced the higher the coefficient of friction. However, Fig. 2.16 shows that the friction associated with the wear surface does not have a large effect on these stresses. The corresponding normal stress distribution is plotted in Fig. 2.17.

This stress component is larger than the shear, and it peaks at a horizontal distance close to the origin where the shear stress is small. The normal stress also changes sign and becomes very slightly positive far behind the contact point. In front of the contact line the normal stress decreases monotonically and becomes of the same order as the shear stress in the region of peak shear stress. The maximum normal stress is found in a similar manner to the maximum shear stress; that is by differentiating eqn (2.104) with respect to Θ and setting the result equal to zero. In the case of shear stress, eqn (2.106) is involved. Thus, for shear stress

$$\tan(\alpha - \Theta^*) = 2 \tan \Theta^* - \cot \Theta^*,$$ (2.108)

where Θ^* corresponds to the position of largest shear. When eqn (2.108) is evaluated numerically, Θ^* is found to be very insensitive to the friction coefficient $\tan \alpha$, only varying between $30°$ and $45°$ as α varies from $0°$ to $90°$.

For normal stress, the critical angle is given by

$$\tan \Theta^* = \tfrac{1}{3} \tan(\Theta^* - \alpha)$$ (2.109)

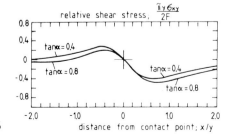

Figure 2.16

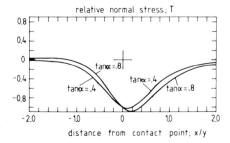

Figure 2.17

and also varies slightly with the coefficient of friction (Θ^* varies from $0°$ to $15°$ as $\tan \alpha$ goes from 0 to 1.37). The stress intensity associated with the crack is obtained from a weighted average of the stresses calculated previously. The stress intensity corresponding to a combined uniform shear-compression stress on the crack can be expressed as

$$K_{\text{II}} = (\sigma_{xy} - \tan \beta \sigma_{yy})(\pi a)^{\frac{1}{2}}, \tag{2.110}$$

where $\tan \beta$ is the coefficient of friction between the opposing faces of the crack and $2a$ is the crack length. According to eqn (2.110) the crack is driven by the shear stress and retarded by the friction forces arising from the compressive stresses. It is suggested that eqn (2.110) can be adopted to a non-uniform stress field by evaluating the quantity $\sigma_{xy} - \tan \beta \sigma_{yy}$ along the crack and integrating according to the procedure described below.

2.12.1. Estimation of stress intensity under non-uniform applied loads

There are situations where one needs to know the stress intensity associated with cracks in non-uniform stress fields, for example when there is a delamination type of wear.

The approximation is derived for a semi-infinite plate containing a crack of length $2a$. The applied stress $\sigma(x)$ can be either tensile or shear so that Mode I, II or III stress intensities can be approximated. If $\sigma(x)$ is the stress that would be acting along the crack plane if the crack were not there

$$K = \left(\frac{1}{\pi a}\right)^{\frac{1}{2}} \int_{-a}^{a} \left(\frac{a+x}{a-x}\right)^{\frac{1}{2}} \sigma(x)\,\mathrm{d}x, \tag{2.111}$$

where K is the stress intensity and $2a$ is the crack length. Equation (2.111) evaluates the stress intensity at $x = a$. When $x = a\cos \Theta$, eqn (2.111) becomes

$$K = \left(\frac{a}{\pi}\right)^{\frac{1}{2}} \int_{0}^{\pi} (1 + \cos \Theta)\sigma(\Theta)\,\mathrm{d}\Theta. \tag{2.112}$$

If a term σ_{eff} given by

$$K = \sigma_{\text{eff}}(\pi a)^{\frac{1}{2}} \tag{2.113}$$

is introduced, eqn (2.112) becomes

$$\sigma_{\text{eff}} = \frac{1}{\pi} \int_{0}^{\pi} (1 + \cos \Theta)\sigma(\Theta)\,\mathrm{d}\Theta. \tag{2.114}$$

Equation (2.114) can be evaluated by the Simpson rule. If the crack length is divided into two intervals, the Simpson rule approximation is

$$\sigma_{\text{eff}} \approx \frac{\sigma_T + 2\sigma_0}{3}, \tag{2.115}$$

where σ_T is the applied stress at $x=a$ and σ_0 is the applied stress at the crack point. For four intervals

$$\sigma_{\text{eff}} \approx \frac{\sigma_T + 3.414\sigma_m^+ + \sigma_0 + 0.586\sigma_m^-}{6}, \tag{2.116}$$

where σ_m^+ and σ_m^- are the applied stresses at $x=a/\sqrt{2}$ and $x=-a/\sqrt{2}$ respectively.

If the effective stress on the crack calculated in this way is denoted by $\bar{\sigma}_{xy}$

$$K_{11} = \bar{\sigma}_{yy}(\pi a)^{\frac{1}{2}} \tag{2.117}$$

or

$$K_{11} = \frac{2W}{\pi y^{\frac{1}{2}}} \bar{S} \left(\frac{\pi a}{y} \right)^{\frac{1}{2}}, \tag{2.118}$$

where $\bar{S}$ is the non-dimensional form of $\bar{\sigma}_{xy}$. Values of $\bar{S}$ were calculated for two different friction coefficients, $\tan \beta$, of the opposing faces of the crack. The case of no friction on the wear surface was used since $\bar{S}$ is relatively insensitive to $\tan \alpha$.

2.13.   Film lubrication

2.13.1.   Coefficient of viscosity

Relative sliding with film lubrication is accompanied by friction resulting from the shearing of viscous fluid. The coefficient of viscosity of a fluid is defined as the tangential force per unit area, when the change of velocity per unit distance at right angles to the velocity is unity.

Referring to Fig. 2.18, suppose AB is a stationary plane boundary and CD a parallel boundary moving with linear velocity, V. AB and CD are separated by a continuous oil film of uniform thickness, h. The boundaries are assumed to be of infinite extent so that edge effects are neglected. The fluid velocity at a boundary is that of the adherent film so that velocity at AB is zero and at CD is V.

Let us denote by v velocity of fluid in the plane EF at a perpendicular distance y from AB and by $v+\delta v$ the velocity in the plane GH at a distance $y+\delta y$ from AB. Then, if the tangential force per unit area at position y is denoted by q

$$\text{coefficient of viscosity} = \mu = q / \frac{\delta v}{\delta y}$$

and in the limit

$$q = \mu \frac{\mathrm{d}v}{\mathrm{d}y}. \tag{2.119}$$

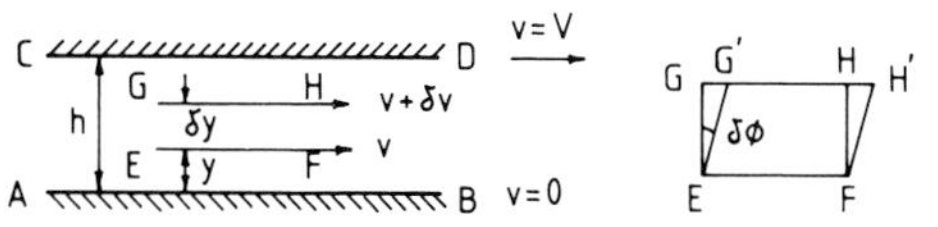

Figure 2.18

Alternatively, regarding q as a shear stress, then, if ϕ is the angle of shear in an interval of time δt, shown by GEG' in Fig. 2.18

$$GG' = \delta v \delta t = \delta y \delta \phi$$

and in the limit

$$\frac{dv}{dy} = \frac{d\phi}{dt}$$

or

$$q = \mu \frac{d\phi}{dt}. \tag{2.120}$$

Hence for small rates of shear and thin layers of fluid, μ may be defined as the shear stress when the rate of shear is one radian per second. Thus the physical dimensions of μ are

$$\left(\frac{\text{force}}{\text{area}}\right) \bigg/ \left(\frac{\text{radian}}{\text{sec}}\right) = \frac{ML}{T^2} \frac{1}{L^2} \; T = \frac{M}{LT}.$$

2.13.2. Fluid film in simple shear

The above considerations have been confined to the simple case of parallel surfaces in relative tangential motion, and the only assumption made is that the film is properly supplied with lubricant, so that it can maintain itself between the surfaces. In Fig. 2.19 the sloping lines represent the velocity distribution in the film so that the velocity at E is $EF = v$, and the velocity at P is $PQ = V$. Thus

$$\frac{dv}{dy} = \frac{v}{y} = \frac{V}{h},$$

$$\text{tangential force per unit area} = \mu \frac{dv}{dy} = \mu \frac{V}{h}.$$

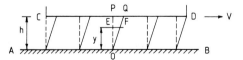

Figure 2.19

It will be shown later that, from considerations of equilibrium, the pressure within a fluid film in simple shear must be uniform, i.e. there can be no pressure gradient.

If the intensity of pressure per unit area of AB or CD is p and f is the virtual coefficient of friction

$$q = fp = \mu \frac{V}{h}$$

or

$$F = \mu A \frac{V}{h},\qquad(2.121)$$

where F is the total tangential force resisting relative motion and A is the area of the surface CD wetted by the lubricant. This is the Petroff law and gives a good approximation to friction losses at high speeds and light loads, under conditions of lubrication, that is when interacting surfaces are completely separated by the fluid film. It does not apply when the lubrication is with an imperfect film, that is when boundary lubrication conditions apply.

2.13.3.   Viscous flow between very close parallel surfaces

Figure 2.20 represents a viscous fluid, flowing between two stationary parallel plane boundaries of infinite extent, so that edge effects can be neglected. The axes Ox and Oy are parallel and perpendicular, respectively, to the direction of flow, and Ox represents a plane midway between the boundaries. Let us consider the forces acting on a flat rectangular element of width δx, thickness δy, and unit length in a direction perpendicular to the plane of the paper. Let

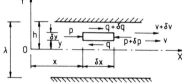

Figure 2.20

tangential drag per unit area at $y = q$,
tangential drag per unit area at $y + \delta y = q + \delta q$,
net tangential drag on the element $= \delta q \delta x$,
normal pressure per unit area at $x = p$,
normal pressure per unit area at $x + \delta x = p + \delta p$,
net normal load on the ends of the element $= \delta p \delta y$.

Hence the surrounding fluid exerts a net forward drag on the element of amount $\delta q \delta x$, which must be equivalent to the net resisting load $\delta p \delta y$ acting on the ends of the element, so that

$$\delta p \delta y = \delta q \delta x$$

and in the limit

$$\frac{\mathrm{d}p}{\mathrm{d}x} = \frac{\mathrm{d}q}{\mathrm{d}y}.\qquad(2.122)$$

Combining this result with the viscosity equation $q = \mu\, \mathrm{d}v/\mathrm{d}y$, we obtain the fundamental equation for pressure

$$\frac{\mathrm{d}p}{\mathrm{d}x} = \mu \frac{\mathrm{d}^2 v}{\mathrm{d}y^2}.\qquad(2.123)$$

Rewriting this equation, and integrating twice with respect to y and keeping x constant

$$\frac{\mathrm{d}^2 v}{\mathrm{d}y^2} = \frac{1}{\mu}\frac{\mathrm{d}p}{\mathrm{d}x},$$

$$v = \frac{y^2}{2\mu}\frac{\mathrm{d}p}{\mathrm{d}x} + Ay + B.$$

The distance between the boundaries is $2h$, so that $v=0$ when $y=\pm h$. Hence the constant A is zero and

$$B = -\frac{h^2}{2\mu}\frac{dp}{dx},$$

$$v = -\frac{1}{2\mu}\frac{dp}{dx}(h^2 - y^2). \tag{2.124}$$

It follows from this equation that the pressure gradient dp/dx is negative, and that the velocity distribution across a section perpendicular to the direction of flow is parabolic. The pressure intensity in the film falls in the direction of flow. Further, if Q represents the volume flowing, per second, across a given section

$$Q = 2\int_0^h v\,dy = -\frac{1}{\mu}\frac{dp}{dx}\int_0^h (h^2 - y^2)\,dy$$

or

$$Q = -\frac{1}{\mu}\frac{dp}{dx}\frac{2}{3}h^3 = -\frac{1}{\mu}\frac{db}{dx}\frac{\lambda^3}{12}, \tag{2.125}$$

where $\lambda = 2h$ is the distance between the boundaries. This result has important applications in lubrication problems.

2.13.4.   Shear stress variations within the film

For the fluid film in simple shear, q is constant, so that

$$\frac{dp}{dx} = \frac{dq}{dy} = 0$$

and p is also constant. In the case of parallel flow between plane boundaries, since Q must be the same for all sections, dp/dx is constant and p varies linearly with x. Further

$$\frac{dq}{dy} = \frac{dp}{dx} = \text{a constant}$$

and so

$$q = y\frac{dp}{dx}. \tag{2.126}$$

2.13.5.   Lubrication theory by Osborne Reynolds

Reynolds' theory is based on experimental observations demonstrated by Tower in 1885. These experiments showed the existence of fluid pressure within the oil film which reached a maximum value far in excess of the mean pressure on the bearing. The more viscous the lubricant the greater was the friction and the load carried. It was further observed that the wear of

properly lubricated bearings is very small and is almost negligible. On the basis of these observations Reynolds drew the following conclusions:
(i) friction is due to shearing of the lubricant;
(ii) viscosity governs the load carrying capacity as well as friction;
(iii) the bearing is entirely supported by the oil film.
He assumed the film thickness to be such as to justify its treatment by the theory of viscous flow, taking the bearing to be of infinite length and the coefficient of viscosity of the oil as constant. Let

$r =$ the radius of the journal,
$f =$ the virtual coefficient of friction,
$F =$ the tangential resisting force at radius, r,
$P =$ the total load carried by the bearing.

Then

frictional moment, $M = Fr = fPr$. $\qquad$ (2.127)

Again, if

$A =$ the area wetted by the lubricant,
$V =$ the peripheral velocity of the journal,
$c =$ the clearance between the bearing and the shaft, when the shaft is placed centrally,

then using eqn (2.121)

$$F = \mu A \left(\frac{V}{c} \right) = fP$$

and

$$M = \mu r A \left(\frac{V}{c} \right). \qquad (2.128)$$

This result, given by Petroff in 1883, was the first attempt to relate bearing friction with the viscosity of the lubricant.

In 1886 Osborne Reynolds, without any knowledge of the work of Petroff, published his treatise, which gave a deeper insight into the hydrodynamic theory of lubrication. Reynolds recognized that the journal cannot take up a central position in the bearing, but must so find a position according to its speed and load, that the conditions for equilibrium are satisfied. At high speeds the eccentricity of the journal in the bearing decreases, but at low speeds it increases. Theoretically the journal takes up a position, such that the point of nearest approach of the surfaces is in advance of the point of maximum pressure, measured in the direction of rotation. Thus the lubricant, after being under pressure, has to force its way through the narrow gap between the journal and the bearing, so that friction is increased. Two particular cases of the Reynolds theory will be discussed separately.

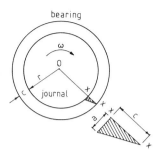

bearing

journal

Figure 2.21

2.13.6. High-speed unloaded journal

Here it can be assumed that eccentricity is zero, i.e. the journal is placed centrally when rotating, and the fluid film is in a state of simple shear. The load P, the tangential resisting force, F and the frictional moment, M are measured per unit length of the bearing.

Referring to Fig. 2.21, ω is the angular velocity of the journal, so that

$$V = \omega r$$

$$\text{tangential drag per unit area at radius, } r = q = \mu \frac{\mathrm{d}\phi}{\mathrm{d}t},$$

where

$$\mathrm{d}\phi/\mathrm{d}t = \text{ratio of shearing} = V/c,$$

$$\text{shear stress at radius, } r = \mu\frac{V}{c},$$

$$\text{tangential resisting force } F = \frac{AV}{c} = 2\pi r\mu\frac{V}{c},$$

$$\text{frictional moment, } M = 2\pi r^2\mu\frac{V}{c},$$

$$\text{power absorbed} = 2\pi r\mu\frac{V^2}{c}. \tag{2.129}$$

This result may be obtained directly from the Petroff eqn (2.128). Theoretically, the intensity of normal pressure on the journal is uniform, so that the load carried must be zero. It should be noted that the shaded area in Fig. 2.21 represents the volume of lubricant passing the section X–X in time δt, where $a = V\delta t$, so that

$$\text{volume of lubricant passing per second per unit length} = \tfrac{1}{2}Vc. \tag{2.130}$$

The effect of the load is to produce an eccentricity of the journal in the bearing and a pressure gradient in the film. The amount of eccentricity is determined by the condition, that the resultant of the fluid action on the surface of the journal, must be equal and opposite to the load carried.

2.13.7. Equilibrium conditions in a loaded bearing

Figure 2.22 shows a journal carrying a load P per unit length of the bearing acting vertically downwards through the centre O. If O' is the centre of the bearing then it follows from the conditions of equilibrium that the eccentricity OO' is always perpendicular to the line of action of P. The journal is in equilibrium under the load P, acting through O, the normal pressure intensity, p, the friction force, q, per unit area and an externally applied couple, M', per unit length equal and opposite to the frictional

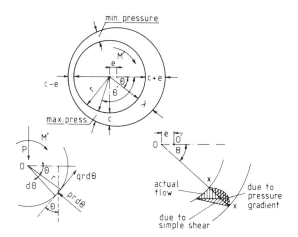

Figure 2.22

moment M. Resolving p and q parallel and perpendicular to P respectively, the equations of equilibrium become

$$\int_0^{2\pi} pr \sin \Theta \, d\Theta + \int_0^{2\pi} qr \cos \Theta \, d\Theta = P, \tag{2.131}$$

$$\int_0^{2\pi} pr \cos \Theta \, d\Theta - \int_0^{2\pi} qr \sin \Theta \, d\Theta = O \tag{2.132}$$

and

$$\int_0^{2\pi} qr^2 \, d\Theta = M'. \tag{2.133}$$

These equations are similar to those used in determining the frictional moment of the curved brake shoe to be discussed later. The evaluation of the integrals in any particular case depends upon the variation of p and q with respect to Θ.

2.13.8.   Loaded high-speed journal

In a film of uniform thickness and in simple shear, the pressure gradient is zero. When the journal is placed eccentrically, the wedge-like character of the film introduces a pressure gradient, and the flow across any section then depends upon pressure changes in the direction of flow in addition to the simple shearing action. Let

$c =$ the clearance between bearing and journal,
$r + c =$ the radius of the bearing,
$e =$ the eccentricity when under load,
$c - e =$ the film thickness at the point of nearest approach,
$c + e =$ the maximum film thickness,
$\lambda =$ the film thickness at a section X–X.

Referring to Fig. 2.22

$$r + \lambda = r + c + e \cos \Theta \qquad \text{(approximately)},$$

i.e.

$$\lambda = c + e \cos \Theta. \tag{2.134}$$

The flow across section $X - X$ is then

$$\tfrac{1}{2} V \lambda \qquad \text{(due to simple shear)}$$

$$-\frac{1}{\mu} \frac{dp}{dx} \frac{\lambda^3}{12} \qquad \text{(due to pressure gradient)}$$

and writing $x = r\Theta$

$$\text{flow across } X - X = Q = \tfrac{1}{2} V \lambda - \frac{1}{r\mu} \frac{dp}{d\Theta} \frac{\lambda^3}{12}. \tag{2.135}$$

For continuity of flow, Q must be the same for all sections. Suppose Θ' is the value of Θ at which maximum pressure occurs. At this section $dp/d\Theta = 0$ and the film is in simple shear, so that

$$Q = \tfrac{1}{2} V \lambda = \tfrac{1}{2} V (c + e \cos \Theta').$$

Equating these values of Q

$$\tfrac{1}{2} v (c + e \cos \Theta') = \tfrac{1}{2} V (c + e \cos \Theta) - \frac{1}{r\mu} \frac{dp}{d\Theta} \frac{\lambda^3}{12}$$

so that

$$\frac{dp}{d\Theta} = \frac{6\mu V r e}{\lambda^3} (\cos \Theta - \cos \Theta'). \tag{2.136}$$

Again, taking into account the results obtained earlier, the friction force, q per unit area at the surface of the journal is

$$\frac{\mu V}{\lambda} \qquad \text{(due to simple shear)},$$

$$y \frac{dp}{dx} = \frac{\lambda}{2r} \frac{dp}{d\Theta} \qquad \text{(due to pressure gradient)},$$

hence

$$q = \frac{\mu V}{\lambda} + \frac{\lambda}{2r} \frac{dp}{d\Theta}.$$

Substituting for $dp/d\Theta$, this becomes

$$q = \frac{\mu V}{\lambda} \left[1 + \frac{3e}{\lambda} (\cos \Theta - \cos \Theta') \right]. \tag{2.137}$$

These expressions for q and $dp/d\Theta$, together with the conditions of equilibrium, are the basis of the theory of fluid-film lubrication or hydrodynamic lubrication.

The solution of the indefinite integral $\int_0^\theta (\mathrm{d}p/\mathrm{d}\Theta)\,\mathrm{d}\Theta$ is given in an abbreviated form as follows: using the method of substitution, integrate eqn (2.136) in terms of a new variable ϕ such that

$$\lambda = c + e\cos\Theta = \frac{c^2 - e^2}{c - e\cos\phi},$$

where

$$\frac{\mathrm{d}\Theta}{\mathrm{d}\phi} = \frac{\sqrt{(c^2 - e^2)}}{c - e\cos\phi} = \frac{\lambda}{\sqrt{c^2 - e^2}},$$

$$\sin\phi = \sqrt{(c^2 - e^2)}\,\frac{\sin\Theta}{\lambda},$$

$$\cos\phi = \frac{c\cos\Theta + e}{\lambda},$$

hence

$$\frac{\sqrt{(c^2 - e^2)}}{6\mu Vr}\frac{\mathrm{d}p}{\mathrm{d}\phi} = \frac{1}{\lambda} - \frac{1}{\lambda^2}(c + e\cos\Theta'). \tag{2.138}$$

This equation can readily be integrated in terms of ϕ. Further, since the pressure equation must give the same value of p when $\Theta = 0$ or $\Theta = 2\pi$, i.e. when $\phi = 0$ or $\phi = 2\pi$, the sum of the terms involving ϕ must vanish. This condition determines the angle Θ' and leads to the result

$$\cos\Theta' = -\frac{3ec}{2c^2 + e^2}. \tag{2.139}$$

The integral then becomes

$$\frac{(p - p_0)(c^2 - e^2)^{\frac{3}{2}}(2c^2 + e^2)}{6\mu Vr} = e\sin\phi(2c^2 - e^2 - ce\cos\phi).$$

Substituting for $\sin\phi$ and $\cos\phi$ in terms of Θ, the solution becomes

$$\frac{(p - p_0)(2c^2 + e^2)}{6\mu Vr} = e\sin\Theta\frac{\lambda + c}{\lambda^2}$$

or

$$p = p_0 + \left(\frac{\mu Vr}{c^2}\right)k, \tag{2.140}$$

where

$$k = -2c\frac{\lambda + c}{\lambda^2}\cos\Theta'\sin\Theta. \tag{2.141}$$

In these equations p_0 is the arbitrary uniform pressure of simple shear. The constant $\varepsilon = e/c$ is called the attitude of the journal, so that

$$\frac{\lambda}{c} = 1 + \varepsilon \cos \Theta \tag{2.142}$$

The variation of p around the circumference, for the value of $\varepsilon = e/c = 0.2$, is very close to the sine curve. For small values of e/c we can write $\lambda = c$ and $\cos \Theta' = -(3/2)(e/c)$, so that $k = 6(e/c) \sin \Theta$ and the pressure closely follows the sine law

$$p = p_0 + \frac{\mu V r}{c^2} 6 \frac{e}{c} \sin \Theta. \tag{2.143}$$

For the value of $e/c = 0.7$, maximum pressure occurs at the angle $\Theta' = 147.2°$ and $k_{max} = 7.62$. Also at an angle $\Theta' = 212.8°$, the pressure is minimum and, if p_0 is small, the pressure in the upper half of the film may fall below the atmospheric pressure. It is usual in practice to supply oil under slight pressure at a point near the top of the journal, appropriate to the assumed value of e/c. This ensures that p_{min} shall have a small positive value and prevents the possibility of air inclusion in the film and subsequent cavitation.

2.13.9. Equilibrium equations for loaded high-speed journal

Referring now to the equilibrium equations discussed earlier, the uniform pressure p_0 will have no effect upon the value of the load P and many be neglected. In addition it can be shown that the effect of the tangential drag or shear stress, q, upon the load is very small when compared with that of the normal pressure intensity, p, and therefore may also be neglected. The error involved is of the order c/r, i.e. less than 0.1 per cent. Hence

$$P = \int_0^{2\pi} pr \sin \Theta \, d\Theta, \tag{2.144}$$

$$M' = \int_0^{2\pi} qr^2 \, d\Theta, \tag{2.145}$$

where

$$p = p_0 - \left(\frac{\mu V r}{c^2}\right) 2c \frac{\lambda + c}{\lambda^2} \cos \Theta' \sin \Theta$$

and

$$q = \frac{\mu V}{\lambda} \left[1 + \frac{3e}{\lambda} (\cos \Theta - \cos \Theta') \right].$$

The integrals arising from $pr \cos \Theta$ and $qr \sin \Theta$ in eqn (2.132) will vanish separately, proving P is the resultant load on the journal, and that the eccentricity e is perpendicular to the line of action of P. The remaining integrals required for the determination of P and M' are tabulated below

film thickness, $\lambda = c + e \cos \Theta$,

where e is the eccentricity and c is the radial clearance.

$$I_1 = \int_0^{2\pi} \frac{d\Theta}{\lambda} = \frac{2\pi}{\sqrt{c^2 - e^2}},$$

$$I_2 = \int_0^{2\pi} \frac{d\Theta}{\lambda^2} = \frac{2\pi c}{(c^2 - e^2)^{\frac{3}{2}}},$$

$$I_3 = \int_0^{2\pi} \frac{\cos \Theta \, d\Theta}{\lambda^2} = -\frac{2\pi e}{(c^2 - e^2)^{\frac{3}{2}}},$$

$$I_4 = \int_0^{2\pi} \frac{\sin^2 \Theta \, d\Theta}{\lambda} = \frac{2\pi}{e^2}[c - \sqrt{(c^2 - e^2)}],$$

$$I_5 = \int_0^{2\pi} \frac{\sin^2 \Theta \, d\Theta}{\lambda^2} = \frac{2\pi}{e^2}\left[\frac{c}{\sqrt{(c^2 - e^2)}} - 1\right].$$

Using the given notation and substituting for p in eqn (2.144), the load per unit length of journal is given by

$$P = -\frac{\mu V r^2}{c^2} 2c \cos \Theta' (I_4 + c I_5)$$

$$= -\frac{\mu V r^2}{c^2} 2c \cos \Theta' \frac{2\pi}{\sqrt{(c^2 - e^2)}}.$$

Writing $\cos \Theta' = -3ec/(2c^2 + e^2)$ and $e/c = \varepsilon$, this becomes

$$P = \mu V r^2 \frac{12\pi e}{(2c^2 + e^2)\sqrt{(c^2 - e^2)}} \tag{2.146}$$

or

$$P = \frac{\mu V r^2}{c^2} \frac{12\pi\varepsilon}{(2 + \varepsilon^2)\sqrt{1 - \varepsilon^2}} \tag{2.147}$$

Proceeding in a similar manner the applied couple M' becomes

$$M' = \mu V r^2 (I_1 + 3eI_3 - 3e \cos \Theta' I_2)$$

$$= \mu V r^2 \frac{2\pi}{(c^2 - e^2)^{\frac{3}{2}}} (c^2 - 4e^2 - 3ec \cos \Theta')$$

$$= \mu V r^2 \frac{4\pi}{\sqrt{(c^2 - e^2)}} \frac{c^2 + 2e^2}{2c^2 + e^2} \tag{2.148}$$

or

$$M' = \frac{\mu V r^2}{c} \frac{4\pi}{\sqrt{(1 - \varepsilon^2)}} \frac{1 + 2\varepsilon^2}{2 + \varepsilon^2}. \tag{2.149}$$

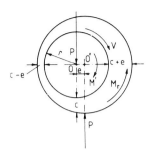

Figure 2.23

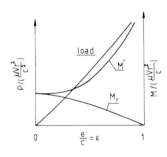

Figure 2.24

2.13.10. Reaction torque acting on the bearing

The journal and the bearing as a whole are in equilibrium under the action of a downward force P at the centre of the journal, an upward reaction P through the centre of the bearing, the externally applied couple M' and a reaction torque on the bearing, M_r acting in opposite sense to M' as shown in Fig. 2.23. For equilibrium, it follows that

$$M' = M_r + Pe$$

or

$$M_r = M' - Pe.$$

Substituting for M' and P from eqns (2.146) and (2.148), the reaction torque (on the bearing) is given by

$$M_r = \mu V r^2 \frac{4\pi\sqrt{(c^2 - e^2)}}{2c^2 + e^2} \tag{2.150}$$

$$= \frac{\mu V r^2}{c} \frac{4\pi\sqrt{(1 - \varepsilon^2)}}{2 + \varepsilon^2}. \tag{2.151}$$

The variation of the load P together with the applied and reaction torques, for varying values of $\varepsilon = e/c$ is shown in Fig. 2.24. It should be noted that the friction torque on the journal is equal and opposite to M'. Similarly, the friction torque on the bearing is equal and opposite to M_r. Theoretically, when $e/c = 0$, P is zero and $M' = M_r$. Alternatively, when e/c approaches unity, M' and P approach infinity and M_r tends to zero, since the flow of the lubricant is prevented by direct contact of the bearing surfaces. It will be remembered, however, that this condition is one of boundary lubrication and the foregoing theory no longer applies.

2.13.11. The virtual coefficient of friction

If f is the virtual coefficient of friction for the journal under a load, P per unit length, the frictional moment, M per unit length is given by eqn (2.127),

$$M = fPr.$$

The magnitudes of M and P are also given by eqns (2.146) and (2.148), so that

$$fr = \frac{M'}{P} = \frac{1}{3e}(c^2 + 2e^2)$$

or putting $\varepsilon = e/c$

$$\frac{fr}{c} = \frac{1 + 2\varepsilon^2}{3\varepsilon} = \frac{1}{3\varepsilon} + \frac{2\varepsilon}{3}. \tag{2.152}$$

Differentiating with respect to ε and equating to zero, the minimum value of

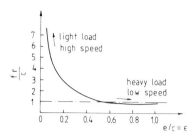

Figure 2.25

fr/c occurs when $\varepsilon^2 = \frac{1}{2}$ or

$$\varepsilon = e/c = \frac{1}{\sqrt{2}} = 0.707 \qquad (2.153)$$

and the virtual coefficient of friction is then given by

$$f_{min} = \frac{2\sqrt{2}}{3}\frac{c}{r}. \qquad (2.154)$$

The graph of Fig. 2.25 shows the variation of fr/c for varying values of e/c. The value of f when $\varepsilon = e/c = 0.7$ occurs at the transition point from film to boundary lubrication, and at this point f may reach an abnormally low value. Thus, if $c/r = 1/1000$; $f_{min} = 0.0009$. It should be noted that this value of f_{min} relates to the journal. If f_r is the virtual coefficient of friction resulting from the reaction torque M_r on the bearing, then

$$f_r r = \frac{Mr}{P} = \frac{1}{3e}(c^2 - e^2),$$

or

$$f_r\frac{r}{c} = \frac{1}{3\varepsilon} - \frac{\varepsilon}{3}, \qquad (2.155)$$

so that,

$$f - f_r = \varepsilon\frac{c}{r}. \qquad (2.156)$$

Equation (2.156) is the result of the subtraction of eqn (2.155) from eqn (2.152). For the value $\varepsilon = e/c = 0.7$ the virtual coefficient of friction for the bearing is then

$$f_r = 0.00024.$$

2.13.12. The Sommerfeld diagram

Rewriting eqn (2.147) the load per unit length of journal is

$$P = \frac{\mu V r^2}{c^2}\frac{12\pi\varepsilon}{(2 + \varepsilon^2)\sqrt{1 - \varepsilon^2}} \qquad \text{where} \quad \varepsilon = e/c.$$

Using the Sommerfeld notation

$$Z = \left(\frac{r}{c}\right)^2\frac{\mu V}{P}, \qquad (2.157)$$

where

$$Z = \frac{(2 + \varepsilon^2)\sqrt{(1 - \varepsilon^2)}}{12\pi\varepsilon} \qquad (2.158)$$

and

$$\frac{fr}{c} = \frac{1}{3\varepsilon} + \frac{2\varepsilon}{3}.$$ (2.159)

If Z is plotted against fr/c, the diagram shown in Fig. 2.26 results. Line OA represents the Petroff line and is given by

$$\frac{fr}{c} = 2\pi \left(\frac{r}{c}\right)^2 \frac{\mu V}{P} = 2\pi Z$$

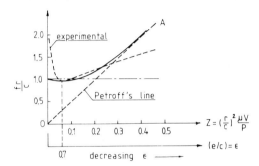

Figure 2.26

for the transition point where $f_{\min}$ occurs, i.e. $\varepsilon = 1/\sqrt{2}$, $Z = 5/24\pi$. The theoretical curve closely follows the experimental curve for values of $\varepsilon = e/c$ from 0.25 to 0.7. For smaller values of e/c (approaching high-speed conditions) the experimental curve continues less steeply. This is explained by the rise in temperature and the decrease in viscosity of the lubricant, so that the increase of frictional moment is less than that indicated by the theoretical curve.

Alternatively, for values of $e/c > 0.7$, the experimental curve rises steeply and fr/c ultimately attains a value corresponding to static conditions. The theory indicates that, although M and P both approach infinity, the ratio $fr/c = M/Pc$ approaches unity.

It must be remembered, however, that Reynolds assumed μ to be constant for all values of e/c, whereas for most lubricants μ increases strongly with pressure. It follows, therefore, that μ is a variable increasing with e/c and varying also within the film itself. This variation results in a tilting of the theoretical curve as shown by the experimental curve. The generally accepted view, however, is that the rapidly increasing value of fr/c under heavy load and low speed, is due to the interactions of surface irregularities, when the film thickness becomes very small.

The conclusion is, that, so long as μ remains constant and the hydrodynamic lubrication conditions are fulfilled, the virtual coefficient of friction is independent of the properties of the lubricant and depends only upon the value of e/c, and the clearance and radius of the journal. For design calculations a value of e/c somewhat less than that corresponding to

f_{min} is recommended. Thus, if $\varepsilon = e/c = 0.4$ and $\Theta' = 123.7°$, then

$$\frac{fr}{c} = \frac{1}{3\varepsilon} + \frac{2\varepsilon}{3} = 1.1, \qquad (2.160)$$

$$\frac{\mu V}{P}\left(\frac{r}{c}\right)^2 = Z = \frac{(2+\varepsilon^2)\sqrt{(1-\varepsilon^2)}}{12\pi\varepsilon} = \frac{0.4125}{\pi} = 0.1312,$$

so that

$$P = 7.616\mu V\left(\frac{r}{c}\right)^2$$

and

$$M = fPr = 8.378\mu V\frac{r^2}{c}.$$

If p' denotes the load per unit of projected area of the bearing surface and N is the speed in r.p.s. then $P = 2p'r$ and $V = 2\pi rN$, so that

$$Z = \frac{\mu N}{p'}\pi\left(\frac{r}{c}\right)^2$$

and

$$p' = 23.93\mu N\left(\frac{r}{c}\right)^2.$$

In using these expressions care must be taken to ensure that the units are consistent.

Numerical example

A journal and complete bearing, of nominal diameter 100 mm and length 100 mm, operates with a clearance of $c = 0.05$ mm. The speed of rotation is 600 r.p.m. and $\mu = 20$ cP. Determine:
  (i) the frictional moment when rotating without load;
 (ii) the maximum permissible load and specific pressure (i.e. load per unit of projected area of bearing surface) knowing that the eccentricity ratio e/c must not exceed 0.4;
(iii) the frictional moment under load.

Solution

The virtual coefficient of friction when $e/c = 0.4$ is $1.1(c/r)$ and the load per unit length is

$$P = 7.62\mu V\left(\frac{r}{c}\right)^2.$$

 (i) when rotating without load, eqn (2.128) will apply, namely

$$M = 2\pi\mu V\frac{r^2}{c} \qquad \text{per unit length.}$$

To convert from centipoise to $(Ns)/m^2$ we have to multiply $\mu = 20\,cP$ by 10^{-3}. Thus

$$\mu = 20 \times 10^{-3}\,Nsm^{-2},$$
$$V = 2\pi rN = 2 \times 3.1416 \times 0.05 \times (600/60) = 3.1416\,m\,s^{-1},$$

$$r^2/c = \frac{(0.05)^2}{0.00005} = 50\,m,$$

and so

$$M = 2\pi 20 \times 10^{-3} \times 3.1416 \times 50 = 19.73\,Nm \text{ per meter length,}$$

but length of the bearing $= 0.1\,m$, thus

$$\text{frictional moment} = 19.73 \times 0.1 = 1.973\,Nm.$$

(ii) load per unit length

$$P = 7.62\mu V \left(\frac{r}{c}\right)^2$$

$$\frac{\mu V r^2}{c} = 20 \times 10^{-3} \times 50 = 3.1416\,N$$

and so

$$P = 7.62 \times 3.1416(1/0.00005) = 478\,779.8\,N/m$$

$$\begin{aligned}\text{maximum permissible load} &= P \times \text{length of bearing}\\ &= 478\,779.8 \times 0.1 = 47\,877.98\,N\end{aligned}$$

$$\text{specific pressure} = p' = \frac{P}{2r} = \frac{478\,779.8}{0.1} = 4.78\,MPa$$

(iii) frictional moment under load, $M = fPr$ per unit length, where

$$fr = 1.1c = 1.1 \times 0.00005 = 0.000055\,m$$
$$\text{total frictional moment} = 478\,779.8 \times 0.000055 = 2.63\,Nm.$$

References to Chapter 2

1. E. Rabinowicz. *Friction and Wear of Materials.* New York: Wiley, 1965.
2. J. Halling (ed.). *Principles of Tribology.* Macmillan Education Ltd., 1975.
3. D. F. Moore. *Principles and Applications of Tribology.* Pergamon Press, 1975.
4. N. P. Suh. *Tribophysics.* Prentice-Hall, 1986.
5. F. P. Bowden and D. Tabor. *The Friction and Lubrication of Solids.* Parts I & II. Oxford: Clarendon Press, 1950, 1964.
6. I. V. Kragelskii, *Friction and Wear.* Washington: Butterworth, 1965.

3 Elements of contact mechanics

3.1. Introduction

There is a group of machine components whose functioning depends upon rolling and sliding motion along surfaces while under load. Both surfaces are usually convex, so that the area through which the load is transferred is very small, even after some surface deformation, and the pressures and local stresses are very high. Unless logically designed for the load and life expected of it, the component may fail by early general wear or by local fatigue failure. The magnitude of the damage is a function of the materials and by the intensity of the applied load or pressure, as well as the surface finish, lubrication and relative motion.

The intensity of the load can be determined from equations which are functions of the geometry of the surfaces, essentially the radii of curvature, and the elastic constants of the materials. Large radii and smaller moduli of elasticity, give larger contact areas and lower pressures. Careful alignment, smoother surfaces, and higher strength and oil viscosity minimize failures.

In this chapter, presentation and discussion of *contact mechanics* is confined, for reasons of space, to the most technically important topics. However, a far more comprehensive treatment of contact problems in a form suitable for the practising engineer is given in the ESDU tribology series. The following items are recommended:

ESDU–78035, Contact phenomena I; stresses, deflections and contact dimensions for normally loaded unlubricated elastic components;

ESDU–84017, Contact phenomena II; stress fields and failure criteria in concentrated elastic contacts under combined normal and tangential loading;

ESDU–85007, Contact phenomena III; calculation of individual stress components in concentrated elastic contacts under combined normal and tangential loading.

Although a fairly comprehensive treatment of thermal effects in surface contacts is given here it is appropriate, however, to mention the ESDU tribology series where thermal aspects of bearings, treated as a system are presented, and network theory is employed in an easy to follow step-by-step procedure. The following items are esentially recommended for the practising designer:

ESDU–78026, Equilibrium temperatures in self-contained bearing assemblies;

Part I – outline of method of estimation;

ESDU-78027, Part II – first approximation to temperature rise;
ESDU-78028, Part III – estimation of thermal resistance of an assembly;
ESDU-78029, Part IV – heat transfer coefficient and joint conductance.

Throughout this chapter, references are made to the appropriate ESDU item number, in order to supplement information on contact mechanics and thermal effects, offer alternative approach or simply to point out the source of technical data required to carry out certain analysis.

3.2. Concentrated and distributed forces on plane surfaces

The theory of contact stresses and deformations is one of the more difficult topics in the theory of elasticity. The usual approach is to start with forces applied to the plane boundaries of semi-infinite bodies, i.e. bodies which extend indefinitely in all directions on one side of the plane. Theoretically this means that the stresses which radiate away from the applied forces and die out rapidly are unaffected by any stresses from reaction forces or moments elsewhere on the body.

A concentrated force acts at point O in case 1 of Table 3.1. At any point Q there is a resultant stress q on a plane perpendicular to OZ, directed through O and of magnitude inversely proportional to $(r^2 + z^2)$, or the

Table 3.1

Loading case	Pictorial	Stress and deflection
1. Point		$q = \sqrt{\sigma_z^2 + \tau_{rz}^2} \quad \dfrac{3}{2\pi} \dfrac{P\cos^2\theta}{(r^2 + z^2)}$ $w = \dfrac{1-v^2}{\pi E} \dfrac{P}{r}$ at surface
2. Line		$\sigma = -\dfrac{2}{\pi} \dfrac{(P/l)\cos\theta}{\sqrt{x^2 + z^2}}$
3. Knife edge or pivot		$\sigma_r = -\dfrac{(P/L)\cos\theta}{r(\alpha + \frac{1}{2}\sin 2\alpha)}$
4. Uniform distributed load p over circle of radius a		with $v = 0.3$ at point O $\sigma_z = -p, \sigma_r = \sigma_\theta = -0.8p$ $w_m = \dfrac{2(1-v^2)pa}{E} = 2\eta pa$ $T_m = 0.33p$ at $z = 0.638a$
5. Rigid cylinder ($E_1 \gg E_2$)		$(\sigma_z)_{z=0} = -p = -\dfrac{P}{2\pi a\sqrt{a^2 - r^2}}$ $w = \dfrac{(1-v^2)P}{2E_2 a} = \dfrac{\eta_2 P}{2a}$

square of the distance OQ from the point of load application. This is an indication of the rate at which stresses die out. The deflection of the surface at a radial distance r is inversely proportional to r, and hence, is a hyperbola asymptotic to axes OR and OZ. At the origin, the stresses and deflections theoretically become infinite, and one must imagine the material near O cut out, say, by a small hemispherical surface to which are applied distributed forces that are statistically equivalent to the concentrated force P. Such a surface is obtained by the yielding of the material.

An analogous case is that of concentrated loading along a line of length l (case 2). Here, the force is P/l per unit length of the line. The result is a normal stress directed through the origin and inversely proportional to the first power of distance to the load, not fading out as rapidly. Again, the stress approaches infinite values near the load. Yielding, followed by work-hardening, may limit the damage. Stresses in a knife or wedge, which might be used to apply the foregoing load, are given under case 3. The solution for case 2 is obtained when $2\alpha = \pi$, or when the wedge becomes a plane.

In the deflection equation of case 1, we may substitute for the force P, an expression that is the product of a pressure p, and an elemental area, such as the shaded area in Fig. 3.1. This gives a deflection at any point, M, on the surface at a distance $r = s$ away from the element, namely

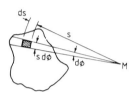

Figure 3.1

$$dw = \frac{(1-v^2)}{\pi E}\frac{p(s\,d\phi)\,ds}{s} = \frac{(1-v^2)p\,ds\,d\phi}{\pi E},$$

where v is the Poisson ratio. The total deflection at M is the superposition or integration over the loaded area of all the elemental deflections, namely

$$w = \frac{(1-v^2)}{\pi E}\int\int p\,ds\,d\phi = \frac{\eta}{\pi}\int\int p\,ds\,d\phi, \tag{3.1}$$

where η is an elastic constant $(1-v^2)/E$. If the pressure is considered uniform, as from a fluid, and the loaded area is a circle, the resulting deflections, in terms of elliptic integrals, are given by two equations, one for M outside the circle and one for M inside the circle. The deflections at the centre are given under case 4 of Table 3.1. The stresses are also obtained by a superposition of elemental stresses for point loading. Shear stress is at a maximum below the surface.

If a rod in the form of a punch, die or structural column is pressed against the surface of a relatively soft material, i.e. one with a modulus of elasticity much less than that of the rod, the rod may be considered rigid, and the distribution of deflection is initially known. For a circular section, with deflection w constant over the circle, the results are listed in case 5. The pressure p is least at the centre, where it is $0.5p_{\text{avg}}$, and it is infinite at the edges. The resultant yielding at the edges is local and has little effect on the general distribution of pressure. For a given total load, the deflection is inversely proportional to the radius of the circle.

3.3. Contact between two elastic bodies in the form of spheres

When two elastic bodies with convex surfaces, or one convex and one plane surface, or one convex and one concave surface, are brought together in point or line contact and then loaded, local deformation will occur, and the point or line will enlarge into a surface of contact. In general, its area is bounded by an ellipse, which becomes a circle when the contacting bodies are spheres, and a narrow rectangle when they are cylinders with parallel axes. These cases differ from those of the preceding section in that there are two elastic members, and the pressure between them must be determined from their geometry and elastic properties.

The solutions for deformation, area of contact, pressure distribution and stresses at the initial point of contact were made by Hertz. They are presented in ESDU 78035 in a form suitable for engineering application. The maximum compressive stress, acting normal to the surface is equal and opposite to the maximum pressure, and this is frequently called the Hertz stress. The assumption is made that the dimensions of the contact area are small, relative to the radii of curvature and to the overall dimensions of the bodies. Thus the radii, though varying, may be taken as constant over the very small arcs subtending the contact area. Also, the deflection integral derived for a plane surface, eqn (3.1), may be used with very minor error. This makes the stresses and their distribution the same in both contacting bodies.

The methods of solution will be illustrated by the case of two spheres of different material and radii R_1 and R_2. Figure 3.2 shows the spheres before and after loading, with the radius a of the contact area greatly exaggerated for clarity. Distance $z = R - R \cos \gamma \approx R - R(1 - \gamma^2/2 + \cdots) \approx R\gamma^2 2 \approx r^2/2R$ because $\cos \gamma$ may be expanded in series and the small angle $\gamma \approx r/R$. If points M_1 and M_2 in Fig. 3.2 fall within the contact area, their approach distance $M_1 M_2$ is

$$z_1 + z_2 = \frac{r^2}{2}\left(\frac{1}{R_1} + \frac{1}{R_2}\right) = Br^2, \tag{3.2}$$

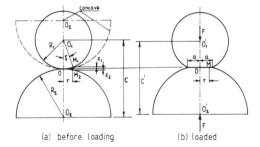

Figure 3.2   (a) before loading   (b) loaded

where B is a constant $(1/2)(1/R_1 + 1/R_2)$. If one surface is concave, as indicated by the dotted line in Fig. 3.2, the distance is $z_1 - z_2 = (r^2/2)(1/R_1 - 1/R_2)$ which indicates that when the contact area is on the inside of a surface the numerical value of its radius is to be taken as negative in all equations derived from eqn (3.1).

The approach between two relatively distant and strain-free points, such as Q_1 and Q_2, consists not only of the surface effect $z_1 + z_2$, but also of the approach of Q_1 and Q_2 relative to M_1 and M_2, respectively, which are the deformations w_1 and w_2 due to the, as yet, undetermined pressure over the contact area. The total approach or deflection δ, with substitution from eqn (3.1) and (3.2), is

$$\delta = (z_1 + z_2) + (w_1 + w_2) = Br^2 + (1/\pi)(\eta_1 + \eta_2) \int \int p \, ds \, d\phi,$$

where

$$\eta_1 = (1 - v_1^2)/E_1 \qquad \text{and} \qquad \eta_2 = (1 - v_2^2)/E_2.$$

With rearrangement,

$$\frac{\eta_1 + \eta_2}{\pi} \int \int p \, ds \, d\phi = \delta - Br^2. \tag{3.3}$$

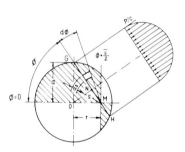

Figure 3.3

For symmetry, the area of contact must be bounded by a circle, say of radius a, and Fig. 3.3 is a special case of Fig. 3.1. A trial will show that eqn (3.3) will be satisfied by a hemispherical pressure distribution over the circular area. Thus the peak pressure at centre O is proportional to the radius a, or $p_0 = ca$. Then, the scale for plotting pressure is $c = p_0/a$. To find w_1 and w_2 at M in eqn (3.3), an integration, $p \, ds$, must first be made along a chord GH, which has the half-length $GN = (a^2 - r^2 \sin^2 \phi)^{\frac{1}{2}}$. The pressure varies as a semicircle along this chord, and the integral equals the pressure scale c times the area A under the semicircle, or

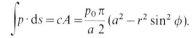

$$\int p \cdot ds = cA = \frac{p_0 \pi}{a \, 2} (a^2 - r^2 \sin^2 \phi).$$

By a rotation of line GH about M from $\phi = 0$ to $\phi = \pi/2$ (half of the contact circle), the shaded area of Fig. 3.3, is covered. Doubling the integral completes the integration in eqn (3.3), namely

$$\frac{p_0(\eta_1 + \eta_2)}{a} \int_0^{\frac{\pi}{2}} (a^2 - r^2 \sin^2 \phi) \, d\phi = \delta - Br^2,$$

whence

$$\frac{p_0 \pi}{4a} (\eta_1 + \eta_2)(2a^2 - r^2) = \delta - Br^2. \tag{3.4}$$

Now the approach δ of centres Q_1 and Q_2, is independent of the particular points M and radius r, chosen in the representation by which eqn (3.4) was obtained. To make the equation independent of r, the two r^2 terms must be equal, whence it follows that the two constant terms are equal. The r^2 terms, equated and solved for a, yield the radius of the contact area

$$a = \frac{p_0 \pi (\eta_1 + \eta_2)}{4B}. \tag{3.5}$$

The two constant terms when equated give

$$\delta = \frac{p_0 \pi (\eta_1 + \eta_2) a}{2}. \tag{3.6}$$

The integral of the pressure over the contact area is equal to the force P by which the spheres are pressed together. This integral is the pressure

Table 3.2

Loading case	Pictorial	Area, Pressure, Approach
1. Spheres or sphere and plane		$a = 0.721[P(\eta_1 + \eta_2)D_1 D_2/(D_1 + D_2)]^{\frac{1}{3}} = c/2$ $P_0 = 1.5P/\pi a^2 = 1.5 p_{avg} = -(\sigma_c)_{max}$ $\max \tau = \frac{1}{3}p_0$, at depth $0.638a$ $\max \sigma_t = (1 - 2v)p_0/3$ at radius a $\delta = 1.04[(\eta_1 + \eta_2)^2 P^2 (D_1 + D_2)/D_1 D_2]^{\frac{1}{3}}$
2. Cylindrical surfaces with parallel axes		$b = 1.13\sqrt{(P/l)(\eta_1 + \eta_2)R_1 R_2/(R_1 + R_2)} = c/2$ $p_0 = 2P/\pi bl = 1.273 p_{avg} = -(\sigma_c)_{max}$ if $v = 0.30$ $\max \tau = 0.304 p_0$ at depth $0.786b$ if $\eta_1 = \eta_2 = \eta$ $\delta = 0.638(P/l)\eta \left[\frac{2}{3} + \ln \frac{2R_1}{b} + \ln \frac{2R_2}{b} \right]$
3. General case		$b = \beta \left[\dfrac{3P(\eta_1 + \eta_2)}{4(B + A)} \right]^{\frac{1}{3}}$ and $a = b/k$ $B + A = \frac{1}{2}\left[\dfrac{1}{R_1} + \dfrac{1}{R_1'} + \dfrac{1}{R_2} + \dfrac{1}{R_2'} \right]$ $B - A = \frac{1}{2}\left[\left(\dfrac{1}{R_1} - \dfrac{1}{R_1'}\right)^2 + \left(\dfrac{1}{R_2} - \dfrac{1}{R_2'}\right)^2 \right.$ $\left. + 2\left(\dfrac{1}{R_1} - \dfrac{1}{R_1'}\right)\left(\dfrac{1}{R_2} - \dfrac{1}{R_2'}\right) \cos 2\psi \right]^{\frac{1}{2}}$ at $x = y = z = 0$ $p_0 = 1.5P/\pi ab = 1.5 p_{avg} = -(\sigma_c)_{max}$ $\sigma_x = -2vp_0 - (1 - 2v)\, p_0 \dfrac{b}{a + b}$ $\sigma_y = -2vp_0 - (1 - 2v)p_0 \dfrac{a}{a + b}$ $\delta = \lambda[P^2(\eta_1 + \eta_2)^2(B + A)]^{\frac{1}{3}}$

β, κ, λ are constants and obtained from appropriate diagrams

$$\eta_1 = \frac{1 - v_1^2}{E_1} \quad \text{and} \quad \eta_2 = \frac{1 - v_2^2}{E_2}$$

scale times the volume under the hemispherical pressure plot, or

$$\frac{p_0}{a}(\tfrac{2}{3}\pi a^3) = P$$

and the peak pressure has the value

$$p_0 = 1.5P/\pi a^2 = 1.5p_{\text{avg}}. \tag{3.7}$$

Substitution of eqn (3.7) and the value of B below eqn (3.2) gives to eqns (3.5) and (3.6) the forms shown for case 1 of Table 3.2. If both spheres have the same elastic modulus $E_1 = E_2 = E$, and the Poisson ratio is 0.30, a simplified set of equations is obtained. With a ball on a plane surface, $R_2 = \infty$, and with a ball in a concave spherical seat, R_2 is negative.

It has taken all this just to obtain the pressure distribution on the surfaces. All stresses can now be found by the superposition or integration of those obtained for a concentrated force acting on a semi-infinite body. Some results are given under case 1 of Table 3.2. An unusual but not unexpected result is that pressures, stresses and deflections are not linear functions of load P, but rather increase at a less rapid rate than P. This is because of the increase of the contact or supporting area as the load increases. Pressures, stresses and deflections from several different loads cannot be superimposed because they are non-linear with load.

3.4. Contact between cylinders and between bodies of general shape

Equations for cylinders with parallel axes may be derived directly, as shown for spheres in Section 3.3. The contact area is a rectangle of width $2b$ and length l. The derivation starts with the stress for line contact (case 2 of Table 3.1). Some results are shown under case 2 of Table 3.2. Inspection of the equations for semiwidth b, and peak pressure p_0, indicates that both increase as the square root of load P. The equations of the table, except that given for δ, may be used for a cylinder on a plane by the substitution of infinity for R_2. The semiwidth b, for a cylinder on a plane becomes $1.13[(P/l)(\eta_1 + \eta_2)R_1]^{\frac{1}{2}}$. All normal stresses are compressive, with σ_y and σ_z equal at the surface to the contact pressure p_0. Also significant is the maximum shear stress τ_{yz}, with a value of $0.304p_0$ at a depth $0.786b$.

Case 3 of Table 3.2 pictures a more general case of two bodies, each with one major and one minor plane of curvature at the initial point of contact. Axis Z is normal to the tangent plane XY, and thus the Z axis contains the centres of the radii of curvature. The minimum and maximum radii for body 1 are R_1 and R_1', respectively, lying in planes Y_1Z and X_1Z. For body 2, they are R_2 and R_2', lying in planes Y_2Z and X_2Z, respectively. The angle between the planes with the minimum radii or between those with the maximum radii is ψ. In the case of two crossed cylinders with axes at $90°$, such as a car wheel on a rail, $\psi = 90°$ and $R_1' = R_2' = \infty$. This general case was solved by Hertz and the results may be presented in various ways. Here, two sums $(B + A)$ and $(B - A)$, obtained from the geometry and defined under case 3 of Table 3.2 are taken as the basic parameters. The area of contact is an ellipse with a minor axis $2b$ and a major axis $2a$. The distribution of pressure is that of an ellipsoid built upon these axes, and the peak pressure is

1.5 times the average value $P/\pi ab$. However, for cylinders with parallel axes, the results are not usable in this form, and the contact area is a rectangle of known length, not an ellipse. The principal stresses shown in the table occur at the centre of the contact area, where they are maximum and compressive. At the edge of the contact ellipse, the surface stresses in a radial direction (along lines through the centre of contact) become tensile. Their magnitude is considerably less than that of the maximum compressive stresses, e.g. only $0.133p_0$ with two spheres and $v = 0.30$ by an equation of case 1, Table 3.2, but the tensile stresses may have more significance in the initiation and propagation of fatigue cracks. The circumferential stress is everywhere equal to the radial stress, but of opposite sign, so there is a condition of pure shear. With the two spheres $\tau = 0.133p_0$. Forces applied tangentially to the surface, such as by friction, have a significant effect upon the nature and location of the stresses. For example, two of the three compressive principal stresses immediately behind the tangential force are changed into tensile stresses. Also, the location of the maximum shear stress moves towards the surface and may be on it when the coefficient of friction exceeds 0.10.

More information on failure criteria in contacts under combined normal and tangential loading can be found in ESDU–84017.

3.5. Failures of contacting surfaces

There are several kinds of surface failures and they differ in action and appearance. Indentation (yielding caused by excessive pressure), may constitute failure in some machine components. Non-rotating but loaded ball-bearings can be damaged in this way, particularly if vibration and therefore inertia forces are added to dead weight and static load. This may occur during shipment of machinery and vehicles on freight cars, or in devices that must stand in a ready status for infrequent and short-life operations. The phenomena is called false brinelling, named after the indentations made in the standard Brinell hardness test.

The term, *surface failure*, is used here to describe a progressive loss of quality by the surface resulting from shearing and tearing away of particles. This may be a flat spot, as when a locked wheel slides on a rail. More generally the deterioration in surface quality is distributed over an entire active surface because of a combination of sliding and rolling actions, as on gear teeth. It may occur in the presence of oil or grease, where a lubricating film is not sufficiently developed, for complete separation of the contacting surfaces. On dry surfaces, it may consist of a flaking of oxides. If pressures are moderate, surface failures may not be noticeable until loose particles develop. The surface may even become polished, with machining and grinding marks disappearing. The generation of large amounts of particles, may result from misalignments and unanticipated deflections, on only a portion of the surface provided to take the entire load. This has been observed on the teeth of gears mounted on insufficiently rigid shafts, particularly when the gear is overhung. Rapid deterioration of surface quality may occur from insufficient lubrication, as on cam shafts, or from negligence in lubrication and protection from dirt.

A type of surface failure, particularly characteristic of concentrated

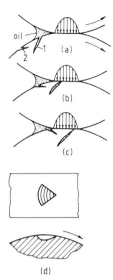

Figure 3.4

contacts, consists of fatigue cracks which progress into and under the surface, and particles which then fall out of the surface. The holes resulting from this process are called pits or spalls. This pitting occurs on convex surfaces, such as gear teeth, rolling element bearings and cams. It is a well-established fact that the maximum shearing stress occurs below the surfaces of bodies which are in contact. Hence, at one time, it was strongly held that the crack forming a pit started at this point of maximum shear stress, then progressed outwards. Data from pure rolling tests disclosed, however, that the cracks commonly started at the surface and progressed only in the presence of oil. A good penetrant, filling any fine cracks present, acted as a hydraulic wedge. Experiments also revealed that only cracks with their lips facing the approaching load would progress to failure.

In Fig. 3.4, a crack, 1, filled with oil, approaches the loading zone and has its lip sealed off. As the full length of the crack comes under the load, oil in the crack cannot escape, and high hydraulic pressure results. After repeated occurrence of this process, high stress from stress concentration along the root of the crack to spread by fatigue. Eventually, the crack will progress towards the surface, favouring the most highly stressed regions. Then, a particle will fall out, exposing a pit with the typical lines of progressive cracking, radiating from the pointed lip. The pit may look much as though it were moulded from a tiny sea shell, with an arrowhead point of origin. Pit depths may vary from a few microns to about 1 mm, with lengths from two to four times their depths.

Cracks facing away from the approaching zone of loading, such as crack 2 (Fig. 3.4), will not develop into pits. The root of the crack first reaches the loaded area and the oil in the crack is squeezed out by the time its lip is sealed off. A more viscous oil reduces or eliminates pitting, either by not penetrating into fine cracks, or by forming an oil film thick enough to prevent contact between asperities.

There are several possible causes for the initial surface cracks, which only need to be microscopic or even submacroscopic. Machining and grinding are known to leave fine surface cracks, either from a tearing action or from thermal stresses. Polishing inhibits pitting, presumably by the removal of these cracks. Along the edges of spherical and elliptical contact areas a small tensile stress is present under static and pure-rolling conditions. Tangential forces caused by sliding combined with rolling, as on gear teeth, add tensile stresses to the above and to the rectangular contact area of cylinders. Surface inclusions at the tensile areas create stress concentrations and add to the chance that the repeated tensile stresses will initiate cracks. Sometimes a piece that has dropped out of a pit passes through the contact zone, making a shallow indentation probably with edge cracks. Sometimes the breaking out of material continues rapidly in a direction away from the arrowhead point of origin, increasing in width and length. It is then called spalling. Spalling occurs more often in rolling-element bearings than in gears, sometimes covering more than half the width of a bearing race. Propagation of the crack from the surface is called a point-surface origin mode of failure. There might be so-called inclusion-origin failure. Inclusions are non-metallic particles that are formed in, and not eliminated from, the melt in the refining process. They may be formed during the deoxidization of steel or by a reaction with the refractory of the container. The inclusion does not bond with the metal, so that essentially a cavity is present with a concentration of stress. The usual way to detect inclusions is by a magnetic particle method. A crack, starting at the inclusion, may propagate through the subsurface region for some distance, or the crack may head for the surface. If cracks on the surface form, further propagation may be by hydraulic action, with a final appearance similar to that from a point-surface origin. The damaged area is often large. It is well known that bearings made from vacuum-melted steel, and therefore a cleaner, more oxide-free steel, are less likely to fail and may be given higher load ratings.

There are three other types of failure which usually occur in heavily loaded roller bearings in test rigs. Geometric stress concentration occurs at the ends of a rectangular contact area, where the material is weaker without side support. Slight misalignment, shaft slope or taper error will move much of the load to one of the ends. In peeling, fatigue cracks propagate over large areas but at depths of 0.005 to 0.01 mm. This has been attributed to loss of hydrodynamic oil film, particularly when the surface finish has many asperities which are greater than the film thickness under the conditions of service. Subcase fatigue occurs on carburized elements where the loads are heavy, the core is weak and the case is thin, relative to the radii of curvature in contact. Cracks initiate and propagate below the effective case depth, and cracks break through to the surface at several places, probably from a crushing of the case due to lack of support.

3.6. Design values and procedures

Previous investigations, some of which are published, have not produced a common basis on which materials, properties, component configuration, operating conditions and theory may be combined to determine dimensions for a satisfactory life of concentrated contacts. The investigations indicate that much progress is being made, and they do furnish a guide to conditions and changes for improvement. Most surface-contact components operate satisfactorily, and their selection is often based on a nominal Hertz pressure determined from experience with a particular component and material, or a selection is made from the manufacturers tables based on tests and experience with their components. The various types of stresses, failures and their postulated causes, including those of subsurface origin, are all closely related to the maximum contact pressure calculated by the Hertz equations. If an allowable maximum Hertz pressure seems large compared with other physical properties of the particular material, it is because it is a compressive stress and the other two principal stresses are compressive. The shear stresses and tensile stresses that may initiate failures are much smaller. Also, the materials used are often hardened for maximum strength. Suggestions for changes in contact-stress components by which their load or life may be increased are:

1. larger radii or material of a lower modulus of elasticity to give larger contact area and lower stress;

2. provision for careful alignment or minimum slope by deflection of parallel surfaces, or the provision of crowned surfaces as has been done for gear teeth, bearing rollers and cam followers;
3. cleaner steels, with fewer entrapped oxides (as by vacuum melting);
4. material and treatments to give higher hardness and strength at and near the surface, and if carburized, a sufficient case depth (at least somewhat greater than the depth to maximum shear stress) and a strong core;
5. smoother surfaces, free of fine cracks, by polishing, by careful running-in or by avoidance of coarse machining and grinding and of nicks in handling;
6. oil of higher viscosity and lower corrosiveness, free of moisture and in sufficient supply at the contacting surfaces. No lubricant on some surfaces with pure rolling and low velocity;
7. provision for increased film thickness of asperity-height ratio, the so-called lambda ratio ($\lambda > 1.5$).

3.7.   Thermal effects in surface contacts

The surface temperature generated in contact areas has a major influence on wear, scuffing, material properties and material degradation. The friction process converts mechanical energy primarily into thermal energy which results in a temperature rise. In concentrated contacts, which may be separated by a full elastohydrodynamic film, thin-film boundary lubricated contacts, or essentially unlubricated contacts, the friction intensity may be sufficiently large to cause a substantial temperature rise on the surface. The methods to estimate surface temperature rise presented in this section are all based on simplifying assumptions but nevertheless can be used in design processes. Although the temperature predicted may not be precise, it will give an indication of the level of temperature to expect and thereby give the designer some confidence that it can be ignored, or it will alert the designer to possible difficulties that may be encountered because of excessive temperatures.

The most significant assumption, involved in calculating a surface temperature, is the actual or anticipated coefficient of friction between the two surfaces where the temperature rise is sought. The coefficient of friction will depend on the nature of the surface and can vary widely depending on whether the surfaces are dry/unlubricated or if they are lubricated by boundary lubricants, solids, greases, hydrodynamic or elastohydrodynamic films. The coefficient of friction enters to the first power and is, in general, relatively unpredictable. If measurements of the coefficient of friction are available for the system under consideration, they frequently show substantial fluctuations. Another assumption is that all the energy is conducted into the solids in contact, which are assumed to be at a bulk temperature some distance away from the contact area. However, the presence of a lubricant in the immediate vicinity of the contact results in convection heat transfer, thereby cooling the surfaces close to the contact. This would generally tend to lower the predicted temperature.

The calculations focus on the flash temperature. That is the temperature rise in the contact area above the bulk temperature of the solid as a result of friction energy dissipation. However, the temperature level, not rise, in the contact area is frequently of major concern in predicting problems associated with excessive local temperatures. The surface temperature rise can influence local surface geometry through thermal expansion, causing high spots on the surface which concentrate the load and lead to severe local wear.

The temperature level, however, can lead to physical and chemical changes in the surface layers as well as the surface of the solid. These changes can lead to transitions in lubrication mechanisms and wear phenomena resulting in significant changes in the wear rate. Therefore, an overall system-heat transfer analysis may be required to predict the local bulk temperature and therefore the local surface temperature. Procedures are available for modelling the system-heat transfer problems by network theory and numerical analysis using commercially available finite element modelling systems. ESDU items 78026 to 78029 are especially recommended in this respect.

There is considerable literature on the subject of surface temperatures, covering both general aspects and specific special situations, but compared to theoretical analysis, little experimental work has been reported.

3.7.1. Analysis of line contacts

Blok proposed a theory for line contacts which will be summarized here. The maximum conjunction temperature, T_c, resulting from frictional heating between counterformal surfaces in a line contact is

$$T_c = T_f + T_b, \qquad [°C], \tag{3.8}$$

where T_b, the bulk temperature, is representative of the fairly uniform level of the part at some distance from the conjunction zone. T_f represents the maximum flash temperature in the conjunction zone resulting from frictional heating. T_f may be calculated from the following formula:

$$T_f = 1.11 \frac{fW|V_1 - V_2|}{b_1\sqrt{V_1} + b_2\sqrt{V_2}} \frac{1}{L\sqrt{w}}, \tag{3.9a}$$

where f is the instantaneous coefficient of friction, w is the instantaneous width of the band shaped conjunction, m, W is the instantaneous load on the conjunction, N, L is the instantaneous length of the conjunction perpendicular to motion, m, V_1, V_2 are the instantaneous velocities of surfaces 1 and 2 tangential to the conjunction zone and perpendicular to the conjunction band length, m/s, b_1, b_2 are the thermal contact coefficients of bodies 1 and 2 and

$$b_i = \sqrt{k_i \rho_i c_i} = k_i / \sqrt{\alpha_{Ti}}, \qquad \left[\frac{J}{m^2 °C s^{\frac{1}{2}}} \right],$$

where k_i, ρ_i, c_i and α_{Ti} are the thermal conductivity, density, specific heat per unit mass and thermal diffusivity of solid i, respectively.

If both bodies are of the same material, the maximum flash temperature can be written in one of the following three forms in commonly available variables:

$$T_f = 1.11 \left(\frac{fW}{L}\right) \left|\sqrt{V_1} - \sqrt{V_2}\right| (b\sqrt{w})^{-1}, \tag{3.9b}$$

$$T_f = 0.62 f \left(\frac{W}{L}\right)^{\frac{3}{4}} \left|\sqrt{V_1} - \sqrt{V_2}\right| \left(\frac{E}{R}\right)^{\frac{1}{4}} b^{-1}, \tag{3.9c}$$

$$T_f = 2.45 f p_H^{\frac{3}{2}} \left|\sqrt{V_1} - \sqrt{V_2}\right| \left(\frac{R}{E}\right)^{\frac{1}{2}} b^{-1}. \tag{3.9d}$$

In addition to the previously defined variables we have

$$b = \sqrt{\rho k c} - \text{thermal contact coefficient} \left[\frac{J}{m^2 \, ^\circ C s^{\frac{1}{2}}}\right]$$

$$E = E_1/(1 - v_1^2)$$

$E_1, v_1 =$ modulus of elasticity and Poisson's ratio of the solid

$$R = \left[\frac{1}{R_1} \pm \frac{1}{R_2}\right]^{-1} \qquad \text{the equivalent radius of undeformed}$$

surfaces, m,

$p_H =$ Hertzian (maximum) pressure in the contact, N/m^2.

The numerical factors, 1.11, 0.62 and 2.45 are valid for a semi-elliptical (Hertzian) distribution of the frictional heat over the contact width, w. This would be expected from a Hertzian contact with a constant coefficient of friction, or an elastohydrodynamic contact for a lubricant with a limiting shear stress proportional to pressure. Obviously, a consistent set of units must be chosen to give T_f in units of degrees centigrade.

The Blok flash temperature formulas apply only to cases for which the surface Peclet numbers, L, are sufficiently high. This is generally true for gear contacts which were the focus of Blok's experiments.

The Peclet number, or dimensionless speed criterion, is defined as

$$L_i = \frac{V_i w}{4\alpha_{Ti}} = \frac{V_i w \rho_i c_i}{4 k_i}, \tag{3.9e}$$

where the variables involved are defined above. An interpretation of the Peclet number can be given in terms of the heat penetration into the bulk of the material.

We now consider the instantaneous generation of energy at the surface of body i at a time zero. At a depth of half the contact width, $w/2$, below the surface, the maximum effect of this heat generation occurs after a time t_1, where

$$t_1 = \frac{\left(\dfrac{w}{2}\right)^2}{2\alpha_{Ti}}. \tag{3.9f}$$

The time for the point on the surface to move through half the contact width, $w/2$, is

$$t_2 = \frac{w}{2V_i} \tag{3.9g}$$

and therefore the Peclet number, eqn (3.9e), can be written as

$$L_i = \frac{t_1}{t_2} = \frac{\dfrac{w}{8\alpha_{Ti}}}{\dfrac{w}{2V_i}} = \frac{V_i w}{4\alpha_{Ti}}. \tag{3.9h}$$

Hence the Peclet number (or dimensionless speed parameter) can be interpreted as the ratio of the time required for the friction heat to penetrate the surface a distance equal to half the contact width, divided by the time in which a point on the surface travels the same distance.

Equations (3.9a) to (3.9d) assume that both surfaces are moving and that the Peclet number of each surface is at least 5 to 10. The analysis prediction accuracy will increase as the Peclet number increases beyond these values. The flash temperature, T_f, is the maximum temperature rise on the surface in the contact region above the bulk temperature. For the assumption underlying the theory and given above, the maximum temperature will be located at about

$$X \approx 0.825w,$$

where X is measured from the contact edge at which the material enters the contact region.

The analysis presented above can be illustrated by the following numerical example. Consider two steel (1 per cent chrome) cylinders each 100 mm diameter and 30 mm long rotating at different speeds such that the surface velocities at the conjunction are $V_1 = 3.0\,\mathrm{m\,s^{-1}}$ and $V_2 = 1.0\,\mathrm{m\,s^{-1}}$. The contact load is $10^5\,\mathrm{N}$ or $3.33 \times 10^6\,\mathrm{N/m}$ and the bulk temperature of each roller is $100\,^\circ\mathrm{C}$.

The thermal properties of the material (see Table 3.3 or ESDU–84041 for a more comprehensive list of data) are

$$k = 55\,\mathrm{W/m\,^\circ C} \quad \text{(at } 100\,^\circ\mathrm{C}),$$
$$\rho = 7865\,\mathrm{kg/m^3},$$
$$c_p = 0.46\,\mathrm{kJ/kg\,^\circ C},$$

therefore

$$\alpha_T = \frac{k}{\rho c_p} = \frac{55}{7865 \cdot 460} = 1.52 \times 10^{-5}\,\mathrm{m^2\,s^{-1}},$$

$$b = \sqrt{k\rho c_p} = 1.41 \times 10^4\,\frac{\mathrm{Nm}}{\mathrm{m^2\,^\circ C\,s^{\frac{1}{2}}}}.$$

The Young modulus $E_1 = 2.068 \times 10^{11}\,\mathrm{N/m^2}$ and $\nu_1 = 0.30$, thus

Table 3.3. *Typical thermal properties of some solids*

Material	Properties at 20°C				Thermal conductivity, k[W/m°C]										
	ρ kg/m³	c_p kJ/kg°C	k W/m°C	α m²/sec ×10	−100°C	0°C	100°C	200°C	300°C	400°C	600°C	800°C	1000°C	1200°C
Aluminium (pure)	2707	0.896	204	8.418	215	202	206	215	228	249				
Steel ($C_{max}=0.5\%$)	7833	0.465	54	1.474		55	52	48	45	42	35	31	29	31
Tungsten steel	7897	0.452	73	2.026										
Copper	8954	0.3831	386	11.234		386	379	369	363	353				
Aluminium bronze	8666	0.410	83	2.330	407									
Bronze	8666	0.343	26	0.859										
Silicon nitride	3200	0.710	30.7	1.35				27		23	20	18		
Titanium carbide	6000	0.543	55	1.69						32			28	
Graphite	1900	0.71	178	13.2						112			62	
Nylon	1140	1.67	0.25	0.013										
Polymide	1430	1.13	0.36	0.023										
PTFE	2200	1.05	0.24	0.010										
Silicon oxide (glass)	2200	0.8	1.25	0.08				1.05		1.25	1.4	1.6	1.8	

$E = 2.27 \times 10^{11}\,\text{N/m}^2$. The equivalent radius of contact is

$$R = \left[\frac{1}{R_1} + \frac{1}{R_2} \right]^{-1} = 25\,\text{mm}.$$

Contact width, based on Hertz theory

$$w = 2b = 2 \times 1.598 \sqrt{\frac{10^5 \times 0.025}{0.030 \times 2.27 \times 10^{11}}}$$

$$= 1.94 \times 10^{-3}\,\text{m} = 1.94\,\text{mm}.$$

Hertzian stress

$$p_{\text{H}} = 2.19\,\text{GPa}.$$

Checking the Peclet number for each surface we get

$$L_1 = \frac{V_1 w}{4\alpha_{T_1}} = \frac{3.0 \times 1.94 \times 10^{-3}}{4 \times 1.52 \times 10^{-5}} = 95.7,$$

$$L_2 = \frac{V_2 w}{4\alpha_{T_2}} = \frac{1.0 \times 1.94 \times 10^{-3}}{4 \times 1.52 \times 10^{-3}} = 31.9.$$

We find that both Peclet numbers are greater than 10. Thus, using eqn (3.9a)

$$T_{\text{f}} = \frac{1.11 \times 0.10 \times 3.33 \times 10^6\,|3-1|}{1.41 \times 10^4 (\sqrt{3} - \sqrt{1})} \times \frac{1}{\sqrt{1.94 \times 10^{-3}}} = 435\,°\text{C}$$

and with equal bulk temperatures of $100\,°\text{C}$ the maximum surface temperature is

$$T_{\text{c}} = 435 + 100 = 535\,°\text{C}.$$

3.7.2. Refinement for unequal bulk temperatures

It has been assumed that the bulk temperature, T_{b}, is the same for both surfaces. If the two bodies have different bulk temperatures, $T_{\text{b}1}$ and $T_{\text{b}2}$, the T_{b} in eqn (3.8) should be replaced with

$$T_{\text{b}} = \tfrac{1}{2}(T_{\text{b}1} + T_{\text{b}2}) + \tfrac{1}{2}\frac{n-1}{n+1}(T_{\text{b}1} - T_{\text{b}2}), \tag{3.10}$$

where

$$n = \frac{b_1 \sqrt{V_1}}{b_2 \sqrt{V_2}}. \tag{3.11}$$

If $0.2 \leqslant n \leqslant 5$, to a good approximation,

$$T_{\text{b}} = \tfrac{1}{2}(T_{\text{b}1} + T_{\text{b}2}). \tag{3.12}$$

3.7.3. Refinement for thermal bulging in the conjunction zone

Thermal bulging relates to the fact that friction heating can cause both thermal stresses and thermoelastic strains in the conjunction region. The thermoelastic strains may result in local surface bulging, which may shift and concentrate the load onto a smaller region, thereby causing higher flash temperatures. A dimensionless thermal bulging parameter, K, has the form

$$K = \tfrac{1}{2}\frac{fW\,|V_1 - V_2|}{p_{\mathrm{H}}} \times \frac{(1+v_1)E_1\varepsilon}{k(1-v_1^2)},\qquad(3.13)$$

where all the variables are as defined above except, ε is the coefficient of linear thermal expansion ($1/^\circ$C). Note: p_{H} is the maximum Hertz pressure that would occur under conditions of elastic contact in the absence of thermal bulging. In other words, it can be calculated using Hertz theory. In general, for most applications

$$0 \leqslant K \leqslant 2$$

and for this range there is a good approximation to the relation between the maximum conjunction pressure resulting from thermal bulging, p_{k}, and the maximum pressure in the absence of thermal bulging, p_{H}, namely

$$\frac{p_{\mathrm{k}}}{p_{\mathrm{H}}} \approx 1 + 0.62K \qquad(3.14a)$$

and the ratio of the contact widths w_{k} and w_{H}, respectively, is

$$\frac{w_{\mathrm{k}}}{w_{\mathrm{H}}} = (1 + 0.62K)^{-1},\qquad(3.14b)$$

which, when substituted into the flash temperature expressions, eqn (3.9a), results simply in a correction factor multiplying the original flash temperature relation

$$T_{\mathrm{f,k}} \approx (1 + 0.62K)^{\frac{1}{2}}T_{\mathrm{f}},\qquad(3.15)$$

where the second subscript, k, refers to the flash temperature value corrected for the thermal bulge phenomena.

The thermal bulging phenomena can lead to a thermoelastic instability in which the bulge wears, relieving the local stress concentration, which then shifts the load to another location where further wear occurs.

3.7.4. The effect of surface layers and lubricant films

The thermal effects of surface layers on surface temperature increase may be important if they are thick and of low thermal conductivity relative to the bulk solid. If the thermal conductivity of the layer is low, it will raise the surface temperature, but to have a significant influence, it must be thick compared to molecular dimensions. Another effect of excessive surface temperature will be the desorption of the boundary lubricating film leading to direct metal–metal contacts which in turn could lead to a further increase of temperature. Assuming the same frictional energy dissipation, at low sliding speeds, the surface temperature is unchanged by the presence of the film. At high sliding speeds, the layer influence is determined by its thickness relative to the depth of heat penetration, x_p, where

$$x_p = \sqrt{2\alpha_T t} = \sqrt{\frac{2\alpha_T w}{V}}$$

α_T = thermal diffusivity of the solid, $(\text{m}^2\,\text{s}^{-1})$ and $t = w/V$ = time of heat application, (sec).

For practical speeds on materials and surface films, essentially all the heat penetrates to the substrate and its temperature is almost the same as without the film. Thus, the thermal effect of the film is to raise the surface temperature and to lower or leave unchanged the temperature of the substrate. The substrate temperature will not be increased by the presence of the film unless the film increases the friction. A more likely mechanism by which the surface film will influence the surface temperature increase, is through the influence the film will have on the coefficient of friction, which results in a change in the amount of energy being dissipated to raise the surface temperature. The case of a thin elastohydrodynamic lubricant film is more complicated because it is both a low thermal conductivity film and may be thick enough to have substantial temperature gradients. It is possible to treat this problem by assuming that the frictional energy dissipation occurs at the midplane of the film, and the energy division between the two solids depends on their thermal properties and the film thickness. This results in the two surfaces having different temperatures as long as they are separated by a film. As the film thickness approaches zero the two surface temperatures approach each other and are equal when the separation no longer exists.

For the same kinematics, materials and frictional energy dissipation, the presence of the film will lower the surface temperatures, but cause the film middle region to have a temperature higher than the unseparated surface temperatures. The case of a thin elastohydrodynamic film can be modelled using the notion of a slip plane. Assuming that in the central region of the film there is only one slip plane, $y = h_1$ (see Fig. 3.5), the heat generated in this plane will be dissipated through the film to the substrates.

Because the thickness of the film is much less than the width of the contact, it can therefore, be assumed that the temperature gradient along the x-axis is small in comparison with that along the y-axis. It is further assumed that the heat is dissipated in the y direction only. Friction-generated heat per unit area of the slip plane is

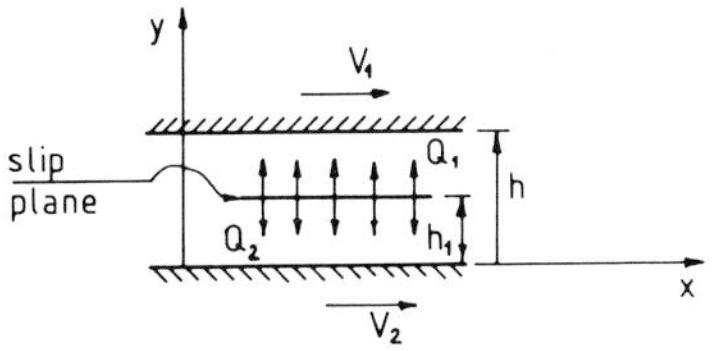

$$Q_0 = \tau_s V, \tag{3.16}$$

where τ_s is the shear stress in the film and V is the relative sliding velocity. If all the friction work is converted into heat, then

$$Q_0 = Q_1 + Q_2.$$

Figure 3.5

The ratio of Q_1 and Q_2 is

$$Q_1/Q_2 = \frac{h - h_1}{h_1}. \tag{3.17}$$

Equation (3.17) gives the relationship between the heat dissipated to the substrates and the location of the slip plane. Temperatures of the substrates will increase as a result of heat generated in the slip plane. Thus, the increase in temperature is given by

$$\Delta T = \frac{1}{\sqrt{\pi k_i \rho_i c_i}} \int_0^t Q(t - \xi) \frac{\mathrm{d}\xi}{\sqrt{\xi}}, \tag{3.18}$$

where $Q(t - \xi)$ is the flow of heat during the time $(t - \xi)$, k_i is the thermal conductivity, c_i is the specific heat per unit mass and ρ_i is the density.

3.7.5.   Critical temperature for lubricated contacts

The temperature rise in the contact zone due to frictional heating can be estimated from the following formula, proposed by Bowden and Tabor

$$T_f = 0.443 \frac{g f V (W p_m)^{\frac{1}{2}}}{J(k_1 + k_2)}, \tag{3.19}$$

where J is the mechanical equivalent of heat and g is the gravitational constant. The use of the fractional film defect is the simplest technique for estimating the characteristic lubricant temperature, T_c, without getting deeply involved in surface chemistry.

The fractional film defect is given by eqn (2.67) and has the following form

$$\beta = 1 - \exp \left\{ - \left[\frac{30.9 \times 10^5 T m^{\frac{1}{2}}}{V M^{\frac{1}{2}}} \right] \exp \left(- \frac{E_c}{R T_c} \right) \right\}.$$

If a closer look is taken at the fractional film defect equation, as affected by the heat of adsorption of the lubricant, E_c, and the surface contact temperature, T_c, it can be seen that the fractional film defect is a measure of the probability of two bare asperital areas coming into contact. It would be far more precise if, for a given heat of adsorption for the lubricant-substrate combination, we could calculate the critical temperature just before encountering $\beta > 0$.

In physical chemistry, it is the usual practice to use the points, T_{c1} and T_{c2}, shown in Fig. 3.6, at the inflection point in the curves. However, even a small probability of bare asperital areas in contact can initiate rather large regenerative heat effects, thus raising the flash temperature T_f. This substantially increases the desorption rate at the exit from the conjunction zone so that almost immediately β is much larger at the entrance to the conjunction zone. It is seen from Fig. 3.6 that when T_c is increased, for a given value E_c, β is also substantially increased. It is proposed therefore, that the critical point on the β-curve will be where the change in curvature

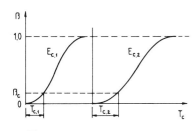

Figure 3.6

first becomes a maximum. Mathematically, this is where $d^2\beta/dT_c^2$ is the first maximum value or the minimum value of β, where $d^3\beta/dT_c^3 = 0$. Thus, starting with eqn (2.67) it is possible to derive the following expression for T_c

$$T_c = \frac{3\left(\dfrac{E_c}{RT_c} - 2\right) - \left[5\left(\dfrac{E_c}{RT_c}\right)^2 - 12\left(\dfrac{E_c}{RT_c}\right) + 12\right]^{\frac{1}{2}}}{\dfrac{30.9 \times 10^5}{V}\left(\dfrac{T_m}{M}\right)^{\frac{1}{2}}\dfrac{E_c}{RT_c^2}\exp\left(-\dfrac{E_c}{RT_c}\right)}. \tag{3.20}$$

Equation (3.20) is implicit and must be solved by using a microcomputer, for instance, in order to obtain values for T_c.

3.7.6. The case of circular contact

Archard has presented a simple formulation for the mean flash temperature in a circular area of real contact of diameter $2a$. The friction energy is assumed to be uniformly distributed over the contact as shown schematically in Fig. 3.7. Body 1 is assumed stationary, relative to the conjunction area and body 2 moves relative to it at a velocity V. Body 1, therefore, receives heat from a stationary source and body 2 from a moving heat source. If both surfaces move (as with gear teeth for instance), relative to the conjunction region, the theory for the moving heat source is applied to both bodies.

Archard's simplified formulation also assumes that the contacting portion of the surface has a height approximately equal to its radius, a, at the contact area and that the bulk temperature of the body is the temperature at the distance, a, from the surface. In other words, the contacting area is at the end of a cylinder with a length-to-diameter ratio of approximately one-half, where one end of the cylinder is the rubbing surface and the other is maintained at the bulk temperature of the body. Hence the model will cease to be valid, or should be modified, as the length-to-diameter ratio of the slider deviates substantially from one-half, and/or as the temperature at the root of the slider increases above the bulk temperature of the system as the result of frictional heating. If these assumptions are kept in mind, Archard's simplified formulation can be of value in estimating surface flash temperature, or as a guide to calculations with modified contact geometries.

For the stationary heat source, body 1, the mean temperature increase above the bulk solid temperature is

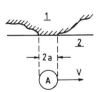

Figure 3.7

$$T_m = \frac{Q_1}{4ak_1}, \tag{3.21}$$

where Q_1 is the rate of frictional heat supplied to body 1, $(\mathrm{Nm\,s^{-1}})$, k_1 is the thermal conductivity of body 1, $(\mathrm{W/m\,^\circ C})$ and a is the radius of the circular contact area, (m).

If body 2 is moving very slowly, it can also be treated as essentially a stationary heat source case. Therefore

$$T_m = \frac{Q_2}{4ak_2},\qquad(3.22)$$

where Q_2 is the rate of frictional heat supplied to body 2 and k_2 is the thermal conductivity of body 2.

The speed criterion used for the analysis is the dimensionless parameter, L, called the Peclet number, given by eqn (3.9e). For $L < 0.1$, eqn (3.22) applies to the moving surface. For larger values of L ($L > 5$) the surface temperature of the moving surface is

$$T = \frac{2Q_2}{a^2}\sqrt{\frac{x}{\pi k\rho c V}},\qquad(3.23)$$

where x is the distance from the leading edge of the contact. The average temperature over the circular contact in this case then becomes

$$T_m = \frac{Q_2}{3.25k_2 a}\sqrt{\frac{1}{2L}} = \frac{0.31Q_2}{k_2 a}\sqrt{\frac{\alpha_{T_2}}{Va}}.\qquad(3.24)$$

The above expression can be simplified if we define:

$$N = \frac{Q}{a^2 \rho c V}.$$

Then, for $L < 0.1$, eqns (3.21) and (3.22) become

$$T_m = 0.5NL\qquad(3.25)$$

and for high speed moving surfaces, ($L > 5$), eqn (3.24) becomes

$$T_m = 0.435NL^{\frac{1}{2}}\qquad(3.26)$$

and for the transformation region ($0.1 \leqslant L \leqslant 5$)

$$T_m = 0.5N\beta L,\qquad(3.27)$$

where it has been shown that the factor β is a function of L ranging from about 0.85 at $L = 0.1$ to about 0.35 at $L = 5$. Equations (3.25–3.27) can be plotted as shown in Fig. 3.8.

To apply the results to a practical problem the proportion of frictional heat supplied to each body must be taken into account. A convenient procedure is to first assume that all the frictional heat available ($Q = fWV$) is transferred to body 1 and calculate its mean temperature rise (T_{m1}) using N_1 and L_1. Then do the same for body 2. The true temperature rise T_m (which must be the same for both contacting surfaces), taking into account the division of heat between bodies 1 and 2, is given by

$$1/T_m = 1/T_{m1} + 1/T_{m2}.\qquad(3.28)$$

To obtain the mean contact surface temperature, T_c, the bulk temperature, T_b, must be added to the temperature rise, T_m.

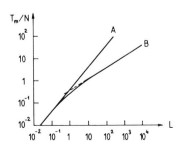

Figure 3.8

Numerical example

Now consider a circular contact 20 mm in diameter with one surface stationary and one moving at $V = 0.5 \, \mathrm{m\,s^{-1}}$. The bodies are both of plain carbon steel ($C \approx 0.5\%$) and at 24 °C bulk temperature. We recall that the assumption in the Archard model implies that the stationary surface is essentially a cylindrical body of diameter 20 mm and length 10 mm with one end maintained at the bulk temperature of 24 °C. The coefficient of friction is 0.1 and the load is $W = 3000 \, \mathrm{N}$ (average contact pressure of 10 MPa). The properties of contacting bodies are (see Table 3.3 or ESDU–84041 for a more comprehensive list of data)

$$k = 54 \, \mathrm{W/m\,°C}$$
$$\alpha_T = 1.474 \times 10^{-5} \, \mathrm{m^2\,s^{-1}}.$$

Therefore

$$L_m = \frac{Va}{2\alpha_T} = \frac{0.5 \times 0.01}{2 \times 1.474 \times 10^{-5}} = 169.$$

If we assume that all the frictional energy is conducted into the moving surface ($L_m = 169 > 5$), we can then use eqn (3.24)

$$T_{m2} = \frac{0.31 \times 0.1 \times 3000 \times 0.5}{54 \times 0.01} \sqrt{\frac{1.474 \times 10^{-5}}{0.5 \times 0.01}} = 4.7 \, \mathrm{°C}$$

and if all the frictional energy went into the stationary surface ($L_s = 0$), then we use eqn (3.21)

$$T_{m1} = \frac{0.1 \times 3000 \times 0.5}{4 \times 0.01 \times 54} = 69 \, \mathrm{°C}.$$

The true temperature rise for the two surfaces is then obtained from eqn (3.28) and is

$$1/T_m = 1/69 + 1/4.7 = 4.4 \, \mathrm{°C}.$$

3.7.7. Contacts for which size is determined by load

There are special cases where the contact size is determined by either elastic or plastic contact deformation.

If the contact is plastic, the contact radius, a, is

$$a = \left[\frac{W}{\pi p_m} \right]^{\frac{1}{2}}, \tag{3.29a}$$

where W is the load and p_m is the flow pressure or hardness of the weaker material in contact.

If the contact is elastic

$$a = 1.1 \left[\frac{WR}{E} \right]^{\frac{1}{3}}, \tag{3.29b}$$

where R is the undeformed radius of curvature and E denotes the elastic modulus of a material.

Employing these contact radii in the low and high speed cases discussed in the previous section gives the following equations for the average increase in contact temperature

- plastic deformation, low speed ($L<0.1$)

$$T_m = \frac{fV\sqrt{\pi p_m W}}{8k};$$

(3.30)

- plastic deformation, high speed ($L>100$),

$$T_m = \frac{f}{3.25}(\pi p_m)^{\frac{3}{4}}W^{\frac{1}{4}}\sqrt{\frac{v}{k\rho c}};$$

(3.31)

- elastic deformation, low speed ($L<0.1$),

$$T_m = \frac{fV}{8.8k}W^{\frac{2}{3}}\left(\frac{E}{R}\right)^{\frac{1}{3}};$$

(3.32)

- elastic deformation, high speed ($L>100$),

$$T_m = \frac{f}{3.8}\left(\frac{EWV}{k\rho cR}\right)^{\frac{1}{2}}.$$

(3.33)

3.7.8.   Maximum attainable flash temperature

The maximum average temperature will occur when the maximum load per unit area occurs, which is when the load is carried by a plastically deformed contact. Under this condition the N and L variables discussed previously become

$$N = f\frac{\pi p_m}{\rho c}; \qquad L = \frac{V}{2\alpha_T}\sqrt{\frac{W}{\pi p_m}}.$$

(3.34)

Then at low speeds ($L<0.1$), the heat supply is equally divided between surfaces 1 and 2, and the surface temperatures are

$$T_{max} = 0.25NL.$$

(3.35)

At moderate speeds ($0.1 \leqslant L \leqslant 5$), less than half the heat is supplied to body 1, and therefore

$$T_{max} = 0.25\beta NL,$$

(3.36)

where β ranges from about 0.95 at $L=0.1$ to about 0.5 at $L=5$. At very high speeds ($L>100$), practically all the heat is supplied to body 2, and then

$$T_{max} = 0.435NL^{\frac{1}{2}}.$$

(3.37)

At lower speeds ($5<L<100$), less heat is supplied to body 2 and

$$T_{max} = 0.435\gamma NL^{\frac{1}{2}},$$

(3.38)

where

$$\gamma = [1 + 0.87 L^{-\frac{1}{2}}]^{-1} \tag{3.39}$$

and γ ranges from 0.72 at $L = 5$ to 0.92 at $L = 100$.

3.8. Contact between rough surfaces

There are no topographically smooth surfaces in engineering practice. Mica can be cleaved along atomic planes to give an atomically smooth surface and two such surfaces have been used to obtain perfect contact under laboratory conditions. The asperities on the surface of very compliant solids such as soft rubber, if sufficiently small, may be squashed flat elastically by the contact pressure, so that perfect contact is obtained through the nominal contact area. In general, however, contact between solid surfaces is discontinuous and the real area of contact is a small fraction of the nominal contact area. It is not easy to flatten initially rough surfaces by plastic deformation of the asperities.

The majority of real surfaces, for example those produced by grinding, are not regular, the heights and the wavelengths of the surface asperities vary in a random way. A machined surface as produced by a lathe has a regular structure associated with the depth of cut and feed rate, but the heights of the ridges will still show some statistical variation. Most man-made surfaces such as those produced by grinding or machining have a pronounced lay, which may be modelled, to a first approximation, by one-dimensional roughness.

It is not easy to produce wholly isotropic roughness. The usual procedure for experimental purposes is to air-blast a metal surface with a cloud of fine particles, in the manner of shot-peening, which gives rise to a randomly cratered surface.

3.8.1. Characteristics of random rough surfaces

The topographical characteristics of random rough surfaces which are relevant to their behaviour when pressed into contact will now be discussed briefly. Surface texture is usually measured by a profilometer which draws a stylus over a sample length of the surface of the component and reproduces a magnified trace of the surface profile. This is shown schematically in Fig. 3.9. It is important to realize that the trace is a much distorted image of the actual profile because of using a larger magnification in the normal than in the tangential direction. Modern profilometers digitize the trace at a suitable sampling interval and send the output to a computer in order to extract statistical information from the data. First, a datum or centre-line is established by finding the straight line (or circular arc in the case of round components) from which the mean square deviation is at a minimum. This implies that the area of the trace above the datum line is equal to that below it. The average roughness is now defined by

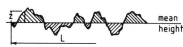

Figure 3.9

$$R_{\mathrm{a}} = \frac{1}{L} \int_0^L |z| \, \mathrm{d}z, \tag{3.40}$$

where $z(x)$ is the height of the surface above the datum and L is the sampling length. A less common but statistically more meaningful measure of average roughness is the root mean square (r.m.s.) or standard deviation σ of the height of the surface from the centre-line, i.e.

$$\sigma^2 = \frac{1}{L} \int_0^L z^2 \, \mathrm{d}z. \tag{3.41}$$

The relationship between σ and R_a depends, to some extent, on the nature of the surface; for a regular sinusoidal profile $\sigma = (\pi/2\sqrt{2})R_a$ and for a Gaussian random profile $\sigma = (\pi/2)^{\frac{1}{2}}R_a$.

The R_a value by itself gives no information about the shape of the surface profile, i.e. about the distribution of the deviations from the mean. The first attempt to do this was by devising the so-called bearing area curve. This curve expresses, as a function of the height z, the fraction of the nominal area lying within the surface contour at an elevation z. It can be obtained from a profile trace by drawing lines parallel to the datum at varying heights, z, and measuring the fraction of the length of the line at each height which lies within the profile (Fig. 3.10). The bearing area curve, however, does not give the true bearing area when a rough surface is in contact with a smooth flat one. It implies that the material in the area of interpenetration vanishes and no account is taken of contact deformation.

An alternative approach to the bearing area curve is through elementary statistics. If we denote by $\phi(z)$ the probability that the height of a particular point in the surface will lie between z and $z + \mathrm{d}z$, then the probability that the height of a point on the surface is greater than z is given by the cumulative probability function: $\Phi(z) = \int_z^\infty \phi(z') \mathrm{d}z'$. This yields an S-shaped curve identical to the bearing area curve.

It has been found that many real surfaces, notably freshly ground surfaces, exhibit a height distribution which is close to the normal or Gaussian probability function:

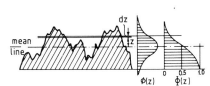

Figure 3.10

$$\phi(z) = \sigma(2\pi)^{-\frac{1}{2}} \exp(-z^2/2\sigma^2), \tag{3.42}$$

where σ is that standard (r.m.s.) deviation from the mean height. The cumulative probability, given by the expression

$$\Phi(z) = \frac{1}{2} - \frac{1}{(2\pi)^{\frac{1}{2}}} \int_0^{z/\sigma} \exp(-z'^2/2\sigma^2) \mathrm{d}(z'/\sigma) \tag{3.43}$$

can be found in any statistical tables. When plotted on normal probability graph paper, data which follow the normal or Gaussian distribution will fall on a straight line whose gradient gives a measure of the standard deviation. It is convenient from a mathematical point of view to use the normal probability function in the analysis of randomly rough surfaces, but it must be remembered that few real surfaces are Gaussian. For example, a ground surface which is subsequently polished so that the tips of the higher asperities are removed, departs markedly from the straight line in the upper height range. A lathe turned surface is far from random; its peaks are nearly all the same height and its valleys nearly all the same depth.

So far only variations in the height of the surface have been discussed. However, spatial variations must also be taken into account. There are several ways in which the spatial variation can be represented. One of them uses the r.m.s. slope σ_m and r.m.s. curvature σ_k. For example, if the sample length L of the surface is traversed by a stylus profilometer and the height z is sampled at discrete intervals of length h, and if z_{i-1} and z_{i+1} are three consecutive heights, the slope is then defined as

$$m = (z_{i+1} - z_i)/h \tag{3.44}$$

and the curvature by

$$k = (z_{i+1} - 2z_i + z_{i-1})/h^2. \tag{3.45}$$

The r.m.s. slope and r.m.s. curvature are then found from

$$\sigma_m^2 = (1/n) \sum_{i=1}^{i=n} m^2, \tag{3.46}$$

$$\sigma_m^2 = (1/n) \sum_{i=1}^{i=n} k^2, \tag{3.47}$$

where $n = L/h$ is the total number of heights sampled.

It would be convenient to think of the parameters σ, σ_m and σ_k as properties of the surface which they describe. Unfortunately their values in practice depend upon both the sample length L and the sampling interval h used in their measurements. If a random surface is thought of as having a continuous spectrum of wavelengths, neither wavelengths which are longer than the sample length nor wavelengths which are shorter than the sampling interval will be recorded faithfully by a profilometer. A practical upper limit for the sample length is imposed by the size of the specimen and a lower limit to the meaningful sampling interval by the radius of the profilometer stylus. The mean square roughness, σ, is virtually independent of the sampling interval h, provided that h is small compared with the sample length L. The parameters σ_m and σ_k, however, are very sensitive to sampling interval; their values tend to increase without limit as h is made smaller and shorter, and shorter wavelengths are included. This fact has led to the concept of function filtering. When rough surfaces are pressed into contact they touch at the high spots of the two surfaces, which deform to bring more spots into contact. To quantify this behaviour it is necessary to know the standard deviation of the asperity heights, σ_s, the mean curvature of their peaks, $\bar{k}_s$, and the asperity density, η_s, i.e. the number of asperities per unit area of the surface. These quantities have to be deduced from the information contained in a profilometer trace. It must be kept in mind that a maximum in the profilometer trace, referred to as a peak does not necessarily correspond to a true maximum in the surface, referred to as a summit since the trace is only a one-dimensional section of a two-dimensional surface.

The discussion presented above can be summarized briefly as follows:
(i) for an isotropic surface having a Gaussian height distribution with

standard deviation, σ, the distribution of summit heights is very nearly Gaussian with a standard deviation

$$\sigma_s \approx \sigma.$$

The mean height of the summits lies between 0.5σ and 1.5σ above the mean level of the surface. The same result is true for peak heights in a profilometer trace. A peak in the profilometer trace is identified when, of three adjacent sample heights, z_{i-1} and z_{i+1}, the middle one z_i is greater than both the outer two.

(ii) the mean summit curvature is of the same order as the r.m.s. curvature of the surface, i.e.

$$\bar{k}_s \approx \sigma_k.$$

(iii) by identifying peaks in the profile trace as explained above, the number of peaks per unit length of trace η_p can be counted. If the wavy surface were regular, the number of summits per unit area η_s would be η_p^2. Over a wide range of finite sampling intervals

$$\eta_s \approx 1.8\eta_p^2.$$

Although the sampling interval has only a second-order effect on the relationship between summit and profile properties it must be emphasized that the profile properties themselves, i.e. σ_k and σ_p are both very sensitive to the size of the sampling interval.

3.8.2.   Contact of nominally flat rough surfaces

Although in general all surfaces have roughness, some simplification can be achieved if the contact of a single rough surface with a perfectly smooth surface is considered. The results from such an argument are then reasonably indicative of the effects to be expected from real surfaces. Moreover, the problem will be simplified further by introducing a theoretical model for the rough surface in which the asperities are considered as spherical cups so that their elastic deformation characteristics may be defined by the Hertz theory. It is further assumed that there is no interaction between separate asperities, that is, the displacement due to a load on one asperity does not affect the heights of the neighbouring asperities.

Figure 3.11 shows a surface of unit nominal area consisting of an array of identical spherical asperities all of the same height z with respect to some reference plane XX'. As the smooth surface approaches, due to the

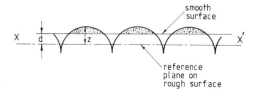

Figure 3.11

application of a load, it is seen that the normal approach will be given by $(z-d)$, where d is the current separation between the smooth surface and the reference plane. Clearly, each asperity is deformed equally and carries the same load W_i so that for η asperities per unit area the total load W will be equal to ηW_i. For each asperity, the load W_i and the area of contact A_i are known from the Hertz theory

$$W=\tfrac{4}{3}E'R^{\frac{1}{2}}\delta^{\frac{3}{2}}$$

and

$$A=\pi R\delta,$$

where δ is the normal approach and R is the radius of the sphere in contact with the plane. Thus if β is the asperity radius, then

$$N_i=\tfrac{4}{3}E'\beta^{\frac{1}{2}}(z-d)^{\frac{3}{2}} \tag{3.48}$$

and

$$A_i=\pi\beta(z-d) \tag{3.49}$$

and the total load will be given by

$$W=\tfrac{4}{3}\eta E'\beta^{\frac{1}{2}}\left(\frac{A_i}{\pi\beta}\right)^{\frac{3}{2}},$$

that is the load is related to the total real area of contact, $A=\eta A_i$, by

$$W=\frac{4E'}{3\pi^{\frac{3}{2}}\eta^{\frac{1}{2}}\beta}A^{\frac{3}{2}}. \tag{3.50}$$

This result indicates that the real area of contact is related to the two-thirds power of the load, when the deformation is elastic.

If the load is such that the asperities are deformed plastically under a constant flow pressure H, which is closely related to the hardness, it is assumed that the displaced material moves vertically down and does not spread horizontally so that the area of contact A' will be equal to the geometrical area $2\pi\beta\delta$. The individual load, W_i', will be given by

$$W_i'=HA_i'=2H\pi\beta(z-d).$$

Thus

$$W'=\eta N_i'=\eta HA_i'=HA'=2HA, \tag{3.51}$$

that is, the real area of contact is linearly related to the load.

It must be pointed out at this stage that the contact of rough surfaces should be expected to give a linear relationship between the real area of contact and the load, a result which is basic to the laws of friction. From the simple model of rough surface contact, presented here, it is seen that while a plastic mode of asperity deformation gives this linear relationship, the elastic mode does not. This is primarily due to an oversimplified and hence unrealistic model of the rough surface. When a more realistic surface model is considered, the proportionality between load and real contact area can in fact be obtained with an elastic mode of deformation.

It is well known that on real surfaces the asperities have different heights indicated by a probability distribution of their peak heights. Therefore, the simple surface model must be modified accordingly and the analysis of its contact must now include a probability statement as to the number of the asperities in contact. If the separation between the smooth surface and that reference plane is d, then there will be a contact at any asperity whose height was originally greater than d (Fig. 3.12). If $\phi(z)$ is the probability density of the asperity peak height distribution, then the probability that a particular asperity has a height between z and $z + dz$ above the reference plane will be $\phi(z)\,dz$. Thus, the probability of contact for any asperity of height z is

$$\mathrm{prob}(z > d) = \int_d^\infty \phi(z)\,dz.$$

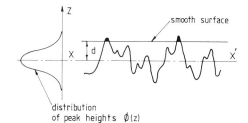

Figure 3.12

distribution
of peak heights $\phi(z)$

If we consider a unit nominal area of the surface containing asperities, the number of contacts n will be given by

$$n = \eta \int_d^\infty \phi(z)\,dz. \tag{3.52}$$

Since the normal approach is $(z - d)$ for any asperity and N_i and A_i are known from eqns (3.48) and (3.49), the total area of contact and the expected load will be given by

$$A = \pi\eta\beta \int_d^\infty (z - d)\phi(z)\,dz \tag{3.53}$$

and

$$N = \tfrac{4}{3}\eta\beta^{\frac{1}{2}}E' \int_d^\infty (z - d)^{\frac{3}{2}}\phi(z)\,dz. \tag{3.54}$$

It is convenient and usual to express these equations in terms of standardized variables by putting $h = d/\sigma$ and $s = z/\sigma$, σ being the standard deviation of the peak height distribution of the surface. Thus

$$n = \eta F_0(h),$$
$$A = \pi\eta\beta\sigma F_1(h),$$
$$N = \tfrac{4}{3}\eta\beta^{\frac{1}{2}}\sigma^{\frac{3}{2}}E'F_{\frac{3}{2}}(h),$$

where

$$F_m(h) = \int_h^\infty (s-h)^m \phi^*(s)\,ds$$

$\phi^*(s)$ being the probability density standardized by scaling it to give a unit standard deviation. Using these equations one may evaluate the total real area, load and number of contact spots for any given height distribution.

An interesting case arises where such a distribution is exponential, that is,

$$\phi^*(s) = e^{-s}.$$

In this case

$$F_n(h) = m!e^{-h}$$

so that

$$n = \eta e^{-h},$$
$$A = \pi \eta \beta \sigma e^{-h},$$
$$N = \pi^{\frac{1}{2}} \eta \beta^{\frac{1}{2}} \sigma^{\frac{3}{2}} E' e^{-h}.$$

These equations give

$$N = C_1 A \qquad \text{and} \qquad N = C_2 n,$$

where C_1 and C_2 are constants of the system. Therefore, even though the asperities are deforming elastically, there is exact linearity between the load and the real area of contact. For other distributions of asperity heights, such a simple relationship will not apply, but for distributions approaching an exponential shape it will be substantially true. For many practical surfaces the distribution of asperity peak heights is near to a Gaussian shape.

Where the asperities obey a plastic deformation law, eqns (3.53) and (3.54) are modified to become

$$A' = 2\pi\eta\beta \int_d^\infty (z-d)\phi(z)\,dz, \tag{3.55}$$

$$N' = 2\pi\eta\beta H \int_d^\infty (z-d)\phi(z)\,dz. \tag{3.56}$$

It is immediately seen that the load is linearly related to the real area of contact by $N' = HA'$ and this result is totally independent of the height distribution $\phi(z)$, see eqn (3.51).

The analysis presented has so far been based on a theoretical model of the rough surface. An alternative approach to the problem is to apply the concept of profilometry using the surface bearing-area curve discussed in Section 3.8.1. In the absence of the asperity interaction, the bearing-area curve provides a direct method for determining the area of contact at any given normal approach. Thus, if the bearing-area curve or the all-ordinate distribution curve is denoted by $\psi(z)$ and the current separation between the smooth surface and the reference plane is d, then for a unit nominal surface area the real area of contact will be given by

$$A = \int_d^\infty \psi(z)\,\mathrm{d}z \tag{3.57}$$

so that for an ideal plastic deformation of the surface, the total load will be given by

$$N = H \int_d^\infty \psi(z)\,\mathrm{d}z. \tag{3.58}$$

To summarize the foregoing it can be said that the relationship between the real area of contact and the load will be dependent on both the mode of deformation and the distribution of the surface profile. When the asperities deform plastically, the load is linearly related to the real area of contact for any distribution of asperity heights. When the asperities deform elastically, the linearity between the load and the real area of contact occurs only where the distribution approaches an exponential form and this is very often true for many practical engineering surfaces.

3.9. Representation of machine element contacts

Many contacts between machine components can be represented by cylinders which provide good geometrical agreement with the profile of the undeformed solids in the immediate vicinity of the contact. The geometrical errors at some distance from the contact are of little importance.

For roller-bearings the solids are already cylindrical as shown in Fig. 3.13. On the inner race or track the contact is formed by two convex

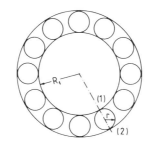

Figure 3.13

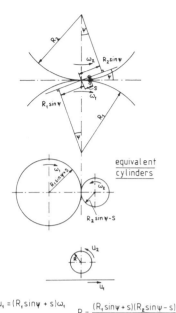

equivalent
cylinders

$J_1 = (R_1 \sin\psi + s)\omega_1$

$J_2 = (R_2 \sin\psi - s)\omega_2$

$R = \dfrac{(R_1 \sin\psi + s)(R_2 \sin\psi - s)}{(R_1 + R_2)\sin\psi}$

Figure 3.14

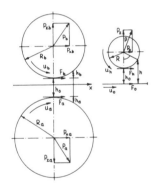

Figure 3.15

cylinders of radii r and R_1, and on the outer race the contact is between the roller of radius r and the concave surface of radius $(R_1 + 2r)$.

For involute gears it can readily be shown that the contact at a distance s from the pitch point can be represented by two cylinders of radii, $R_{1,2} \sin\psi \pm s$, rotating with the angular velocity of the wheels. In this expression R represents the pitch radius of the wheels and ψ is the pressure angle. The geometry of an involute gear contact is shown in Fig. 3.14. This form of representation explains the use of disc machines to simulate gear tooth contacts and facilitate measurements of the force components and the film thickness.

From the point of view of a mathematical analysis the contact between two cylinders can be adequately described by an equivalent cylinder near a plane as shown in Fig. 3.15. The geometrical requirement is that the separation of the cylinders in the initial and equivalent contact should be the same at equal values of x. This simple equivalence can be adequately satisfied in the important region of small x, but it fails as x approaches the radii of the cylinders. The radius of the equivalent cylinder is determined as follows:

$$h = h_0 + h_a + h_b. \tag{3.59}$$

Using approximations

$$h \approx h_0 + \frac{x^2}{2R_a} + \frac{x^2}{2R_b}$$

and

$$h \approx h_0 + \frac{x^2}{2}\left(\frac{1}{R_a} + \frac{1}{R_b}\right).$$

For the equivalent cylinder

$$h \approx h_0 + \frac{x^2}{2R}.$$

Hence, the separation of the solids at any given value of x will be equal if

$$1/R = 1/R_a + 1/R_b.$$

The radius of the equivalent cylinder is then

$$R = \frac{R_a R_b}{R_a + R_b}. \tag{3.60}$$

If the centres of the cylinders lie on the same side of the common tangent at the contact point and $R_a > R_b$, the radius of the equivalent cylinder takes the form

$$R = \frac{R_a R_b}{R_a - R_b}. \tag{3.61}$$

From the lubrication point of view the representation of a contact by an equivalent cylinder near a plane is adequate when pressure generation is considered, but care must be exercised in relating the force components on the original cylinders to the force components on the equivalent cylinder. The normal force components along the centre-lines as shown in Fig. 3.15 are directly equivalent since, by definition

$$P_{za} = P_{zb} = P_z = \int p \, dx.$$

The normal force components in the direction of sliding are defined as

$$P_{xa} = -\int p \, dh_a = -\frac{1}{R_a} \int px \, dx,$$

$$P_{xb} = -\int p \, dh_b = -\frac{1}{R_b} \int px \, dx,$$

$$P_x = -\int p \, dh = -\frac{1}{R} \int px \, dx.$$

Hence

$$P_{xa} = \frac{R}{R_a} P_x$$

and

$$P_{xb} = \frac{R}{R_b} P_x.$$

For the friction force components it can also be seen that

$$F_a = F_0 = \int \tau_0 \, dx,$$

$$F_b = F_h = \int \tau_h \, dx,$$

where $\tau_{o,h}$ represents the tangential surface stresses acting on the solids.

References to Chapter 3

1. S. Timoshenko and J. N. Goodier. *Theory of Elasticity*. New York: McGraw-Hill, 1951.
2. D. Tabor. *The Hardness of Metals*. Oxford: Oxford University Press, 1951.
3. J. A. Greenwood and J. B. P. Williamson. Contact of nominally flat surfaces. *Proc. Roy. Soc.*, **A295** (1966), 300.
4. J. F. Archard. The temperature of rubbing surfaces. *Wear*, **2** (1958-9), 438.
5. K. L. Johnson. *Contact Mechanics*. Cambridge: Cambridge University Press, 1985.
6. H. S. Carslaw and J. C. Jaeger. *Conduction of Heat in Solids*. London: Oxford University Press, 1947.
7. H. Blok. *Surface Temperature under Extreme Pressure Conditions*. Paris: Second World Petroleum Congress, 1937.
8. J. C. Jaeger. Moving sources of heat and the temperature of sliding contacts. *Proc. Roy. Soc. NSW*, **10**, (1942), 000.

4 Friction, lubrication and wear in lower kinematic pairs

4.1. Introduction

Every machine consists of a system of pieces or lines connected together in such a manner that, if one is made to move, they all receive a motion, the relation of which to that of the first motion, depends upon the nature of the connections. The geometric forms of the elements are directly related to the nature of the motion between them. This may be either:

(i) sliding of the moving element upon the surface of the fixed element in directions tangential to the points of restraint;

(ii) rolling of the moving element upon the surface of the fixed element; or

(iii) a combination of both sliding and rolling.

If the two profiles have identical geometric forms, so that one element encloses the other completely, they are referred to as a closed or lower pair. It follows directly that the elements are then in contact over their surfaces, and that motion will result in sliding, which may be either in curved or rectilinear paths. This sliding may be due to either turning or translation of the moving element, so that the lower pairs may be subdivided to give three kinds of constrained motion:

(a) a turning pair in which the profiles are circular, so that the surfaces of the elements form solids of revolution;

(b) a translating pair represented by two prisms having such profiles as to prevent any turning about their axes;

(c) a twisting pair represented by a simple screw and nut. In this case the sliding of the screw thread, or moving element, follows the helical path of the thread in the fixed element or nut.

All three types of constrained motion in the lower pairs might be regarded as particular modifications of the screw; thus, if the pitch of the thread is reduced indefinitely so that it ultimately disappears, the motion becomes pure turning. Alternatively, if the pitch is increased indefinitely so that the threads ultimately become parallel to the axis, the motion becomes a pure translation. In all cases the relative motion between the surfaces of the elements is by sliding only.

It is known that if the normals to three points of restraint of any plane figure have a common point of intersection, motion is reduced to turning about that point. For a simple turning pair in which the profile is circular, the common point of intersection is fixed relatively to either element, and continuous turning is possible.

4.2. The concept of friction angle

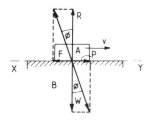

Figure 4.1

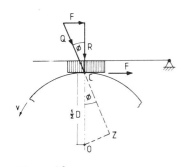

Figure 4.2

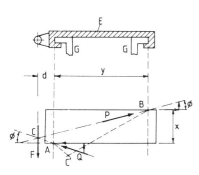

Figure 4.3

Figure 4.1 represents a body A supporting a load W and free to slide on a body B bounded by the stationary horizontal surface $X-Y$. Suppose the motion of A is produced by a horizontal force P so that the forces exerted by A on B are P and the load W. Conversely, the forces exerted by B on A are the frictional resistance F opposing motion and the normal reaction R. Then, at the instant when sliding begins, we have by definition

$$\text{static coefficient of friction} = f = F/R. \tag{4.1}$$

We now combine F with R, and P with W, and then, since $F = P$ and $R = W$, the inclination of the resultant force exerted by A and B, or vice versa, to the common normal NN is given by

$$\tan\phi = F/R = P/W = f. \tag{4.2}$$

The angle $\phi = \tan^{-1} f$ is called the angle of friction or more correctly the *limiting angle of friction*, since it represents the maximum possible value of ϕ at the commencement of motion. To maintain motion at a constant velocity, V, the force P will be less than the value when sliding begins, and for lubricated surfaces such as a crosshead slipper block and guide, the minimum possible value of ϕ will be determined by the relation

$$\phi_{\min} = \tan^{-1} f_{\min}. \tag{4.3}$$

In assessing a value for f, and also ϕ, for a particular problem, careful distinction must be made between kinetic and static values. An example of dry friction in which the kinetic value is important is the brake block and drum shown schematically in Fig. 4.2. In this figure

$R =$ the normal force exerted by the block on the drum,
$F =$ the tangential friction force opposing motion of the drum,
$Q = F/\sin\phi =$ the resultant of F and R,
$D =$ the diameter of the brake drum.

The retarding or braking is then given by

$$\tfrac{1}{2}FD = \tfrac{1}{2}QD\sin\phi = Q \times OZ. \tag{4.4}$$

The coefficient of friction, f, usually decreases with increasing sliding velocity, which suggests a change in the mechanism of lubrication. In the case of cast-iron blocks on steel tyres, the graphitic carbon in the cast-iron may give rise to adsorbed films of graphite which adhere to the surface with considerable tenacity. The same effect is produced by the addition of colloidal graphite to a lubricating oil and the films, once developed, are generally resistant to conditions of extreme pressure and temperature.

4.2.1. Friction in slideways

Figure 4.3 shows the slide rest or saddle of a lathe restrained by parallel guides G. A force F applied by the lead screw will tend to produce clockwise rotation of the moving element and, assuming a small side clearance, rotation will be limited by contact with the guide surfaces at A and B. Let P

and Q be the resultant reactions on the moving element at B and A respectively. These will act at an angle ϕ with the normal to the guide surface in such a manner as to oppose the motion. If ϕ is large, P and Q will intersect at a point C' to the left of F and jamming will occur. Alternatively, if ϕ is small, as when the surfaces are well lubricated or have intrinsically low-friction properties, C' will lie to the right of F so that the force F will have an anticlockwise moment about C' and the saddle will move freely. The limiting case occurs when P and Q intersect at C on the line of action of F, in which case

$$x = (y+d)\tan\phi + d\tan\phi$$
$$x = \tan\phi(y+2d)$$

and

$$f = \tan\phi = \frac{x}{y+2d}. \tag{4.5}$$

Hence, to ensure immunity from jamming f must not exceed the value given by eqn (4.5). By increasing the ratio $x:y$, i.e. by making y small, the maximum permissible value of f greatly exceeds any value likely to be attained in practice.

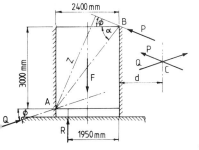

Figure 4.4

Numerical example

A rectangular sluice gate, 3 m high and 2.4 m wide, can slide up and down between vertical guides. Its vertical movement is controlled by a screw which, together with the weight of the gate, exerts a downward force of 4000 N in the centre-line of the sluice. When it is nearly closed, the gate encounters an obstacle at a point 460 mm from one end of the lower edge. If the coefficient of friction between the edges of the gate and the guides is $f = 0.25$, calculate the thrust tending to crush the obstacle. The gate is shown in Fig. 4.4.

Solution

A.   Analytical solution

Using the notion of Fig. 4.4, P and Q are the constraining reactions at B and A. R is the resistance due to the obstacle and F the downward force in the centre-line of the sluice.
Taking the moment about A,

$$R \times 0.45 + P \times z = F \times 1.2.$$

Resolving vertically

$$(P+Q)\sin\phi + R = F.$$

Resolving horizontally

$$P\cos\phi = Q\cos\phi$$

and so

$$P = Q.$$

To calculate the perpendicular distance z we have

$$AB = \sqrt{2.4^2 + 3^2} = 3.84 \text{ m}$$

$$\tan \alpha = \frac{3}{2.4} = 1.25; \qquad \alpha = 51°34'$$

$$\tan \phi = f = 0.25; \qquad \phi = 14°2'$$

and

$$(\alpha + \phi) = 65°36'$$

and so

$$z = AB \sin(\alpha + \phi) = 3.84 \sin 65°36' = 3.48 \text{ m}.$$

Substituting, the above equations become

$$0.45R + 3.48P = 4800$$
$$R + 7.73P = 10\,666.7$$

from this

$$R + 7.73P = 10\,666.7; \qquad R = 10\,666.7 - 7.73P$$

because $P = Q$

$$2P \sin \phi + R = 4000$$
$$2P \sin \phi + 10\,666.7 - 7.73P = 4000$$
$$0.48P + 10\,666.7 - 7.73P = 4000$$
$$-7.25P = -6666.7$$
$$P = 919.5 \text{ N}$$
$$R = 3559 \text{ N}.$$

B.   Graphical solution

We now produce the lines of action of P and Q to intersect at the point C, and suppose the distance of C from the vertical guide through B is denoted by d. Then, taking moment about C

$$F(d + 1.2) = R(d + 1.95).$$

By measurement (if the figure is drawn to scale) $d = 4.8$ m

$$R = \frac{4000(4.8 + 1.2)}{(4.8 + 1.95)} = 3555.5 \text{ N}.$$

4.2.2.   Friction stability

A block B, Fig. 4.5, rests upon a plate A of uniform thickness and the plate is caused to slide over the horizontal surface C. The motion of B is prevented

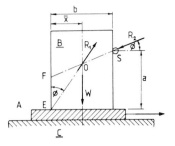

Figure 4.5

by a fixed stop S, and ϕ is the angle of friction between the contact surfaces of B with A and S. Suppose the position of S is such that tilting of B occurs; the resultant reaction R_1 between the surfaces of A and B will then be concentrated at the corner E. Let R_2 denote the resultant reaction between S and B, then, taking moments about E

$$\text{tilting couple} = R_2 \cos \phi a - R_2 \sin \phi b - W\bar{x}.$$

The limiting case occurs when this couple is zero, i.e. when the line of action of R_2 passes through the intersection O of the lines of action of W and R_1. The three forces are then in equilibrium and have no moment about any point. Hence

$$W\bar{x} = R_2(a \cos \phi - b \sin \phi). \tag{4.6}$$

But

$$R_1 \sin \phi = R_2 \cos \phi$$

and

$$W = R_1 \cos \phi - R_2 \sin \phi$$

from which

$$R_1 = W\frac{\cos \phi}{\cos 2\phi} \quad \text{and} \quad R_2 = W\frac{\sin \phi}{\cos 2\phi}.$$

Substituting these values of R_1 and R_2 in eqn (4.6) gives

$$\bar{x} = \frac{\sin \phi}{\cos 2\phi}(a \cos \phi - b \sin \phi)$$

$$= \tfrac{1}{2}\tan 2\phi(a - b \tan \phi)$$

or

$$a = 2\bar{x} \cot 2\phi + b \tan \phi. \tag{4.7}$$

If a exceeds this value tilting will occur.

The above problem can be solved graphically. The triangle of forces is shown by OFE, and the limiting value of a can be determined directly by drawing, since the line of action of R_2 then passes through O. For the particular case when the stop S is regarded as frictionless, SF will be horizontal, so that

$$\tan \phi = \frac{\bar{x}}{a} = f$$

or

$$a = \bar{x}/f. \tag{4.8}$$

Now suppose that A is replaced by the inclined plane or wedge and that B moves in parallel guides. The angle of friction is assumed to be the same at all rubbing surfaces. The system, shown in Fig. 4.6, is so proportioned that,

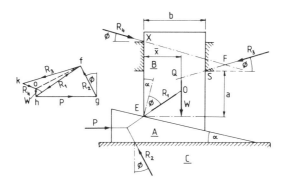

Figure 4.6

as the wedge moves forward under the action of a force P, the reaction R_3 at S must pass above O, the point of intersection of R_1 and W. Hence, tilting will tend to occur, and the guide reactions will be concentrated at S and X as shown in Fig. 4.6.

The force diagram for the system is readily drawn. Thus hgf is the triangle of forces for the wedge (the weight of the wedge is neglected). For the block B, oh represents the weight W; hf the reaction R_1 at E; and, since the resultant of R_3 and R_4 must be equal and opposite to the resultant of R_1 and W, of must be parallel to OF, where F is the point of intersection of R_3 and R_4. The diagram $ohgfk$ can now be completed.

Numerical example

The 5×10^4 N load indicated in Fig. 4.7 is raised by means of a wedge. Find the required force P, given that $\tan \alpha = 0.2$ and that $f = 0.2$ at all rubbing surfaces.

Solution

In this example the guide surfaces are so proportioned that tilting will not occur. The reaction R_4 (of Fig. 4.6) will be zero, and the reaction R_3 will adjust itself arbitrarily to pass through O.

Hence, of in the force diagram (Fig. 4.7) will fall along the direction of R_3 and the value of W for a given value of P will be greater than when tilting occurs. Tilting therefore diminishes the efficiency as it introduces an additional frictional force. The modified force diagram is shown in Fig. 4.7. From the force diagram

$$\text{mechanical advantage} = \frac{W}{P} = \left(\frac{W}{R_1}\right) \Big/ \left(\frac{P}{R_1}\right)$$

$$= \frac{\sin[90° - (\alpha + 2\phi)]}{\sin(90° + \phi)} \frac{\sin(90° - \phi)}{\sin(\alpha + 2\phi)}$$

$$= \frac{\cos(\alpha + 2\phi)}{\sin(\alpha + 2\phi)} = \cot(\alpha + 2\phi). \tag{4.9}$$

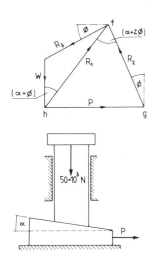

Figure 4.7

Equation (4.9) is derived using the law of sines. Also

$$\text{velocity ratio} = \frac{\text{distance moved by } P}{\text{distance moved by } W} = \cot \alpha \qquad (4.10)$$

and so

$$\text{efficiency} = \frac{\text{mechanical advantage}}{\text{velocity ratio}} = \frac{\cot(\alpha + 2\phi)}{\cot \alpha}$$

$$= \frac{\tan \alpha}{\tan(\alpha + 2\phi)}. \qquad (4.11)$$

In the example given; $\tan \alpha = 0.2$, therefore $\alpha = 11° 18'$ and since $\tan \alpha = \tan \phi$ and $\tan \phi = f = 0.2$, therefore $\phi = 11° 18'$

$$\alpha + 2\phi = 33° 54',$$
$$\tan(\alpha + 2\phi) = 0.672,$$

and thus

$$P = W \tan(\alpha + 2\phi) = 5 \times 10^4 \times 0.672 = 3.36 \times 10^4 \, \text{N}.$$

4.3. Friction in screws with a square thread

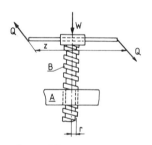

Figure 4.8

Figure 4.8 shows a square threaded screw B free to turn in a fixed nut A. The screw supports an axial load W, which is free to rotate, and the load is to be lifted by the application of forces Q which constitute a couple. This is the ideal case in which no forces exist to produce a tilting action of the screw in the nut. Assuming the screw to be single threaded, let

p = the pitch of the screw,
r = the mean radius of the threads,
α = the slope of the threads at radius r,

then

$$\tan \alpha = \frac{p}{2\pi r}. \qquad (4.12)$$

The reactions on the thread surfaces may be taken as uniformly distributed at radius r. Summing these distributed reactions, the problem becomes analogous to the motion of a body of weight W up an inclined plane of the same slope as the thread, and under the action of a horizontal force P. For the determination of P

$$\text{couple producing motion} = Qz = Pr$$

thus

$$P = \frac{Qz}{r}. \qquad (4.13)$$

It will be seen that the forces at the contact surfaces are so distributed as to give no side thrust on the screw, i.e. the resultant of all the horizontal

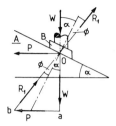

Figure 4.9

components constitute a couple of moment Pr. For the inclined plane, Fig. 4.9, Oab is the triangle for forces from which

$$P = W\tan(\alpha + \phi), \tag{4.14}$$

where ϕ is the angle of friction governed by the coefficient of friction. During one revolution of the screw the load will be lifted a distance p, equal to the pitch of the thread, so that

$$\text{useful work done} = Wp,$$
$$\text{energy exerted} = 2\pi rP,$$

$$\text{efficiency } \eta = \frac{W}{P}\frac{p}{2\pi r},$$

$$\eta = \frac{\tan\alpha}{\tan(\alpha + \phi)}. \tag{4.15}$$

Because $W/P = W/[W\tan(\alpha + \phi)]$ according to eqn (4.14), one full revolution results in lifting the load a distance p, so that

$$\tan\alpha = \frac{p}{2\pi r}.$$

If the limiting angle of friction ϕ for the contacting surfaces is assumed constant, then it is possible to determine the thread angle α which will give maximum efficiency; thus, differentiating eqn (4.15) with respect to α and equating to zero

$$\tan(\alpha + \phi)\sec^2\alpha - \tan\alpha\sec^2(\alpha + \phi) = 0$$

that is

$$\sin 2(\alpha + \phi) = \sin 2\alpha$$

or

$$2\alpha + 2\phi = \pi - 2\alpha$$

and finally

$$\alpha = \tfrac{1}{4}\pi - \tfrac{1}{2}\phi$$

so that

$$\text{maximum efficiency} = \frac{\tan(\tfrac{1}{4}\pi - \tfrac{1}{2}\phi)}{\tan(\tfrac{1}{4}\pi + \tfrac{1}{2}\phi)} = \frac{(1 - \tan\tfrac{1}{2}\phi)^2}{(1 + \tan\tfrac{1}{2}\phi)^2}$$

$$= \frac{1 - \sin\phi}{1 + \sin\phi}. \tag{4.17}$$

For lubricated surfaces, assume $f = 0.1$, so that $\phi = 6°$ approximately. Hence for maximum efficiency $\alpha = (\tfrac{1}{4}\pi - \tfrac{1}{2}\phi) = 42°$ and the efficiency is then 81 per cent.

There are two disadvantages in the use of a large thread angle when the screw is used as a lifting machine, namely low mechanical advantage and

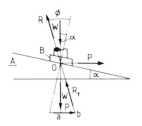

Figure 4.10

the fact that when $\alpha > \phi$ the machine will not sustain the load when the effort is removed. Thus, referring to the inclined plane, Fig. 4.9, if the motion is reversed the reaction R_1 will lie on the opposite side of the normal ON in such a manner as to oppose motion. Hence, reversing the sign of ϕ in eqn (4.14)

$$P = W \tan(\alpha - \phi) \tag{4.18}$$

and if $\alpha > \phi$ this result gives the value of P which will just prevent downward motion. Alternatively, if $\alpha < \phi$, the force P becomes negative and is that value which will just produce downward motion. In the latter case the system is said to be self-locking or self-sustaining and is shown in Fig. 4.10. When $\alpha = \phi$ the system is just self-sustaining. Thus, if $\alpha = \phi = 6°$, corresponding to the value of $f = 0.1$, then when the load is being raised

$$\text{efficiency} = \frac{\tan \alpha}{\tan(\alpha + \phi)} = \frac{0.1051}{0.2126} = 49.5\%$$

and

$$W/P = \cot(\alpha + \phi) = 4.705.$$

On the other hand, for the value $\alpha = 42°$, corresponding to the maximum efficiency given above

$$W/P = \cot 48° = 0.9$$

and the mechanical advantage is reduced in the ratio $4.75 : 0.9 = 5.23 : 1$.

In general, the following is approximately true: a machine will sustain its load, if the effort is removed, when its efficiency, working direct, is less than 50 per cent.

4.3.1. Application of a threaded screw in a jack

The screw jack is a simple example of the use of the square threaded screw and may operate by either:
(i) rotating the screw when the nut is fixed; or
(ii) rotating the nut and preventing rotation of the screw.
Two cases shall be considered.

Case (i) – The nut is fixed

A schematic representation of the screw-jack is shown in Fig. 4.11. The effort is applied at the end of a single lever of length L, and a swivel head is provided at the upper end of the screw. Assuming the jack to be used in such a manner that rotation and lateral movement of the load are prevented, let C denote the friction couple between the swivel head and the upper end of the screw. Then

$$\text{applied couple} = QL$$
$$\text{effective couple on the screw} = Pr = QL - C$$

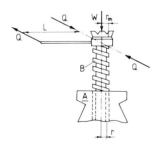

Figure 4.11

so that

$$QL = Pr + C, \tag{4.19}$$

where r = the mean radius of the threads. For the thread, eqn (4.14) applies, namely

$$P = W\tan(\alpha + \phi) \tag{4.20}$$

thus,

$$\text{efficiency} = \frac{Wp}{QL2\pi} = \frac{Wr\tan\alpha}{QL}$$

$$= \frac{Wr\tan\alpha}{Pr + C}$$

or

$$\eta = \frac{Wr\tan\alpha}{Wr\tan(\alpha + \phi) + C}. \tag{4.21}$$

Thus, the efficiency is less than $\tan\alpha/\tan(\alpha + \phi)$, since it is reduced by the friction at the contact surfaces of the swivel head. If the effective mean radius of action of these surfaces is r_{m}, it may be written

$$C = fWr_{\mathrm{m}}$$

and eqn (4.21) becomes

$$\eta = \frac{\tan\alpha}{\tan(\alpha + \phi) + fr_{\mathrm{m}}/r}, \tag{4.22}$$

where fr_{m}/r may be regarded as the virtual coefficient of friction for the swivel head if the load is assumed to be distributed around a circle of radius r.

Case (ii) – The nut rotates

In the alternative arrangement, Fig. 4.12, a torque T applied to the bevel pinion G drives a bevel wheel integral with the nut A, turning on the block C. Two cases are considered:

(1) When rotation of the screw is prevented by a key D. Neglecting any friction loss in the bevel gearing, the torque applied to the nut A is kT, where k denotes the ratio of the number of teeth in A to the number in the pinion G. Again, if P is the equivalent force on the nut at radius r, then

$$kT = Pr. \tag{4.23}$$

Let

r_{m} = the effective mean radius of action of the
      bearing surface of the nut,
d = the distance of the centre-line of the bearing
      surface of the key from the axis of the screw,

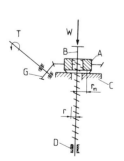

Figure 4.12

then, reducing the frictional effects to radius r

$$\text{the virtual coefficient of friction for the nut} = \tan\phi_2 = f\frac{r_m}{r}$$

$$\text{the virtual coefficient of friction for the key} = \tan\phi_3 = f\frac{d}{r}.$$

The system is now analogous to the problem of the wedge as in Section 4.2. The force diagrams are shown in Fig. 4.13, where $\phi_1 = \tan^{-1} f$ is the true angle of friction for all contact surfaces.

It is assumed that tilting of the screw does not occur; the assumption is correct if turning of the screw is restrained by two keys in diametrically opposite grooves in the body of the jack. Hence

$$\text{mechanical advantage} = \frac{W}{P} = \frac{W}{R_1}\bigg|\frac{P}{R_1} = \frac{oh}{hf}\frac{hf}{hg}$$

$$= \frac{\sin[90° - (\alpha + \phi_1 + \phi_3)]}{\sin(90° + \phi_3)} \frac{\sin(90° - \phi_2)}{\sin(\alpha + \phi_1 + \phi_2)}$$

$$= \frac{\cos(\alpha + \phi_1 + \phi_3)}{\sin(\alpha + \phi_1 + \phi_2)} \frac{\cos\phi_2}{\cos\phi_3}. \tag{4.24}$$

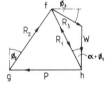

Figure 4.13

Equation (4.24) is derived with the use of the law of sines. The efficiency is given by the expression:

$$\text{efficiency} = \frac{Wp}{T2\pi k} = \frac{W}{P}\frac{p}{2\pi r} = \frac{W}{P}\tan\alpha. \tag{4.25}$$

(2) When rotation of the screw is prevented at the point of application of the load. This method has a wider application in practice, and gives higher efficiency since guide friction is removed. The modified force diagram is shown in Fig. 4.14, where of is now horizontal. Hence, putting $\phi_3 = 0$ in eqn (4.24)

$$\text{mechanical advantage} = \frac{W}{P} = \frac{\cos(\alpha + \phi_1)\cos\phi_2}{\sin(\alpha + \phi_1 + \phi_2)}$$

$$= \frac{\cos(\alpha + \phi_1)\cos\phi_2}{\sin(\alpha + \phi_1)\cos\phi_2 + \cos(\alpha + \phi_1)\sin\phi_2}$$

$$= \frac{1}{\tan(\alpha + \phi_1) + \tan\phi_2} \tag{4.26}$$

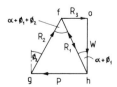

Figure 4.14

Efficiency

$$\eta = \frac{W}{P} \tan \alpha$$

$$= \frac{\tan \alpha}{\tan(\alpha + \phi_1) + \tan \phi_2}. \tag{4.27}$$

Writing $\tan \phi_2 = fr_m/r$, the efficiency becomes

$$\eta = \frac{\tan \alpha}{\tan(\alpha + \phi_1) + fr_m/r}, \tag{4.28}$$

where $f = \tan \phi_1$ is the true coefficient of friction for all contact surfaces. This result is of a similar form to eqn (4.22), and can be deduced directly in the same manner.

In the case of the rotating nut, $C = fWr_m$ is the friction couple for the bearing surface of the nut and, if the pressure is assumed uniformly distributed:

$$r_m = \frac{2}{3} \frac{r_1^3 - r_2^3}{r_1^2 - r_2^2},$$

where r_1 and r_2 are the external and internal radii respectively of the contact surface.

Comparing eqn (4.19) with eqn (4.23), it will be noticed that, in the former, P is the horizontal component of the reaction at the contact surfaces of the nut and screw, whereas in the latter, P is the horizontal effort on the nut at radius r, i.e. in the latter case

$$Pr - fWr_m = Wr \tan(\alpha + \phi_1)$$

or

$$P = W \left[\tan(\alpha + \phi_1) + f \frac{r_m}{r} \right] \tag{4.29}$$

which is another form of eqn (4.26).

Numerical example

Find the efficiency and the mechanical advantage of a screw jack when raising a load, using the following data. The screw has a single-start square thread, the outer diameter of which is five times the pitch of the thread, and is rotated by a lever, the length of which, measured to the axis of the screw, is ten times the outer diameter of the screw; the coefficient of friction is 0.12. The load is free to rotate.

Solution

Assuming that the screw rotates in a fixed nut, then, since the load is free to rotate, friction at the swivel head does not arise, so that $C = 0$. Further, it must be remembered that the use of a single lever will give rise to side friction due to the tilting action of the screw, unless the load is supported laterally. For a single-start square thread of pitch, p, and diameter, d,

$$\text{depth of thread} = \tfrac{1}{2}p$$

so

$$\text{mean radius} = r\tfrac{1}{2}d - \tfrac{1}{4}p$$

but $d = 5p$

$$r = \tfrac{1}{2}d - \frac{1}{20}d = \frac{9}{20}d$$

$$\tan\alpha = \frac{p}{2\pi r} = \frac{1}{2\pi}\cdot\frac{4}{9} = 0.0707$$

$$\alpha = 4° \, 3'$$
$$f = \tan\phi = 0.12; \qquad \phi = 6° \, 51'; \qquad \alpha + \phi = 10° \, 54'$$

now

$$P = W\tan(\alpha + \phi) = \frac{QL}{r}$$

$$\text{mechanical advantage} = \frac{\text{load}}{\text{effort}} = \frac{W}{Q} = \frac{L}{r\tan(\alpha + \phi)}$$

taking into account that $L = 10d$

$$\text{mechanical advantage} = \frac{200}{9 \times 0.1926} = 115.4$$

$$\text{velocity ratio} = \frac{\text{distance moved by the effort}}{\text{distance moved by the load}} = \frac{2\pi L}{p} = 100\pi$$

$$\text{efficiency} = \frac{\text{mechanical advantage}}{\text{velocity ratio}} = \frac{115.4}{100\pi} = 36.7\%$$

alternatively

$$\text{efficiency} = \frac{\tan\alpha}{\tan(\alpha + \phi)} = \frac{0.0707}{0.1226} = 36.7\%.$$

4.4. Friction in screws with a triangular thread

The analogy between a screw thread and the inclined plane applies equally to a thread with a triangular cross-section. Figure 4.15 shows the section of a V-thread working in a fixed nut under an axial thrust load W. In the figure

$\alpha = $ the helix angle at mean radius, r

$\psi_a = $ the semi-angle of the thread measured on a section through the axis of the screw,

$\psi_n = $ the semi-angle on a normal section perpendicular to the helix,

$\phi = $ the true angle of friction, where $f = \tan\phi$.

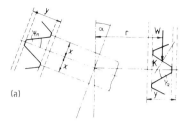

(a)

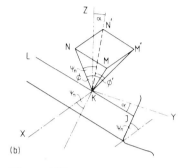

(b)

Figure 4.15

Referring to Fig. 4.15, JKL is a portion of a helix on the thread surface at mean radius, r, and KN is the true normal to the surface at K. The resultant reaction at K will fall along KM at an angle ϕ to KN. Suppose that KN and KM are projected on to the plane YKZ. This plane is vertical and tangential to the cylinder containing the helix JKL. The angle $M'KN' = \phi'$ may be regarded as the virtual angle of friction, i.e. if ϕ' is used instead of ϕ, the thread reaction is virtually reduced to the plane YKZ and the screw may be treated as having a square thread. Hence

$$\text{efficiency} = \frac{\tan \alpha}{\tan(\alpha + \phi')} \tag{4.30}$$

as for a square thread. The relation between ϕ and ϕ' follows from Fig. 4.15

$$\tan \phi' = \frac{M'N'}{KN'} = \frac{MN}{KN \cos \psi_n} = \tan \phi \sec \psi_n. \tag{4.31}$$

Further, if the thread angle is measured on the section through the axis of the screw, then, using the notation of Fig. 4.15, we have

$$\tan \psi_n = \frac{x}{y}$$

so that

$$\tan \psi_a = \frac{x \sec \alpha}{y} = \tan \psi_n \sec \alpha. \tag{4.32}$$

These three equations taken together give the true efficiency of the triangular thread. If $f' = \tan \phi'$ is the virtual coefficient of friction then

$$f' = f \sec \psi_n$$

according to eqn (4.31). Hence, expanding eqn (4.30) and eliminating ϕ',

$$\text{efficiency} = \frac{\sin \alpha - f \sin^2 \alpha \, \dfrac{\sec \psi_n}{\cos \alpha}}{\sin \alpha + f \cos^2 \alpha \, \dfrac{\sec \psi_n}{\cos \alpha}}. \tag{4.33}$$

But from eqn (4.32)

$$\tan \psi_n = \frac{\tan \psi_a}{\sec \alpha}$$

and

$$\sec \psi_n = \frac{\sqrt{(\sec^2 \alpha + \tan^2 \psi_n)}}{\sec \alpha}$$

and eliminating ψ_n

$$\text{efficiency} = \frac{\sin \alpha - f \sin^2 \alpha \sqrt{(\sec^2 \alpha + \tan^2 \psi_a)}}{\sin \alpha + f \cos^2 \alpha \sqrt{(\sec^2 \alpha + \tan^2 \psi_a)}}. \tag{4.34}$$

4.5. Plate clutch – mechanism of operation

A long line of shafting is usually made up of short lengths connected together by couplings, and in such cases the connections are more or less permanent. On the other hand, when motion is to be transmitted from one section to another for intermittent periods only, the coupling is replaced by a clutch. The function of a clutch is twofold: first, to produce a gradual increase in the angular velocity of the driven shaft, so that the speed of the latter can be brought up to the speed of the driving shaft without shock; second, when the two sections are rotating at the same angular velocity, to act as a coupling without slip or loss of speed in the driven shaft.

Referring to Fig. 4.16, if A and B represent two flat plates pressed together by a normal force R, the tangential resistance to the sliding of B over A is $F = fR$. Alternatively, if the plate B is gripped between two flat plates A by the same normal force R, the tangential resistance to the sliding of B between the plates A is $F = 2fR$. This principle is employed in the design of disc and plate clutches. Thus, the plate clutch in its simplest form consists of an annular flat plate pressed against a second plate by means of a spring, one being the driver and the other the driven member. The motor-car plate clutch comprises a flat driven plate gripped between a driving plate and a presser plate, so that there are two active driving surfaces.

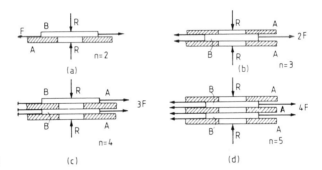

Figure 4.16

Multiple-plate clutches, usually referred to as disc clutches have a large number of thin metal discs, each alternate disc being free to slide axially on splines or feathers attached to the driving and driven members respectively (Fig. 4.17). Let n = the total number of plates with an active driving surface, including surfaces on the driving and driven members, if active, then; $(n-1)$ = the number of pairs of active driving surfaces in contact.

If F is the tangential resistance to motion reduced to a mean radius, r_m, for each pair of active driving surfaces, then

$$\text{total driving couple} = (n-1)Fr_m \qquad (4.35)$$

The methods used to estimate the friction couple Fr_m, for each pair of active surfaces are precisely the same as those for the other lower kinematic pairs, such as flat pivot and collar bearings. For new clutch surfaces the pressure intensity is assumed uniform. On the other hand, if the surfaces become worn the pressure distribution is determined from the conditions of uniform wear, i.e. the intensity of pressure is inversely proportional to the

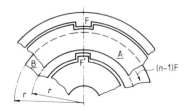

Figure 4.17

radius. Let r_1 and r_2 denote the maximum and minimum radii of action of the contact surfaces, $R =$ the total axial force exerted by the clutch springs and $n_a = (n-1) =$ the number of pairs of active surfaces.

Case A, uniform pressure intensity, p

$$R = \pi(r_1^2 - r_2^2)p \tag{4.36}$$

driving couple for each pair of active surfaces $= \tfrac{2}{3}\pi f p(r_1^3 - r_2^3)$

$$= \tfrac{2}{3}fRr\frac{r_1^3 - r_2^3}{r_1^2 - r_2^2}$$

$$\text{total driving couple} = fRn_a\frac{2(r_1^3 - r_2^3)}{3(r_1^2 - r_2^2)}$$

$$\tag{4.37}$$

Case B, uniform wear; pr = C

If p_2 is the greatest intensity of pressure on the friction surfaces at radius r_2, then

$$p_2 r_2 = C$$
$$R = 2\pi C(r_1 - r_2)$$

or

$$R = 2\pi p_2 r_2 (r_1 - r_2) \tag{4.38}$$

Driving couple for each pair of active surfaces $= fR\tfrac{1}{2}(r_1 - r_2)$

$$\text{total driving couple} = fRn_a\tfrac{1}{2}(r_1 - r_2) \tag{4.39}$$

Comparing eqns (4.37) and (4.39), it is seen that the tangential driving force $F = fR$ can be reduced to a mean radius, r_m, namely

$$\text{for uniform pressure, } r_m = \tfrac{2}{3}\frac{r_1^3 - r_2^3}{r_1^2 - r_2^2} \tag{4.40}$$

$$\text{for uniform wear, } r_m = \tfrac{1}{2}(r_1 + r_2) \tag{4.41}$$

Numerical example

A machine is driven from a constant speed shaft rotating at 300 r.p.m. by means of a friction clutch. The moment of inertia of the rotating parts of the machine is 4.6 kgm². The clutch is of the disc type, both sides of the disc being effective in producing driving friction. The external and internal diameters of the discs are respectively 0.2 and 0.13 m. The axial pressure applied to the disc is 0.07 MPa. Assume that this pressure is uniformly distributed and that the coefficient of friction is 0.25.

If, when the machine is at rest, the clutch is suddenly engaged, what length of time will be required for the machine to attain its full speed.

Solution

For uniform pressure, $p = 0.07$ MPa; the total axial force is

$$R = \pi(r_1^2 - r_2^2)p$$
$$R = 3.14(0.1^2 - 0.065^2)70\,000 = 1270 \text{ N}.$$

Effective radius

$$r_m = \frac{2}{3}\frac{r_1^3 - r_2^3}{r_1^2 - r_2^2} = \frac{2}{3}\frac{0.1^3 - 0.065^3}{0.1^2 - 0.065^2} = 0.084 \text{ m}.$$

Number of pairs of active surfaces $n_a = 2$, then

$$\text{friction couple} = fRn_a r_m = 0.25 \times 1270 \times 2 \times 0.084 = 53.34 \text{ Nm}.$$

Assuming uniform acceleration during the time required to reach full speed from rest

$$\text{angular acceleration} = \frac{\text{couple}}{\text{moment of inertia}}$$

$$= \frac{53.34}{4.6}$$

$$\alpha = 11.6 \text{ rad/s}^2$$

$$\text{Full speed} = 300 \text{ r.p.m.} = \frac{2\pi \times 300}{60} = 31.4 \text{ rad s}^{-1}$$

$$\text{time} = \frac{\omega}{\alpha} = \frac{31.4}{11.6} = 2.71 \text{ s}.$$

It should be noted that energy is dissipated due to clutch slip during the acceleration period. This can be shown as follows:

the angle turned through by the constant speed driving shaft during the period of clutch slip is

$$\omega t = 31.4 \times 2.71 = 85.1 \text{ radn}.$$

the angle turned through by the machine shaft during the same period $= \frac{1}{2}\alpha t^2 = \frac{1}{2}\,11.6 \times 2.71^2 = 42.6$ radn, thus

$$\text{angle of slip} = 85.1 - 42.6 = 42.5 \text{ radn}.$$
$$\text{energy dissipated due to clutch slip} = \text{friction couple} \times \text{angle of slip}$$
$$= 53.34 \times 42.5 = 2267 \text{ Nm}$$
$$\text{kinetic energy developed in machine shaft} = \tfrac{1}{2}I\omega^2 = \tfrac{1}{2}4.6 \times 31.4^2 = 2267 \text{ Nm}$$

thus

$$\text{total energy supplied during the period of clutch slip}$$
$$= \text{energy dissipated} + \text{kinetic energy}$$
$$= 2267 + 2267 = 4534 \text{ Nm}.$$

Numerical example

If, in the previous example, the clutch surfaces become worn so that the intensity of pressure is inversely proportional to the radius, compare the power that can be transmitted with that possible under conditions of uniform pressure, and determine the greatest intensity of pressure on the friction surfaces. Assume that the total axial force on the clutch, and the coefficient of friction are unaltered.

Solution

In the case of uniform wear $pr = C$, the total axial force is

$$R = 2\pi C(r_1 - r_2) = 2\pi p_2 r_2 (r_1 - r_2)$$

and so greatest intensity of pressure at radius r_2 is

$$p_2 = \frac{R}{2\pi r_2(r_1 - r_2)} = \frac{1270}{2 \times 3.14 \times 0.065(0.1 - 0.065)}$$

$$= 0.089 \, \text{MPa}.$$

Effective radius $r_m = \frac{1}{2}(r_1 + r_2) = 0.0825 \, \text{m}$,
friction couple $= fRn_a r_m = 0.25 \times 1270 \times 2 \times 0.0825 = 52.39 \, \text{Nm}$

$$\text{power transmitted} = \frac{\text{couple} \times 2\pi N}{60} = \frac{52.39 \times 2 \times 3.14 \times 300}{60} = 1645 \, \text{Watts}$$

under conditions of uniform pressure $p = 0.07 \, \text{MPa}$, thus

$$\text{power transmitted} = \frac{53.34 \times 2 \times 3.14 \times 300}{60} = 1675 \, \text{Watts}$$

4.6.   Cone clutch – mechanism of operation

The cone clutch depends for its action upon the frictional resistance to relative rotation of two conical surfaces pressed together by an axial force. The internal cone W, Fig. 4.18, is formed in the engine fly-wheel rim keyed to the driving shaft. The movable cone, C faced with friction lining material, is free to slide axially on the driven shaft and, under normal driving conditions, contact is maintained by the clutch spring S. The cone C is disengaged from frictional contact by compression of the clutch spring through a lever mechanism. During subsequent re-engagement the spring force must be sufficient to overcome the axial component of friction between the surfaces, in addition to supplying adequate normal pressure for driving purposes.

Referring to Fig. 4.19, let

Q_e = the total axial force required to engage the clutch,
p = the permissible normal pressure on the lining,
α = the semi-angle of the cone,
f_e = the coefficient of friction for engagement.

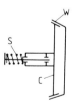

Figure 4.18

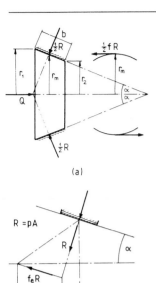

Figure 4.19

Thus, for an element of area δ_a

$$\delta Q_e = p\delta_a \sin\alpha + f_e p\delta_a \cos\alpha$$

or

$$Q_e = R\sin\alpha + f_e R\cos\alpha, \tag{4.42}$$

where $R = pA$ is the total normal load between the bearing surfaces.

Under driving conditions, the normal load R can be maintained by a spring force

$$Q = pA\sin\alpha \tag{4.43}$$

as the friction to be overcome during engagement is then no longer operative. Further, the spring force could be reduced to a value, $R\sin\alpha - f_e R\cos\alpha$, without reduction of the normal load, R, but below this value the clutch would disengage. This conclusion assumes that $\sin\alpha > f_e\cos\alpha$ or $\tan\alpha > f_e$. Alternatively, if $\tan\alpha < f_e$, a reversed axial force will be necessary to disengage the clutch.

One disadvantage of this wedge action resulting from a small cone angle is that clutches of the cone type do not readily respond to disengagement at frequent intervals and, in consequence, are not suited to a purpose where smooth action is desirable. On the other hand, the flat-plate clutch, although requiring a relatively larger axial spring force, is much more sensitive and smooth in action, and is replacing the cone clutch in modern design.

4.6.1. Driving torque

Referring to Fig. 4.19, let r_1 and r_2 denote the radii at the limits of action of the contact surfaces. In the case of uniform pressure

$$\text{torque transmitted} = \tfrac{2}{3}\frac{\pi f p}{\sin\alpha}(r_1^3 - r_2^3)$$

Under driving conditions, however, we must assume

$$Q = pA\sin\alpha,$$

where

$$A = \frac{\pi(r_1^2 - r_2^2)}{\sin\alpha}.$$

Combining these equations, we have

$$\text{torque transmitted} = \tfrac{2}{3}\frac{fQ}{\sin\alpha}\frac{r_1^3 - r_2^3}{r_1^2 - r_2^2}. \tag{4.44}$$

Equation (4.44) can be written in another form, thus

$$\text{mean radius of action, } r_m = \tfrac{2}{3}\frac{r_1^3 - r_2^3}{r_1^2 - r_2^2}$$

and

$$\frac{Q}{\sin \alpha} = pA = R,$$

hence,

$$\text{torque transmitted} = fpAr_{\mathrm{m}}, \tag{4.45}$$

where f is the coefficient of friction for driving conditions. This result is illustrated in Fig. 4.19, where,

$$\text{torque transmitted} = \tfrac{1}{2}fR2r_{\mathrm{m}} = fRr_{\mathrm{m}}.$$

Numerical example

A cone clutch has radii of 127 mm and 152 mm, the semicone angle being 20°. If the coefficient of friction is 0.25 and the allowable normal pressure is 0.14 MPa, find:
(a) the necessary axial load;
(b) the power that can be transmitted at 1000 r.p.m.

Solution

$$\text{mean radius of action} = \tfrac{2}{3}\frac{r_1^3 - r_2^3}{r_1^2 - r_2^2} = \tfrac{2}{3}\frac{0.152^3 - 0.127^3}{0.0152^2 - 0.127^2} = 0.14 \text{ m}$$

$$\text{area of bearing surface, } A = \frac{\pi(r_1^2 - r_2^2)}{\sin \alpha} = \frac{3.14}{0.342} \times 0.007 = 0.064 \text{ m}^2$$

axial load under driving conditions, $Q = pA \sin \alpha$
$$= 0.14 \times 10^6 \times 0.064 \times 0.342$$
$$= 3064 \text{ N}$$

torque transmitted $= fpAr_{\mathrm{m}}$
$$= 0.25 \times 0.14 \times 10^6 \times 0.064 \times 0.14 = 313.6 \text{ Nm}$$

power transmitted $=$ couple $\times$ angular speed

$$= 313.6\frac{2\pi N}{60} = 313.6\frac{3.14 \times 10^3}{30} = 32.8 \text{ kW}$$

4.7. Rim clutch – mechanism of operation

A general purpose clutch, suitable for heavy duty or low speed, as in a line of shafting, is the expanding rim clutch shown in Fig. 4.20. The curved clutch plates, A, are pivoted on the arms, B, which are integral with the boss keyed to the shaft, S. The plates are expanded to make contact with the outer shell C by means of multiple-threaded screws which connect the opposite ends of the two halves of the ring. Each screw has right- and left-hand threads of fast pitch, and is rotated by the lever L, by means of the toggle link E connected to the sliding collar J. The axial pressure on the clutch is provided by a forked lever, the prongs of which enter the groove on the collar, and, when the clutch is disengaged, the collar is in the position marked 1.

Suppose that, when the collar is moved to the position marked 2, the

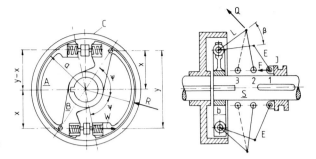

Figure 4.20

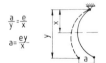

Figure 4.21

axial force F is sufficient to engage the clutch fully. As the screws are of fast pitch, the operating mechanism will not sustain its load if the effort is removed. If, however, the collar is jumped to position 3, the pressure on the clutch plates will tend to force the collar against the boss keyed to the shaft S, and the clutch will remain in gear without continued effort at the sleeve. To avoid undue strain on the operating mechanism, the latter is so designed that the movement of the collar from position 2 to position 3 is small in relation to its total travel. The ends of the operating screw shafts turn in adjusting nuts housed in the arms B and the ends of the clutch plates A. This provides a means of adjustment during assembly and for the subsequent wear of the clutch plate surfaces.

With fabric friction lining the coefficient of friction between the expanding ring and the clutch casing may be taken as 0.3 to 0.4, the allowable pressure on the effective friction surface being in the region of 0.28 to 0.56 MPa. Let e = the maximum clearance between the expanding ring and the outer casing C on the diameter AA, when disengaged. Total relative movement of the free end of the clutch plate in the direction of the screw axis = ey/x (Fig. 4.21).

Hence, if

$$l = \text{the lead of each screw thread}$$
$$\beta = \text{angle turned through by the screw}$$

then

$$\frac{\beta}{2\pi} 2l = e\frac{y}{x},$$

i.e.

$$e = \frac{\beta}{2\pi} \frac{2lx}{y}. \tag{4.46}$$

4.7.1. Equilibrium conditions

It is assumed that the curved clutch plate, A, is circular in form of radius a and that, when fully engaged, it exerts a uniform pressure of intensity p on the containing cylinder. The problem is analogous to that of the hinged

Figure 4.22

brake shoe considered later. Thus, referring to Fig. 4.22

> b = the width of the clutch plate surface,
> 2ψ = the angle subtended at the centre by the effective arc of contact.

Then, length of arc of contact = $2a\psi$, length of chord of contact = $2a\sin\psi$ and the resultant R of the normal pressure intensity, p, on the contact surfaces is given by

$$R = 2pab\sin\psi. \tag{4.47}$$

For an element of length $a \times \mathrm{d}\Theta$ of the clutch surface

$$\text{tangential friction force} = fpab \times \mathrm{d}\Theta.$$

This elementary force can be replaced by a parallel force of the same magnitude, acting at the centre O, together with a couple of moment $fpa^2b \times \mathrm{d}\Theta$. Integrating between the limits $\pm\psi$, the frictional resistance is then equivalent to

(i) a force at O in a direction perpendicular to the line of action of R given by

$$2fpab \int_0^\psi \cos\Theta\,\mathrm{d}\Theta = 2fpab\sin\psi = fR; \tag{4.48}$$

(ii) a couple of moment:

$$M = 2fpa^2b \int_0^\psi \mathrm{d}\Theta = 2fpa^2b\psi. \tag{4.49}$$

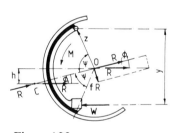

Figure 4.23

The equivalent system of forces and the couple M acting on each curved plate are shown in Fig. 4.23, where W is the axial thrust load in the screw. Taking moments about the hinge and using the notation shown in the figures, we have

$$Wy = Rz\sin\psi + M - fRz\cos\psi$$

so that

$$W = \frac{M}{y} + \frac{Rz}{y}(\sin\psi - f\cos\psi)$$

or

$$W = \frac{M}{y} + \frac{Rz}{y}\frac{\sin(\psi - \phi_1)}{\cos\phi_1}, \tag{4.50}$$

where z = the distance of the centre of the hinge from O, and $\phi_1 = \tan^{-1}f$ is the angle of friction for the clutch plate surface.

An alternative approach is to assume that the resultant of the forces R and fR at O is a force $R_1 = R\sec\phi_1$ at an angle ϕ_1 to the line of action of R. Writing

$$M = R_1 h \tag{4.51}$$

it follows that the couple M and the force R_1 at O may be replaced by a force R_1 acting through the point C on the line of action of R as shown in Fig. 4.23. This force is the resultant reaction on the clutch plate and, taking moments as before

$$Wy = R_1[h + z\sin(\psi - \phi_1)]$$

or

$$W = \frac{M}{y} + \frac{R_1 z}{y}\sin(\psi - \phi_1)$$

i.e.

$$W = \frac{M}{y} + \frac{Rz}{y}\sec\phi_1 \sin(\psi - \phi_1), \tag{4.52}$$

which agrees with eqn (4.50).

4.7.2. Auxiliary mechanisms

If

> r = the mean radius of the operating screw threads,
> α = the slope of the threads at radius r,
> P = the equivalent force on the screw at radius r,

then, since both ends of the screw are in action simultaneously

$$P = 2W\tan(\alpha + \phi), \tag{4.53}$$

where ϕ is the angle of friction for the screw thread surfaces.

The equivalent force at the end of each lever of length L, is then

$$Q = \frac{Pr}{L} = \frac{2Wr}{L}\tan(\alpha + \phi) \tag{4.54}$$

and if k is the velocity ratio of the axial movement of the collar to the circumferential movement of Q, in the position 2 when the clutch plates are initially engaged, then

$$\text{total axial force, } 2F = \frac{2Q}{k} = \frac{4Wr}{kL}\tan(\alpha + \phi). \tag{4.55}$$

In passing from the position 2 to position 3, this axial force will be momentarily exceeded by an amount depending partly upon the elasticity of the friction lining, together with conditions of wear and clearance in the joints of the operating mechanism. Theoretically, the force Q will pass through an instantaneous value approaching infinity, and for this reason, the movement of 2 to 3 should be as small as is possible consistent with the object of sustaining the load when the axial force is removed.

4.7.3. Power transmission rating

The friction torque transmitted by both clutch plates is

$$2M = 4fpa^2b\psi. \tag{4.56}$$

If M is expressed in Nm and N is the speed of the clutch in revolutions per minute, then

$$\text{power transmitted} = \frac{4fpa^2b\psi 2\pi N}{60}. \tag{4.57}$$

4.8. Centrifugal clutch – mechanism of operation

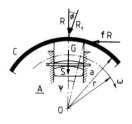

Figure 4.24

In the analysis of the preceding section, inertia effects due to the mass of the clutch plates were neglected. An alternative type of rim clutch operating by centrifugal action is shown in Fig. 4.24. Here, the frictional surfaces are formed on heavy blocks or shoes, A, contained within the cylindrical clutch case, C. The driving member consists of a spider carrying the four shoes which are kept from contact with the clutch case by means of the flat springs until an increase in the centrifugal force overcomes the resistance of the springs, and power is transmitted by friction between the surfaces of the shoes and the case. If

$M =$ the friction couple due to each shoe,
$R =$ the resultant radial pressure on each shoe,

and the angle subtended at the centre O by the arc of contact is assumed to be small, then the uniform pressure intensity between the contact surfaces becomes

$$R = 2pab\psi \qquad \text{very nearly, as} \quad \sin \psi \approx \psi$$

and

$$M = 2fpa^2b\psi = fRa. \tag{4.58}$$

The assumption of uniform pressure is not strictly true, since, due to the tangential friction force, the tendency to tilt in the radial guides will throw the resultant pressure away from the centre-line of the shoe.

Numerical example

Determine the necessary weight of each shoe of the centrifugal friction clutch if 30 kW is to be transmitted at 750 r.p.m., with the engagement beginning at 75 per cent of the running speed. The inside diameter of the drum is 300 mm and the radial distance of the centre of gravity of each shoe from the shaft is 126 mm. Assume a coefficient of friction of 0.25.

Solution

The following solution neglects the tendency to tilt in the parallel guides and assumes uniform pressure intensity on the contact surfaces. Let

$S =$ the radial force in each spring after engagement,
$R =$ the resultant radial pressure on each shoe,

then

$$R + S = \frac{W}{g}\omega^2 r$$

where W is the weight of each shoe and r is the radial distance of the centre of gravity of each shoe from the axis.

At the commencement of engagement $R = 0$ and the angular velocity of rotation is

$$\omega = 0.75 \times 750 \frac{2\pi}{60} = \frac{75\pi}{4}, \text{rad s}^{-1}$$

so that

$$S = W\left(\frac{75\pi}{4}\right)^2 \times 0.126 = 436.7W.$$

At a speed of 750 r.p.m.,

$$R + S = W(78.5)^2 \times 0.126 = 776.4W$$

so

$$R = 776.4W - S = 776.4W - 436.7W = 339.7W.$$

The couple due to each shoe

$$= fRa, \text{ very nearly,}$$
$$= 0.25 \times 339.7W \times 0.150 = 12.74W, \text{Nm}$$

$$\text{power transmitted} = \frac{4 \times 12.74W \times 2\pi \times 750}{60} = 30\,000 \text{ Watts}$$

and finally,

$$W = \frac{30\,000 \times 60}{4 \times 12.74 \times 2\pi \times 750} = 7.5 \text{ kg.}$$

4.9. Boundary lubricated sliding bearings

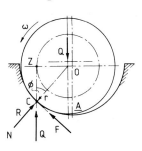

Figure 4.25

Under boundary lubrication conditions the surfaces are considered to be technically dry or only slightly lubricated, so that the resistance to relative motion is due to the interaction between the highest asperities covered by the boundary film. Then, frictional force $F = fR$, where f is the kinetic coefficient of friction. The magnitude of the friction couple retarding the motion of the journal is determined by the assumed geometric conditions of the bearing surface.

Case A. Journal rotating in a loosely fitting bush

Figure 4.25 represents a cross-section of a journal supporting a load Q at the centre of the section. When the journal is at rest the resultant from pressure will be represented by the point A on the line of action of the load Q, i.e. contact is then along a line through A perpendicular to the plane of the section. When rotating commences, we may regard the journal as mounting the bush until the line of contact reaches a position C, where slipping occurs at a rate which exactly neutralizes the rolling action. The resultant reaction at C must be parallel to the line of action of Q at 0, and the two forces will constitute a couple of moment $Q \times OZ$ retarding the motion of the journal. Further, Q at C must act at an angle ϕ to the common normal CN and, if r is the radius of the journal

$$OZ = r \sin \phi$$

hence,

$$\text{Friction couple} = Qr \sin \phi \qquad (4.59)$$

The circle drawn with radius $OZ = r \sin \phi$ is known as the friction circle for the bearing.

Case B. Journal rotating in a closely fitting bush

A closely fitted bearing may be defined as one having a uniform distribution of radial pressure over the complete area of the lower part of the bush (Fig. 4.26). Let

p = the radial pressure per unit area of the bearing surface,
Q = the vertical load on the journal,
l = the length of the bearing surface.

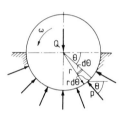

Figure 4.26

Then,

$$Q = \int_0^\pi plr \, d\Theta \sin \Theta = plr \int_0^\pi \sin \Theta \, d\Theta$$

$$= 2plr \qquad (4.60)$$

$$\text{friction couple} = \int_0^\pi fplr \, d\Theta r = fplr^2 \int_0^\pi d\Theta = \pi fplr^2 \qquad (4.61)$$

and substituting for Q,

$$\text{friction couple} = \tfrac{1}{2}\pi frQ. \qquad (4.62)$$

For the purpose of comparison take case A as the standard, and assume boundary conditions of lubrication $f = 0.1$, so that

$$f = \tan \phi = \sin \phi \qquad \text{very nearly}$$

and

$$Qr \sin \phi = frQ \qquad \text{very nearly.}$$

In general, we may then express the friction couple in the form $f'rQ$, where f' is defined as the virtual coefficient of friction, and for the closely fitting bush

$$\text{friction couple} = \tfrac{1}{2}\pi frQ = f'rQ$$

and

virtual coefficient of friction, $f' = \frac{1}{2}\pi f = 1.57f$.

Case C. Journal rotating in a bush under ideal conditions of wear

Let us be assumed that the journal remains circular and unworn and that, after the running-in process, any further wear in the bush reduces the metal in such a way that vertical descent is uniform at all angles. The volume of metal worn away at different angles is proportional to the energy expanded in overcoming friction, so that the pressure will vary over the bearing surface. For vertical displacement, δ, the thickness worn away at angle Θ is $\delta \sin \Theta$, where δ is constant (Fig. 4.27).

Hence, since frictional resistance per unit area is proportional to the intensity of normal pressure p, and the relative velocity of sliding over the circle of radius r is constant, it follows that:

$$p = k \sin \Theta \tag{4.63}$$

where k is a constant,

$$\text{vertical load, } Q = lr \int_0^\pi p \sin \Theta \, d\Theta$$

$$= klr \int_0^\pi \sin^2 \Theta \, d\Theta = \tfrac{1}{2}\pi klr \tag{4.64}$$

$$\text{friction couple} = flr^2 \int_0^\pi p \, d\Theta$$

$$= fklr^2 \int_0^\pi \sin \Theta \, d\Theta = 2fklr^2. \tag{4.65}$$

Hence

$$\text{virtual coefficient} = \frac{\text{friction couple}}{Qr}$$

i.e.

$$f' = \frac{2fklr^2}{\tfrac{1}{2}\pi klr^2} = 1.275f.$$

Summarizing the results of the above three cases

virtual coefficient, $f' = f$ in a loose bearing,
$= 1.57f$ in a new well-fitted bearing,
$= 1.275f$ in a well-worn bearing.

4.9.1.   Axially loaded bearings

Figure 4.28 shows a thrust block or pivot designed on the principle of uniform displacement outlined in case C. In other words, we have the case of a journal rotating in a bush under ideal conditions of wear.

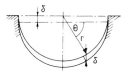

Figure 4.27

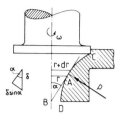

Figure 4.28

The object is to ensure that the thrust block and the collar or rotating pivot maintain an unchanged form after wear. At any radius, r, where the intensity of pressure per unit area of bearing surface is p, work expended in friction is proportional to fpV, volume per unit area worn away by a vertical displacement, $\delta = \delta \sin \alpha$, so that fpV is proportional to $\delta \sin \alpha$. Since f and δ are constant, we have

$$\frac{pV}{\sin \alpha} = \text{const}, \qquad (4.66)$$

where $V = r\omega$ is the circumferential velocity of the pivot surface at radius r, and ω is the angular velocity of the pivot in radians per second. If it is desired that the pressure intensity p should be constant, then, writing $V = r\omega$, eqn (4.66) becomes

$$\frac{r}{\sin \alpha} = \text{const}. \qquad (4.67)$$

Referring to Fig. 4.28, CD is a half-section through the axis of the bearing surface and AB is the tangent to the profile at radius r, where $AB = r/\sin \alpha$. Hence for uniform pressure and uniform wear the profile must be such that the length AB of the tangent is the same for all values of r. If the bearing is of any other shape it will tend to approach this condition after a lapse of time. Equation (4.66) may be applied to any profile. Thus if α is constant and equal to $90°$, then for uniform wear:

$$pV = pr\omega = \text{const},$$

so that the pressure intensity p is proportional to $1/r$ and becomes infinite at the centre where $r = 0$.

4.9.2. Pivot and collar bearings

Two alternative methods of calculation are given below, based on the following assumptions:
 (i) for a new well-fitted bearing the distribution of pressure is uniform;
 (ii) for a well-worn bearing under conditions of uniform wear

$$\frac{pV}{\sin \alpha} = \text{const}$$

or since $V = r\omega$, and α is constant for the bearing surfaces,

$$pr = C. \qquad (4.68)$$

(A). Flat pivot or collar – uniform pressure

Figure 4.29, cases (b) and (c), represent a flat collar and pivot in which the external and internal radii of the bearing surfaces are r_1 and r_2 respectively. Under an axial load Q the bearing pressure is assumed uniform and of

(a)

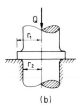

(b)

(c)

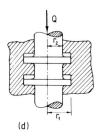

(d)

Figure 4.29

intensity p per unit area, so that

$$Q = \pi(r_1^2 - r_2^2)p. \tag{4.69}$$

Load on an elementary ring of radius $r = 2\pi r p\,dr$, moment of friction due to the elementary ring $= 2\pi f p r^2\,dr$,

$$\text{friction couple} = 2\pi f p \int_{r2}^{r1} r^2\,dr$$

$$= \tfrac{2}{3}\pi f p(r_1^3 - r_2^3),$$

and eliminating p,

$$\text{friction couple} = \tfrac{2}{3}fQ\frac{r_1^3 - r_2^3}{r_1^2 - r_2^2}. \tag{4.70}$$

For the solid pivot (case (a), Fig. 4.29), r_2 is zero, hence

$$\text{friction couple} = \tfrac{2}{3}fQr_1. \tag{4.71}$$

For the thrust block bearing of the type shown in Fig. 4.29, case (d), the thrust is taken on a number of collars, say n, and the pressure intensity p is then given by

$$p = \frac{Q}{n\pi(r_1^2 - r_2^2)} \tag{4.72}$$

so that

$$\text{friction couple} = \tfrac{2}{3}n\pi f p(r_1^3 - r_2^3)$$

$$= \tfrac{2}{3}fQ\frac{r_1^3 - r_2^3}{r_1^2 - r_2^2}$$

as for the single flat collar bearing.

(B). Flat pivot or collar – uniform wear

In this case, the intensity of the bearing pressure at radius r is determined by the condition

$$pr = C,$$

so that

$$\text{normal load on elementary ring} = 2\pi r p\,dr = 2\pi C\,dr$$

$$\text{total load on the pivot or collar} = 2\pi C \int_{r2}^{r1} dr$$

or

$$Q = 2\pi C(r_1 - r_2)$$

and

$$C = \frac{Q}{2\pi(r_1 - r_2)}.$$

Moment of friction due to elementary ring $= 2\pi f p r^2 \, \mathrm{d}r$
$$= 2\pi f C r \, \mathrm{d}r$$

$$\text{friction couple} = 2\pi f C \int_{r_2}^{r_1} r \, \mathrm{d}r$$

$$= \pi f C (r_1^2 - r_2^2)$$

$$= f Q \frac{r_1 + r_2}{2}. \tag{4.73}$$

Again, for the solid pivot (Fig. 4.29, case (a)), writing $r_2 = 0$ as before

$$\text{friction couple} = f Q \frac{r_1}{2}. \tag{4.74}$$

(C). *Conical pivot – uniform pressure*

The system analysed is shown in Fig. 4.30. Proceeding as before, the intensity of the bearing pressure at radius r is determined by the condition $p r = C$, so that

$$\text{normal load on elementary ring} = 2\pi p r \frac{\mathrm{d}r}{\sin \alpha}$$

$$= \frac{2\pi C \, \mathrm{d}r}{\sin \alpha}$$

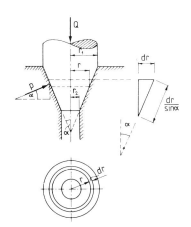

Figure 4.30

and so

$$Q = \frac{2\pi C (r_1 - r_2)}{\sin \alpha} \sin \alpha$$

$$= 2\pi C (r_1 - r_2) \tag{4.75}$$

$$\text{friction couple} = \int_{r_2}^{r_1} 2\pi f p r^2 \frac{\mathrm{d}r}{\sin \alpha}$$

$$= \frac{2\pi f C}{\sin \alpha} \int_{r_2}^{r_1} r \, \mathrm{d}r = \frac{\pi f C}{\sin \alpha} (r_1^2 - r_2^2)$$

and substituting for C

$$\text{friction couple} = \tfrac{1}{2} \frac{f Q}{\sin \alpha} (r_1 + r_2). \tag{4.76}$$

Numerical example

Show that the virtual coefficient of friction for a shaft rotating in a V-groove of semi-angle α, and loaded symmetrically with respect to the groove, is given by

$$f' = \frac{\sin 2\phi}{2 \sin \alpha},$$

where ϕ is the angle of friction for the contact surfaces.

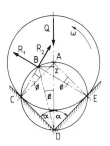

Figure 4.31

Solution

Referring to Fig. 4.31, EA and CA are the common normals to the contact surfaces at the points of contact E and C respectively. R_1 and R_2 are the resultant reactions at E and C inclined at an angle ϕ to the common normals in such a manner as to oppose rotation of the shaft.

If R_1 and R_2 intersect at B, it follows that the points A, B, C, D, E lie on a circle of diameter AD. Hence

$$\text{angle } BDA = \phi$$
$$\text{angle } ABD = \tfrac{1}{2}\pi.$$

The resultant of R_1 and R_2 must be parallel to the line of action of the load Q, so that

$$\text{friction couple} = Q \times BZ,$$

where $BZ = BD \sin \phi = AD \cos \phi \sin \phi$ and $r = AD \sin \alpha$. Thus

$$\text{friction couple} = Q \frac{r}{\sin \alpha} \cos \phi \sin \phi = f'rQ,$$

where f' is the virtual coefficient of friction defined previously, and so

$$f' = \frac{\sin 2\phi}{2 \sin \alpha}.$$

4.10.   Drives utilizing friction force

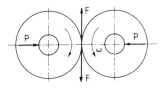

Figure 4.32

In higher pairs of elements there is incomplete restraint of motion. Therefore, force closure is necessary if the motion of one element relative to the other is to be completely constrained. In higher pairing, friction may be a necessary counterpart of the closing force as in the case of two friction wheels (Fig. 4.32). Here, the force P not only holds the cylinders in contact, but must be sufficient to prevent relative sliding between the circular elements if closure is to be complete.

Now, consider the friction drive between two pulleys connected by a belt, Fig. 4.33, then for the pair of elements represented by the driven pulley and the belt (case (b) in Fig. 4.33), the belt behaves as a rigid body in tension only. If the force T_1 were reversed, or the belt speed V were to fall momentarily below $r\omega$, this rigidity would be lost. Hence, force closure is incomplete, and the pulley is not completely restrained since a degree of freedom may be introduced. A pulley and that portion of the belt in contact with it, together constitute an incompletely constrained higher pair which is kinematically equivalent to a lower pair of elements.

Assuming that the pulleys are free to rotate about fixed axes, complete kinematic closure is obtained when an endless flexible belt is stretched tightly over the two pulleys. The effects of elasticity are for a moment neglected, so that the belt behaves as a rigid body on the straight portions and the motion can then be reversed. This combination of two incomplete

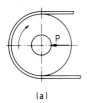

(a)

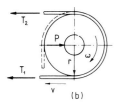

(b)

Figure 4.33

higher pairs is the kinematic equivalent of two lower pairs, and gives the same conditions of motion as the higher pair of elements represented by the two friction wheels in direct contact.

4.10.1. Belt drive

When two pulleys connected by a belt are at rest, the tensions in the two straight portions are equal, and will be referred to as the initial tension. If a torque is applied to the driving pulley, and the initial pressure between the contact surfaces is sufficient, slipping will be prevented by friction, with the result that the tensions in the straight portions will no longer be equal. The difference between the two tensions will be determined by the resistance to motion of the driven pulley and, if limiting friction is reached, slipping will occur. Let

$$T_1 = \text{the tension on the tight side of the belt,}$$
$$T_2 = \text{the tension on the slack side of the belt,}$$
$$f = \text{the limiting coefficient of friction, assumed constant.}$$

Consider the equilibrium of an element of length $r\delta\Theta$ when slipping is about to commence and Θ being measured from the point of tangency of T_2 with the pulley surface (Fig. 4.34). Thus T is the belt tension at angle Θ, $T + \delta T$ is the belt tension at angle $\Theta + \delta\Theta$, R is the normal reaction exerted by the pulley on the element passing through the intersection of T and $T + \delta T$ and fR is the tangential friction force.

If motion of the driven pulley occurs it is assumed that the speed is low, so that centrifugal effects may be neglected. The polygon of forces for the element is shown in Fig. 4.34, and to the first order of small we may write

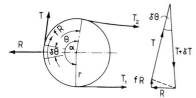

Figure 4.34

$$\delta T = fR,$$
$$R = T\delta\Theta,$$

and in the limit

$$\frac{\mathrm{d}T}{\mathrm{d}\Theta} = fT, \tag{4.77}$$

hence

$$\int \frac{\mathrm{d}T}{T} = f\Theta,$$

$$\log_e T = f\Theta + \text{const.}$$

When $\Theta = 0$, $T = T_2$, and so

$$\log_e \frac{T}{T_2} = \ln \frac{T}{T_2} = f\Theta$$

or

$$\frac{T}{T_2} = e^{f\Theta}. \tag{4.78}$$

Alternatively, if α denotes the total angle of lap for the driven pulley, this result may be written as

$$\frac{T_1}{T_2} = e^{f\alpha}. \tag{4.79}$$

The integration is based on the assumption that f is constant over the contact surface. Under conditions of boundary friction this is not strictly true as f may vary with the intensity of pressure on the bearing surface. Let

> $p =$ the normal pressure per unit area of the contact surface of
> the belt and pulley at position Θ,
> $b =$ the width of the belt,

then, for the element

$$R = pbr\delta\Theta = T\delta\Theta$$

so that,

$$p = \frac{T}{br}. \tag{4.80}$$

This pressure intensity is therefore directly proportional to the tensile stress in the belt at the point considered. If g is the tensile stress in the belt at position Θ and t is the belt thickness

$$g = \frac{T}{bt} \tag{4.81}$$

and so

$$\frac{p}{g} = \frac{t}{r}. \tag{4.82}$$

4.10.2. Mechanism of action

The effect of elasticity on the frictional action between the belt and the pulley surfaces is a vitally important factor in the solution of problems relating to power transmission by belt drives. For a well-designed belt under driving conditions, slip of the belt over the pulley should not occur, i.e. $T_1/T_2 < e^{f\alpha}$, where f is the limiting coefficient of friction and α is the angle of wrap. There are two possible assumptions:

(i) frictional resistance is uniformly distributed over the arc of contact with a reduced coefficient of friction, f;
(ii) the coefficient of friction, f, reaches its limiting value over an active arc which is less than the actual arc of contact, and that over this arc T_1 falls to T_2. For the remaining portion of the arc of contact, the tension remains constant at either T_1 or T_2, depending upon the direction of frictional action relative to the pulley.

If the former assumption were correct, relative movement of the belt over the pulley would be entirely prevented by friction. The first assumption is correct for an inextensible belt. Bodily slip would then occur if f reaches its limiting value. Investigations into the creeping action of a belt under driving conditions do not support this view, since a certain measure of slip occurs under all conditions of loading. Taking the latter assumption as correct, it follows that the angle β over which a change of tension occurs is measured by the equation

$$\frac{T_1}{T_2} = e^{f\beta}, \tag{4.83}$$

where α is the angle subtended by the arc of contact, β is the angle subtended by the active arc and $\alpha - \beta$ is the angle subtended by the idle arc.

Creep in an extensible belt is measured by elastic extension, or contraction, as the belt passes from the straight path to the pulley surface. Further, any relative movement of the belt over the pulley must be directed towards the point of maximum tension.

To examine these changes in length, first consider the active arc, Fig. 4.35. If Θ is measured from the position where growth of tension commences, and T is the tension at angle Θ

$$\frac{T}{T_2} = e^{f\Theta}.$$

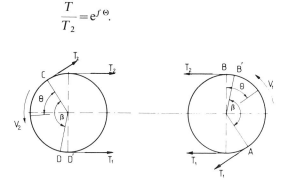

Figure 4.35

The extension of an element of length $r\delta\Theta$ is then $(g/E)r\delta\Theta$.

$$\text{Total extension over the active arc} = \int_0^\beta \frac{T}{btE} r \, d\Theta$$

$$= \frac{T_2 r}{btE} \int_0^\beta e^{f\Theta} \, d\Theta$$

$$= \frac{T_2 r}{btE} \frac{1}{f} (e^{f\beta} - 1)$$

and since $(T_1/T_2) = e^{f\beta}$, this becomes

$$\text{total extension} = \frac{T_1 - T_2}{btE} \cdot \frac{t}{f}, \tag{4.84}$$

where E is the Young modulus of the belt's material.

Referring to Fig. 4.35, suppose AB represents the active arc on the driving pulley. Consider a length of belt $r\beta$ extending backwards from the point A into the straight portion. Friction plays no part over the idle arc; there is no change in tension and no relative movement. Hence for the length $r\beta$:

$$\text{elastic extension} = \frac{T_1}{btE} r\beta.$$

During the time interval in which the point A on the pulley moves to position B, the corresponding point A on the belt will move to B'. The arc BB' is the contraction of a length $r\beta$ of the belt in passing from a condition of uniform tension T_1 to its position AB' on the active arc. Hence

$$BB' = \frac{T_1}{btE} r\beta - \frac{T_1 - T_2}{btE} \frac{r}{f}. \tag{4.85}$$

The second term of this expression follows from eqn (4.84). Similarly for the driven pulley the arc DD' is the increase in elastic extension of a length $r\beta$ in passing from a condition of uniform tension T_2 to its position CD' on the active arc, so that

$$DD' = \frac{T_1 - T_2}{btE} \frac{r}{f} - \frac{T_2}{btE} r\beta. \tag{4.86}$$

Again, let the surface of the driving pulley travel a peripheral distance l, then

$$\text{travel of belt on driving pulley} = \frac{r\beta - BB'}{r\beta} l$$

$$\text{length of belt delivered to driven pulley} = \left[1 - \frac{BB'}{r\beta} \right] l.$$

Corresponding travel of periphery of the driven pulley

$$= \left(1 - \frac{DD'}{r\beta} \right)\left(1 - \frac{BB'}{r\beta} \right) l$$

$$= \left[1 - \frac{DD' + BB^{\circ}}{r\beta} \right] l \quad \text{very nearly.}$$

Hence for pulleys of equal size, if

$V_1 = $ peripheral velocity of surface of the driving pulley,
$V_2 = $ peripheral velocity of surface of the driven pulley,

$$V_2/V_1 = 1 - \frac{DD' + BB'}{r\beta},$$

and substituting for BB' and DD' from eqns (4.85) and (4.86)

$$V_2/V_1 = 1 - \frac{T_1 - T_2}{btE}. \tag{4.87}$$

Also

$$\text{efficiency of transmission} = \frac{(T_1 - T_2)V_2}{(T_1 - T_2)V_1}$$

$$= 1 - \frac{T_1 - T_2}{btE}. \tag{4.88}$$

It should be noted that the idle arc must occupy the earlier portion of the arc of embrace, since contraction of the belt must be directed towards A in the driving pulley and extension of the belt must be directed away-from C on the driven pulley. It follows therefore that

velocity of the surface of the driving pulley
= velocity of the tight side of the belt,

velocity of the surface of the driven pulley
= velocity of the slack side of the belt.

Further, as the power transmitted by the driving pulley increases, the idle arc diminishes in length until $\beta = \alpha$ and the whole arc of contact becomes active. When this condition is reached, the belt commences to slip bodily over the surface of the pulley. This is shown schematically in Fig. 4.36. Thus, eqn (4.87) may be written

$$1 - \frac{V_2}{V_1} = \frac{T_1 - T_2}{btE}$$

or

$$\frac{V_1 - V_2}{V_1} = \frac{T_1 - T_2}{btE} \tag{4.89}$$

and if V_1 is constant, the velocity of slip due to creeping action is proportional to $(T_1 - T_2)$ within the range $\beta \leqslant \alpha$. Since T_1/T_2 is the same for both pulleys, it follows that the angle β subtended by the active arc must be the same for both.

Thus for pulleys of unequal size, the maximum permissible value of β must be less than the angle of lap on the smaller pulley, if the belt is not to slip bodily over the contact surface.

4.10.3. Power transmission rating

In approximate calculations it is usual to assume that the initial belt tension is equal to the mean of the driving tensions, i.e.

$$T_m = T_0 = \tfrac{1}{2}(T_1 + T_2). \tag{4.90}$$

If the belt is on the point of slipping, and the effects of centrifugal action are neglected

$$\frac{T_1}{T_2} = e^{f\alpha},$$

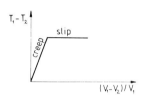

Figure 4.36

where α is the angle of lap of the smaller pulley. Hence

$$T_2 = 2T_0 - T_1 \quad \text{and} \quad T_1 - T_2 = 2(T_1 - T_0).$$

If V = the mean belt speed in $\mathrm{m\,s^{-1}}$,

$$\text{power transmitted} = (T_1 - T_2)V$$
$$= 2(T_1 - T_0)V. \tag{4.91}$$

Alternatively:

$$2T_0 = T_1 + T_2 = T_2(e^{f\alpha} + 1)$$

and

$$T_1 - T_2 = T_2(e^{f\alpha} - 1) = 2T_0 \frac{e^{f\alpha} - 1}{e^{f\alpha} + 1}$$

$$\text{power transmitted} = 2T_0 V \frac{e^{f\alpha} - 1}{e^{f\alpha} + 1}.$$

4.10.4. Relationship between belt tension and modulus

In the foregoing treatment a linear elastic law for the belt material has been assumed. It has already been mentioned that such materials do not in general adhere closely to the simple law of direct proportionality. This is illustrated in Fig. 4.37, which shows the stress-strain curves for samples of leather- and fabric-reinforced rubber belts. Broadly speaking, the curves may be divided into two classes:

(a) those which are approximately linear within the range of stress corresponding to the driving tensions T_1 and T_2 (Fig. 4.37, case (a));
(b) those which are approximately parabolic in form (Fig. 4.37, case (b)).

In the former case we may write

$$e_1 - e_2 = \frac{h_1}{E} - \frac{h_2}{E} = \frac{1}{btE}(T_1 - T_2), \tag{4.92}$$

where e_1 and e_2 are the strains corresponding to the tensions T_1 and T_2, respectively, and E is the slope of the stress-strain curve between these limits. The value of E determined in this way is referred to as the chord modulus of elasticity. If this value of E is used, it readily follows that the expressions for the calculation of creep and initial tension so far obtained are valid when the belt material falls into this group.

In the latter case let h_m and e_m denote a point on the stress-strain curve corresponding to the mean belt tension T_m. Then, if the curve is assumed truly parabolic

$$\text{slope of the tangent at this point} = E = \frac{2h_m}{e_m} \tag{4.93}$$

and for any other point

$$\frac{e_m^2}{h_m} = \frac{e^2}{h}$$

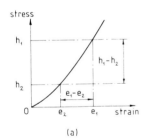

(a)

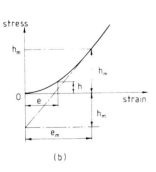

(b)

Figure 4.37

so that

$$e^2 = \frac{h}{h_m} e_m^2 = \frac{4hh_m}{E^2}$$

and

$$e = \frac{2\sqrt{h_m}}{E}\sqrt{h} = \frac{2\sqrt{T_m}}{btE}\sqrt{T} = k\sqrt{T} \tag{4.94}$$

Applying this result to the evaluation of the initial tension, T_0

$$\text{elastic extension due to } T_0 = k\sqrt{T_0}\,[2L + r_1\alpha_1 + r_2\alpha_2] \tag{4.95}$$

Under driving conditions,

$$\text{extension on an active arc } \beta = \int_0^\beta k\sqrt{T}r\,d\Theta = k\int_0^\beta \sqrt{T_2}\,e^{f\,\Theta/2}r\,d\Theta$$

$$= k\frac{2r}{f}\sqrt{T_2}\,[e^{f\,\beta/2} - 1]$$

$$= k\frac{2r}{f}[\sqrt{T_1} - \sqrt{T_2}]. \tag{4.96}$$

Hence, adopting the notation for two unequal pulleys

$$\text{elastic extension on the active arcs} = \frac{2k}{f}(\sqrt{T_1} - \sqrt{T_2})(r_1 + r_2)$$

elastic extension on the straight portions $= kL(\sqrt{T_1} + \sqrt{T_2})$
elastic extension on the idle arc of the larger pulley $= k\sqrt{T_1}r_1(\alpha_1 - \alpha_2)$.

Adding these results and equating to the initial extension due to T_0, we obtain

$$\sqrt{T_0} = \frac{\sqrt{T_1}\left[L + \frac{2}{f}(r_1 + r_2) + r_1(\alpha_1 - \alpha_2)\right] + \sqrt{T_2}\left[L - \frac{2}{f}(r_1 + r_2)\right]}{2L + r_1\alpha_1 + r_2\alpha_2}. \tag{4.97}$$

This is a representation of the maximum permissible value of T_0, if the driving tension T_1 is not to exceed the specified maximum when the belt is transmitting maximum power, and is based on the assumption that the stress-strain curve for the belt is parabolic.

4.10.5.   V-belt and rope drives

An alternative method of increasing the ratio of the effective tensions is by the use of a V-grooved pulley. Figure 4.38, case (a), shows a typical groove section for a cotton driving rope, with groove angle $2\psi = 45°$. The rope does not rest on the bottom of the groove, but only on the sides, the wedge action

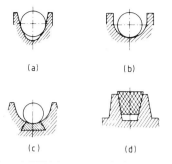

(a) (b)

(c) (d)

Figure 4.38

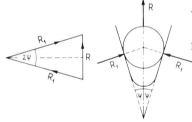

Figure 4.39

giving greater resistance to slipping. If the drive is fitted with a guide pulley, or a tensioning device similar to the belt jockey or idler pulley, the rope rests on the bottom of the groove as shown in Fig. 4.38, case (b). To avoid excessive loss of energy due to bending of the rope as it passes over the pulley, the diameter of the pulley should be greater than 30d, where d is the diameter of the rope.

For wire ropes the groove is as shown in Fig. 4.38, case (c). In this case also, the rope rests on the bottom of the groove, since no jamming action is permissible if increased wear and breakage of the wires is to be avoided. The wedge-shaped recess at the bottom of the groove is packed with a relatively soft material, such as wood or leather. In addition to diminishing the wear on the rope this gives a greater coefficient of friction.

Figure 4.38, case (d), shows the groove for a high speed V-belt machine drive. The belt is of rubber, reinforced with cotton fibre. It consists of a load-carrying core of rubber-impregnated fabric and the surrounding layers are carefully designed to withstand a repeated bending action during driving. The sides of the groove must be prepared to a fine finish and the pulleys placed carefully in alignment if wear of the belt is to be reduced to an acceptable level. The diameter of the smaller pulley should be greater than 7d to 16d, where d is the depth of the V-belt section. The angle of the groove varies from 30 to 40 degrees.

An important feature of rope or V-belt drives is the virtual coefficient of friction. Referring to Fig. 4.39, let

2ψ = the total angle of the groove,

R = the resultant radial force exerted by the pulley on an element of length $r\,d\Theta$ of the rope, where r and Θ have the same meaning as discussed previously in connection with the flat belt action.

The force R is the resultant of the side reactions R_1 where the rope makes contact with the surface of the groove. It is usual to neglect the friction due to wedge action, in which case R_1 is normal to the contact surface.

$$R = 2R_1 \sin \psi. \tag{4.98}$$

Again, as previously shown, the increment of tension in the length $r\delta\Theta$ is

$$T = 2fR_1$$

$$= \frac{fR}{\sin \psi} = f'R, \tag{4.99}$$

where

$$f' = f\,\mathrm{cosec}\,\psi. \tag{4.100}$$

To take account of the V-groove we must therefore replace f by the virtual coefficient of friction f' in the ordinary belt formulae. Thus, eqn (4.79) becomes

$$\frac{T_1}{T_2} = \exp(f\alpha \,\mathrm{cosec}\,\psi), \tag{4.101}$$

where T_1 and T_2 are the effective tensions on the tight and slack sides of the belt or rope respectively.

4.11. Frictional aspects of brake design

The brake-horsepower of an engine is the rate of expenditure of energy in overcoming external resistance or load carried by the engine. The difference between the brake-horsepower and the indicated horsepower represents the rate at which energy is absorbed in overcoming mechanical friction of the moving parts of the engine

$$\text{friction horsepower} = \text{indicated horsepower} - \text{brake-horsepower}$$

and

$$\text{mechanical efficiency} = \frac{\text{brake-horsepower}}{\text{indicated horsepower}}.$$

The operation of braking a machine is a means of controlling the brake-horsepower and so adjusting the output to correspond with variations of indicated horsepower and the external load. A brake may be used either to bring a machine to a state of rest, or to maintain it in a state of uniform motion while still under the action of driving forces and couples.

In engineering practice, the latter alternative is useful as a means of measuring the power that can be transmitted by a machine at a given speed. A brake that is used in this way is termed a dynamometer and is adapted for the purpose simply by the addition of equipment which will measure the friction force or couple retarding the motion of the machine. Dynamometers fall into two classes:

(i) absorption dynamometers, or those which absorb completely the power output of the machine at a given speed;

(ii) transmission dynamometers, which, but for small friction losses in the measuring device itself, transmit power from one machine to another.

Generally speaking, the power developed by an engine may be absorbed by either mechanical, electrical or hydraulic means. In friction brake dynamometers all the power of the engine is absorbed by mechanical friction producing heat.

4.11.1. The band brake

Figure 4.40 shows a mechanical type of friction brake used in a crane. The brake drum D is keyed to the same shaft as the crane barrel E. The flexible band which surrounds the drum, and consists of either a leather or narrow strip of sheet steel with suitable friction material lining is connected to points A and B on the lever pivoted at F. A load P applied to the lever at the point C causes the band to tighten on the drum and friction between the band and drum surface produces the necessary braking torque. Assuming that the drum tends to rotate anticlockwise, let T_1 and T_2 be the effective belt tensions and a the radius of the circle tangential to the lines of action of T_1 and T_2. Then

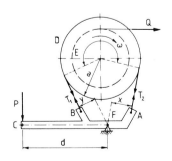

Figure 4.40

$$\text{braking torque, } M = (T_1 - T_2)a. \tag{4.102}$$

Again, suppose x and y are the perpendicular distances of the fulcrum F from the lines of action of T_2 and T_1 respectively. It is assumed that the brake is used in such a manner as to prevent the rotation of the drum when the crane is carrying a load Q attached to a rope passing round the circumference of the barrel. If a force P at leverage d is necessary to support this load, then

$$Pd = T_1 y - T_2 x.$$

The relation between the effective tensions T_1 and T_2 is given by

$$T_1/T_2 = e^{f\Theta}$$

where f is the coefficient of friction for the contact surfaces and Θ is the angle of wrap of the band round the drum. Hence, combining the above expressions

$$M = T_1 a \left[1 - \frac{1}{e^{f\Theta}} \right] \tag{4.103}$$

and

$$P = \frac{T_1}{d} \left[y - \frac{x}{e^{f\Theta}} \right],$$

so that

$$P = \frac{M}{ad} \left[\frac{e^{f\Theta} y - x}{e^{f\Theta} - 1} \right]. \tag{4.104}$$

To study the effect of varying the ratio x/y on the brake action, now consider the following cases:

Case 1. $x = 0$

$$P = \frac{My}{ad} \frac{e^{f\Theta}}{e^{f\Theta} - 1}. \tag{4.105}$$

Here the line of action of T_2 passes through F and a downward movement of the force P produces a tightening effect of the band on the drum.

Case 2. $x = y$

$$P = \frac{My}{ad}. \tag{4.106}$$

In this case there is no tightening effect since the displacements of A and B in the directions of T_2 and T_1 are equal in magnitude. Hence, to maintain the load the band would have to be in a state of initial tension.

Case 3. $x/y = e^{f\Theta}$; i.e.

$$P = 0 \quad \text{and} \quad x/y = T_1/T_2. \tag{4.107}$$

For this ratio a small movement of the lever in the negative direction of P, would have the effect of tightening the band, and the brake would be self-locking.

Case 4. $y = 0$

$$P = -\frac{Mx}{ad} \frac{1}{e^{f\Theta} - 1}.$$ (4.108)

Here, the direction of P must be reversed to tighten the band on the drum.

From the above conclusions it follows that if $e^{f\Theta} > x/y > 1$, downward movement of the force P would tend to slacken the band. Hence for successful action x must be less than y.

When the brake is used in the manner indicated above there is no relative sliding between the friction surfaces, so that f is the limiting coefficient of friction for static conditions. The differential tightening effect of the band brake is used in the design of certain types of friction brake dynamometers.

4.11.2.   The curved brake block

Figure 4.41 represents a brake block A rigidly connected to a lever or hanger LE pivoted at E. The surface of the block is curved to make contact with the rim of the flywheel B, along an arc subtending an angle 2ψ at the centre, and is pressed against the rim by a force P, at the end L of the lever. In general, the normal pressure intensity between the contact surfaces will vary along the length of the arc in a manner depending upon the conditions of wear and the elasticity of the friction lining material of the brake block surface. Let

$p =$ the intensity of normal pressure at position Θ, i.e. p is a function of Θ and varies from $\Theta = 0$ to $\Theta = 2\psi$,
$a =$ the radius of the contact surfaces,
$b =$ the thickness of the brake block,
$R =$ the resultant force on the rim due to the normal pressure intensity p,
$\beta =$ the inclination of the line of action of R to the position $\Theta = 0$.

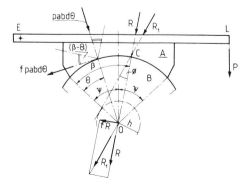

Figure 4.41

Hence, for an element of length $a\,d\Theta$ of the arc of contact

$$\text{normal force} = pab\,d\Theta,$$
$$\text{tangential friction force} = fpab\,d\Theta.$$

The latter elementary force can be replaced by a parallel force of the same magnitude acting at the centre O together with a couple of moment

$$dM = fpa^2b \cdot d\Theta.$$

Proceeding as for the rim clutch and resolving the forces at O in directions parallel and perpendicular to the line of action of R, we have

for the normal force:

$$\text{parallel to } R \quad \int_0^{2\psi} pab\cos(\beta - \Theta)\,d\Theta = R, \tag{4.109}$$

$$\text{perpendicular to } R \quad \int_0^{2\psi} pab\sin(\beta - \Theta)\,d\Theta = 0, \tag{4.110}$$

for the tangential force:

$$\text{parallel to } R \quad \int_0^{2\psi} fpab\sin(\beta - \Theta)\,d\Theta = 0, \tag{4.111}$$

$$\text{perpendicular to } R \quad \int_0^{2\psi} fpab\cos(\beta - \Theta)\,d\Theta = fR. \tag{4.112}$$

If p is given in terms of Θ, the vanishing integral determines the angle β. Further, the resultant force at O is

$$R_1 = R\sqrt{(1 + f^2)} = R\sec\phi \tag{4.113}$$

and is inclined at an angle $\phi = \tan^{-1} f$ to the direction of R. Again, the couple M together with force R_1 at O can be replaced by a parallel force R_1 acting at a perpendicular distance h from O given by

$$h = \frac{M}{R_1} = \frac{\displaystyle\int_0^{2\psi} fpa^2b\,d\Theta}{R\sec\phi}. \tag{4.114}$$

The circle with centre O and radius h is the friction circle for the contact surfaces, and the resultant force on the wheel rim is tangential to this circle.

In the case of symmetrical pressure distribution, $\beta = \psi$, and the line of action of R bisects the angle subtended by the arc of contact at the centre O. The angle Θ is then more conveniently measured from the line of action of R, and the above equations become

$$R = 2\int_0^{\psi} pab\cos\Theta\,d\Theta, \tag{4.115}$$

$$h = \frac{M}{R_1} = \frac{2\displaystyle\int_0^{\psi} fpa^2b\,d\Theta}{R\sec\phi}. \tag{4.116}$$

Now, it is appropriate to consider the curved brake block in action. Three cases shall be discussed.

(i) Uniform pressure

Figure 4.42 represents the ideal case in which the block is pivoted at the point of intersection C of the resultant R_1 and the line of symmetry. Since $\beta = \psi$ and the pressure intensity p is constant, eqns (4.115) and (4.116) apply, so that

$$R = 2pab \int_0^\psi \cos\Theta \, d\Theta$$

$$= 2pab \sin\psi \qquad (4.117)$$

resisting torque

$$M = 2fpa^2b\psi = fRa \frac{\psi}{\sin\psi} \qquad (4.118)$$

and

$$h = \frac{M}{R \sec\phi} = \frac{fa}{\sec\phi} \frac{\psi}{\sin\psi}$$

also

$$\frac{f}{\sec\phi} = \frac{\tan\phi}{\sec\phi} = \sin\phi$$

so that

$$h = a\sin\phi \frac{\psi}{\sin\psi}. \qquad (4.119)$$

(ii) Uniform wear

Referring to Fig. 4.42, it is assumed that the vertical descent δ is constant for all values of Θ. Hence, measuring Θ from the line of symmetry,

normal wear at position $\Theta = \delta\cos\Theta$

and applying the condition for uniform wear, pa is proportional to $\delta\cos\Theta$ or

$$p = k\cos\Theta. \qquad (4.120)$$

Applying the integrals as in the preceding case

$$R = 2kab \int_0^\psi \cos^2\Theta \, d\Theta$$

$$= kab \int_0^\psi (1 + \cos 2\Theta) \, d\Theta$$

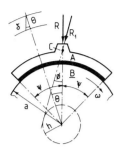

Figure 4.42

$$R = kab(\psi + \sin\psi\cos\psi) \tag{4.121}$$

$$M = 2fka^2b \int_0^\psi \cos\Theta\, d\Theta = 2fka^2b\sin\psi$$

or, resisting torque

$$M = fRa\left[\frac{2\sin\psi}{\psi + \sin\psi\cos\psi}\right] \tag{4.122}$$

and

$$h = \frac{M}{R\sec\phi} = a\sin\phi\left[\frac{2\sin\psi}{\psi + \sin\psi\cos\psi}\right]. \tag{4.123}$$

(iii)   Block pivoted at one extremity

Figure 4.43 shows a brake block or shoe pivoted at or near one extremity of the arc of contact. For a new well-fitted surface, the pressure distribution may be approximately uniform. Wear of the friction lining material will, however, occur more readily at the free end of the shoe, since the hinge may be regarded as being at a constant distance from the centre O.

Taking the radius through the pivot centre as representing the position $\Theta = 0$, let $\delta =$ the angular movement of the shoe corresponding to a given condition of wear.

$$xz = \text{movement at position } \Theta = 2a\sin\tfrac{1}{2}\Theta\delta$$

and

$$\text{normal wear at position } \Theta = 2a\sin\tfrac{1}{2}\Theta\cos\tfrac{1}{2}\Theta\delta$$
$$= \delta a\sin\Theta.$$

Hence, pa is proportional to $\delta a\sin\Theta$ or

$$p = k\sin\Theta \tag{4.124}$$

In this case, eqns (4.109) to (4.112) will apply, and so

$$R = kab\int_0^{2\psi}\sin\Theta\cos(\beta - \Theta)\,d\Theta$$

Expanding the term $\cos(\beta - \Theta)$ and integrating, this becomes

$$R = \tfrac{1}{4}kab[\cos\beta(1 - \cos 4\psi) + \sin\beta(4\psi - \sin 4\psi)].$$

For the angle β we have from eqn (4.110)

$$kab\int_0^{2\psi}\sin\Theta\sin(\beta - \Theta)\,d\Theta = 0.$$

Again, expanding $\sin(\beta - \Theta)$ and integrating

$$\tan\beta = \frac{4\psi - \sin 4\psi}{1 - \cos 4\psi}. \tag{4.125}$$

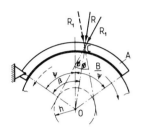

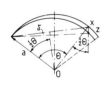

Figure 4.43

Using this value of β the equation for R becomes

$$R = \tfrac{1}{4}kab\cos\beta(1-\cos 4\psi)(1+\tan^2\beta)$$

$$= \tfrac{1}{4}kab\,\frac{1-\cos 4\psi}{\cos\beta}$$

or

$$R = \tfrac{1}{2}kab\sec\beta\sin^2 2\psi. \tag{4.126}$$

For the retarding couple we have

$$M = \int fpa^2 b\,\mathrm{d}\Theta$$

$$= fka^2 b\int_0^{2\psi}\sin\Theta\,\mathrm{d}\Theta$$

$$= fka^2 b(1-\cos 2\psi)$$
$$= 2kfa^2 b\sin^2\psi \tag{4.127}$$

and substituting for k in terms of R this reduces to

$$\text{torque,}\quad M = fRa\cos\beta\sec^2\psi \tag{4.128}$$

so that

$$h = \frac{M}{r\sec\phi} = \frac{fa}{\sec\phi}\cos\beta\sec^2\psi$$

or

$$h = a\sin\phi\cos\beta\sec^2\psi. \tag{4.129}$$

In all three cases, as the angle ψ becomes small, the radius of the friction circle approaches the value

$$h = a\sin\phi$$

and the torque

$$M = fRa.$$

This corresponds to the flat block and the wheel rim. In the general case we may write

$$M = f'Ra,$$

where f' is the virtual coefficient of friction as already applied to friction in journal bearings.
Thus

$$\text{for uniform pressure}\quad f' = \frac{f\psi}{\sin\psi} \tag{4.130}$$

$$\text{for uniform wear}\quad f' = \frac{f2\sin\psi}{\psi+\sin\psi\cos\psi} \tag{4.131}$$

and for zero wear at one extremity

$$f' = f \cos \beta \sec^2 \psi. \tag{4.132}$$

In every case the retarding couple on the flywheel is

$$M = R_1 h = f' R a$$

and

$$h = OC \sin \phi = a \sin \phi \frac{f'}{f}$$

so that

$$OC = a \frac{f'}{f}. \tag{4.133}$$

Numerical example

A brake shoe, placed symmetrically in a drum of 305 mm diameter and pivoted on a fixed fulcrum E, has a lining which makes contact with the drum over an arc as shown in Fig. 4.44. When the shoe is operated by the force F, the normal pressure at position Θ is $p = 0.53 \sin \Theta$ MPa. If the coefficient of friction between the lining and the drum is 0.2 and the width of the lining is 38 mm, find the braking torque required. If the resultant R of the normal pressure intensity p is inclined at an angle β to the position $\Theta = 0$, discuss with the aid of diagrams the equilibrium of the shoe when the direction of rotation is (a) clockwise and (b) anticlockwise.

Solution

Applying eqn (4.127), the braking torque is given by

$$M \int f p a^2 b \, d\Theta$$

$$= f k a^2 b \int_{\pi/4}^{5\pi/6} \sin \Theta \, d\Theta$$

$$= f k a^2 b [\cos \tfrac{1}{4}\pi - \cos \tfrac{5}{6}\pi],$$

where $f = 0.2$, $k = 0.53$ MPa, $a = 1525$ mm and $b = 38$ mm. Thus

$$M = 0.2 \times 0.53 \times 0.1525^2 \times 0.038 \left(\frac{1}{\sqrt{2}} + \tfrac{1}{2}\sqrt{3} \right) \times 10^6$$

$$M = 146.3 \text{ Nm}.$$

Since R is the resultant of the normal pressure intensity, p, the angle β is given by

$$\int p a b \sin(\beta - \Theta) \, d\Theta = 0$$

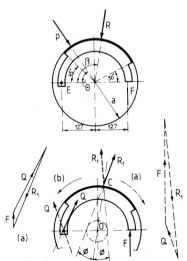

Figure 4.44

i.e.

$$\int_{\pi/4}^{5\pi/6} \sin\Theta \sin(\beta-\Theta)\,d\Theta = 0.$$

Expanding $\sin(\beta-\Theta)$ and integrating, this equation becomes

$$\tan\beta = -11.05 \quad\text{so}\quad \beta = 180 - 84.8 = 95.2 \text{ degrees.}$$

Hence

$$R = kab \int_{\pi/4}^{5\pi/6} \sin\Theta \cos(\beta-\Theta)\,d\Theta$$

and proceeding as follows

$$R = \tfrac{1}{4}kab[-\cos\beta\cos 2\Theta + \sin\beta(2\Theta - \sin 2\Theta)]\Big|_{\pi/4}^{5\pi/6}$$

$$= \tfrac{1}{4}kab[-\tfrac{1}{2}\cos\beta + 5.531\sin\beta],$$

where $\beta = 95.2°$.

Substituting the numerical values

$$R = \frac{0.53 \cdot 10^6 \cdot 0.1525 \cdot 0.038}{4} \, 5.554 = 4264 \text{ N.}$$

For the radius of the friction circle

$$\tan\phi = 0.2 \quad\text{so}\quad \phi = 11.3°$$

$$h = \frac{M}{R\sec\phi} = \frac{146.3}{4264 \cdot 1.02} = 0.034 \text{ m.}$$

Alternatively,

$$f' = \frac{M}{Ra} = \frac{146.3}{4264 \cdot 0.1525} = 0.225$$

so that

$$OC = a\frac{f'}{f} = 1.125a = 0.17 \text{ m.}$$

In Fig. 4.44, $R_1 = R\sec\phi$ is the resultant force opposing the motion of the drum. R_1', equal and opposite to R_1, is the resultant force on the shoe. The reaction Q at the hinge passes through the point of intersection of the lines of action of R_1' and F. As the direction of Q is known, the triangle of forces representing the equilibrium of the shoe can now be drawn. The results are as follows:
(a) clockwise rotation, $F = 1507$ N;
(b) anticlockwise rotation, $F = 2710$ N.

4.11.3. The band and block brake

Figure 4.45 shows a type of brake incorporating the features of both the simple band brake and the curved block. Here, the band is lined with a

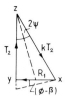

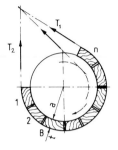

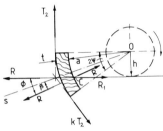

Figure 4.45

number of wooden blocks or other friction material, each of which is in contact with the rim of the brake wheel. Each block, as seen in the elevation, subtends an angle 2ψ at the centre of the wheel. When the brake is in action the greatest and least tensions in the brake strap are T_1 and T_2, respectively, and the blocks are numbered from the point of least tension, T_2.

Let kT_2 denote the band tension between blocks 1 and 2. The resultant force R_1' exerted by the rim on the block must pass through the point of intersection of T_2 and kT_2. Again, since 2ψ is small, the line of action of R_1' will cut the resultant normal reaction R at the point C closely adjacent to the rim, so that the angle between R and R_1' is $\phi = \tan^{-1} f$.

Suppose that the angle between R and the line of symmetry OS is β, then, from the triangle of forces xyz, we have

$$\frac{xz}{zy} = \frac{kT_2}{T_2} = \frac{\sin[(\frac{1}{2}\pi - \psi) + (\phi - \beta)]}{\sin[(\frac{1}{2}\pi - \psi) - (\phi - \beta)]} = k$$

i.e.

$$k = \frac{\cos(\psi - \phi + \beta)}{\cos(\psi + \phi - \beta)}. \tag{4.134}$$

If this process is repeated for each block in turn, the tension between blocks 2 and 3 is k. Hence, if the maximum tension is T_1, and the number of blocks is n, we can write

$$T_1 = k^n T_2 \tag{4.135}$$

i.e.

$$\frac{T_1}{T_2} = \left[\frac{\cos(\psi - \phi + \beta)}{\cos(\psi + \phi - \beta)}\right]^n. \tag{4.136}$$

If the blocks are thin the angle β may be regarded as small, so that

$$k = \frac{\cos(\psi - \phi)}{\cos(\psi + \phi)} = \frac{1 + \tan\psi \tan\phi}{1 - \tan\psi \tan\phi} \tag{4.137}$$

or

$$k = \frac{1 + f \tan\psi}{1 - f \tan\psi} \quad \text{approximately} \tag{4.138}$$

so that

$$\frac{T_1}{T_2} = \left[\frac{1 + f \tan\psi}{1 - f \tan\psi}\right]^n \quad \text{approximately.} \tag{4.139}$$

4.12. The role of friction in the propulsion and the braking of vehicles

The maximum possible acceleration or retardation of a vehicle depends upon the limiting coefficient of friction between the wheels and the track. Thus if

R = the total normal reaction between the track and the driving wheels, or between the track and the coupled wheels in the case of a locomotive,

F = the maximum possible tangential resistance to wheel spin or skidding, then

$$F = fR. \tag{4.140}$$

Average values of f are 0.18 for a locomotive and 0.35 to 0.4 for rubber tyres on a smooth road surface. Here f is called the coefficient of adhesion and F is the traction effort for forward acceleration, or the braking force during retardation. Both the tractive effort and the braking force are proportional to the total load on the driving or braking wheels.

During forward motion, wheel spin will occur when the couple on the driving axle exceeds the couple resisting slipping, neglecting rotational inertia of the wheels. Conversely, during retardation, skidding will occur when the braking torque on a wheel exceeds the couple resisting slipping. The two conditions are treated separately in the following sections.

Case A.   Tractive effort and driving couple when the rear wheels only are driven

Consider a car of total mass M in which a driving couple L is applied to the rear axle. Let

I_1 = the moment of inertia of the rear wheels and axle,
I_2 = the moment of inertia of the front wheels,
F_1 = the limiting force of friction preventing wheel spin due to the couple L,
F_2 = the tangential force resisting skidding of the front wheels.

Also, if $\dot{v}$ is the maximum possible acceleration, α the corresponding angular acceleration of the wheels, and a their effective radius of action, then

$$\dot{v} = a\alpha \tag{4.141}$$

(a)

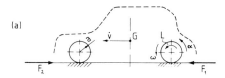

(b)

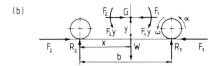

Figure 4.46

Referring to Fig. 4.46, case (a), the following equations can be written

for the rear wheels $L - F_1 a = I_1 \alpha$ $\tag{4.142}$

for the front wheels $F_2 a = I_2 \alpha$ $\tag{4.143}$

for the car $F_1 - F_2 = M\dot{v} = Ma\alpha$ $\tag{4.144}$

Adding eqns (4.142) and (4.143) and eliminating $(F_1 - F_2)$ from eqn (4.144), then

$$\dot{v} = \frac{La}{I_1 + I_2 + Ma^2}.$$ (4.145)

Also, from eqns (4.143) and (4.144)

$$F_1 a = (I_2 + Ma^2)\alpha,$$

and eliminating α

$$F_1 = \frac{L}{a}\left[\frac{I_2 + Ma^2}{I_1 + I_2 + Ma^2}\right].$$ (4.146)

This equation gives the least value of F_1 if wheel spin is to be avoided. For example, suppose $M = 1350\,\mathrm{kg}$, $I_1 = 12.3\,\mathrm{kgm^2}$; $I_2 = 8.1\,\mathrm{kgm^2}$ and $a = 0.33\,\mathrm{m}$, then

$$F_1 = 0.926(L/a)$$

or

$$L = 1.08 F_1 a$$

so that, if L exceeds this value, wheel spin will occur.

The maximum forward acceleration

Equation (4.145) gives the forward acceleration in terms of the driving couple L, which in turn depends upon the limiting friction force F_1 on the rear wheels. The friction force F_2 on the front wheels will be less than the limiting value. Thus, if R_1 and R_2 are the vertical reactions at the rear and front axles, then

$$F_1 = fR_1,$$ (4.147)

$$F_2 = I_2\frac{\alpha}{a} < fR_2.$$ (4.148)

To determine R_1 and R_2, suppose that the wheel base is b and that the centre of gravity of the car is x, behind the front axle and y, above ground level. Since the car is under the action of acceleration forces, motion, for the system as a whole, must be referred to the centre of gravity G. Thus the forces F_1 and F_2 are equivalent to:
 (i) equal and parallel forces F_1 and F_2 at G (Fig. 4.46, case (b)
 (ii) couples of moment $F_1 y$ and $F_2 y$ which modify the distribution of the weight on the springs.
Treating the forces R_1 and R_2 in a similar manner, and denoting the weight of the car by W, we have

$$F_1 - F_2 = \frac{W\dot{v}}{g},$$ (4.149)

$$W = R_1 + R_2,$$ (4.150)

$$(F_1 - F_2)y = R_1(b - x) - R_2 x.$$ (4.151)

Equation (4.151) neglects the inertia couples due to the wheels. For greater accuracy we write

$$(F_1 - F_2)y = R_1(b - x) - R_2 x - (I_1 + I_2)\alpha. \tag{4.152}$$

From eqns (4.150) and (4.151)

$$R_1 = [Wx + (F_1 - F_2)y]/b, \tag{4.153}$$

$$R_2 = [W(b - x) - (F_1 - F_2)y]/b. \tag{4.154}$$

Thus forward acceleration increases the load on the rear wheels and diminishes the load on the front wheels of the car. Again, since $F_1 = fR_1$, eqn (4.153) gives

$$F_1 = \frac{f}{b}[Wx + (F_1 - F_2)y], \tag{4.155}$$

where

$$F_2 = I_2\frac{\alpha}{a} = I_2\frac{\dot{v}}{a^2}.$$

Writing $(F_1 - F_2) = M\dot{v}$ and $W = Mg$, then from eqn (4.155)

maximum forward acceleration, $\dot{v} = \dfrac{f \times g}{b - fy + \dfrac{I_2 b}{Ma^2}}$. $\tag{4.156}$

Case B. Braking conditions

Brakes applied to both rear and front wheels

Proceeding as in the previous paragraph, let L_1 and L_2 represent the braking torques applied to the rear and front axles; F_1 and F_2 denote the tangential resistance to skidding. Referring to Fig. 4.47, $\dot{v}$ is the maximum possible retardation and α the corresponding angular retardation of the wheels, so that, if skidding does not occur, we have:

$$\text{for the rear wheels} \quad L_1 - F_1 a = I_1\alpha \tag{4.157}$$

$$\text{for the front wheels} \quad L_2 - F_2 a = I_2\alpha \tag{4.158}$$

$$\text{for the car} \quad F_1 + F_2 = M\dot{v} = Ma\alpha \tag{4.159}$$

from which

$$\dot{v} = \frac{(L_1 + L_2)a}{I_1 + I_2 + Ma^2}, \tag{4.160}$$

$$F_1 + F_2 = \frac{L_1 + L_2}{a}\left[\frac{Ma^2}{I_1 + I_2 + Ma^2}\right], \tag{4.161}$$

where

$$F_1/F_2 = \frac{L_1(I_2 + Ma^2) - L_2 I_1}{L_2(I_1 + Ma^2) - L_1 I_2}. \tag{4.162}$$

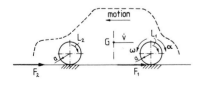

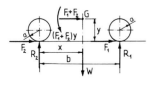

Figure 4.47

Using the same numerical data given in the previous section

$$F_1 + F_2 = 0.875 \frac{L_1 + L_2}{a}$$

or

$$L_1 + L_2 = 1.144(F_1 + F_2)a.$$

If $(L_1 + L_2)$ exceeds this value, skidding will occur.

The maximum retardation

Suppose that limiting friction is reached simultaneously on all the wheels, so that

$$F_1 = fR_1 \quad \text{and} \quad F_2 = fR_2 \tag{4.163}$$

then, referring all the forces, for the system as a whole, to the centre of gravity, G

$$F_1 + F_2 = \frac{W\dot{v}}{g} \tag{4.164}$$

$$W = R_1 + R_2 \tag{4.165}$$

$$(F_1 + F_2)y = R_2 x - R_1(b - x) \tag{4.166}$$

or, more accurately, in accordance with eqn (4.152)

$$(F_1 + F_2)y = R_2 x - R_1(b - x) - (I_1 + I_2)\alpha, \tag{4.167}$$

so that,

$$R_1 = [Wx - (F_1 + F_2)y]/b, \tag{4.168}$$

$$R_2 = [W(b - x) + (F_1 + F_2)y]/b. \tag{4.169}$$

Hence from eqns (4.163) and (4.164)

$$F_1 + F_2 = fW = \frac{W\dot{v}}{g}$$

and

$$\text{maximum retardation} = fg. \tag{4.170}$$

Under running conditions the braking torques on the front and rear axles may be removed in a relatively short time interval, during which the retardation remains sensibly constant. As L_2 is reduced, F_2 diminishes also and passes through a zero value when $L_2 = I_2\alpha$. For smaller values of L_2 it becomes negative; when $L_2 = 0$ the angular retardation of the wheels is due entirely to the tangential friction force.

    If, through varying conditions of limiting friction at either of the front wheels, or because of uneven wear in the brake linings, the braking torques on the two wheels are not released simultaneously, a couple tending to rotate the front axle about a vertical axis will be instantaneously produced, resulting in unsteady steering action. This explains the importance of equal distribution of braking torque between the two wheels of a pair.

Brakes applied to rear wheels only

When $L_2 = 0$, eqns (4.157)–(4.159) become

$$L_1 - F_1 a = I_1 \alpha,$$
$$- F_2 a = I_2 \alpha,$$

$$F_1 - \frac{I_2 \alpha}{a} = Mv = Ma\alpha,$$

from which

$$\text{maximum retardation} = \dot{v} = \frac{L_1 a}{I_1 + I_2 + Ma^2}, \qquad (4.171)$$

$$F_1 = \frac{L_1}{a} \left[\frac{I_2 + Ma^2}{I_1 + I_2 + Ma^2} \right]. \qquad (4.172)$$

These results correspond with eqns (4.145) and (4.146) for driving conditions.

The maximum retardation

In this case limiting friction is reached on the rear wheels only, so that $F_1 = fR_1$ and, applying eqn (4.168),

$$F_1 = \frac{f}{b} [Wx - (F_1 + F_2)y], \qquad (4.173)$$

where

$$F_2 = -I_2 \frac{\alpha}{a} = -I_2 \frac{f}{a^2}.$$

Again, writing $W = Mg$, eqn (4.164) may be written as

$$F_1 + F_2 = M\dot{v}$$

and, eliminating F_1 and F_2 from eqn (4.173)

$$\text{maximum retardation} = \dot{v} = \frac{fx}{b + fy + \dfrac{I_2 b}{Ma^2}} \, g. \qquad (4.174)$$

4.13. Tractive resistance

In the foregoing treatment of driving and braking, the effects of friction in the bearings were neglected. However, friction in the wheel bearings and in the transmission gearing directly connected to the driving wheels is always present and acts as a braking torque. Therefore, for a vehicle running freely on a level road with the power cut off, the retardation is given by eqn (4.160), where L_1 and L_2 may be regarded as the friction torques at the rear and front axle bearings. When running at a constant speed these friction torques will exert a constant tractive resistance as given by eqns (4.157) and (4.158) when $\alpha = 0$, i.e. $F_1 + F_2 = (L_1 + L_2)/a$. This tractive resistance must be deducted from the tractive effort to obtain the effective force for the calculation of acceleration. It must be remembered that there is no loss of energy in a pure rolling action, provided that wheel spin or skidding does not occur. In the ideal case, when friction in a bearing is neglected, so that $L_1 = L_2 = 0$ and $F_1 = F_2 = 0$, the vehicle would run freely without retardation.

4.14. Pneumatic tyres

A pneumatic tyre fitted on the wheel can be modelled as an elastic body in rolling contact with the ground. As such, it is subjected to creep and micro-slip. Tangential force and twisting arising from the lateral creep and usually referred to as the cornering force and the self-aligning torque, play, in fact, a significant role in the steering process of a vehicle. For obvious reasons, the analysis which is possible for solid isotropic bodies cannot be done in the case of a tyre. Simple, one-dimensional models, however, have been proposed to describe the experimentally observed behaviour. An approximately elliptically shaped contact area is created when a toroidal membrane with internal pressure is pressed against a rigid plane surface. The size of the contact area can be compared with that created by the intersection of the plane with the undeformed surface of the toroid, at such a location as to give an area which is sufficient to support the applied load by the pressure inside the toroid. The apparent dimensions of the contact ellipse x and y (see Fig. 4.48) are a function of the vertical deflection of the tyre

$$x = [(2R - \delta)\delta]^{\frac{1}{2}} \quad \text{and} \quad y = [(b - \delta)\delta]^{\frac{1}{2}}. \tag{4.175}$$

The apparent contact area is

$$xy = \pi\delta[(b - \delta)(2R - \delta)]^{\frac{1}{2}} 2\pi bR\delta. \tag{4.176}$$

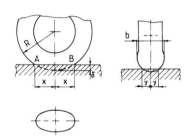

Figure 4.48

It is known, however, that the tyre is tangential to the flat surface at the edge of the contact area and therefore the true area is only about 80 per cent of the apparent area given by eqn (4.176). It has been found that approximately 80 to 90 per cent of the external load is supported by the inflation pressure. On the other hand, an automobile tyre having a stiff tread on its surface forms an almost rectangular contact zone when forced into contact with the road. The external load is transmitted through the walls to the rim. Figure 4.49 shows, schematically, both unloaded and loaded automobile tyres in contact with the road. As a result of action of the external load, W, the tension in the walls decreases and as a consequence of that the curvature of the walls increases. An effective upthrust on the hub is created in this way. In the ideal case of a membrane model the contact pressure is uniformly distributed within the contact zone and is equal to the pressure inside the membrane. The real tyre case is different because the contact pressure tends to be concentrated in the centre of the contact zone. This is mainly due to the tread.

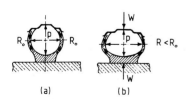

Figure 4.49

4.14.1. Creep of an automobile tyre

An automobile tyre will tend to creep longitudinally if the circumferential strain in the contact patch is different from that in the unloaded periphery. In accordance with the theory of the membrane, there is a shortening in the contact patch of the centre-line of the running surface. This is equal to the difference between the chord AB and the arc AB (see Fig. 4.48). This leads to a strain and consequently to a creep given here as a creep ratio:

$$\kappa = -\frac{\delta}{3R}. \tag{4.177}$$

The silent assumption regarding eqn (4.177) is that the behaviour of the contact is controlled by the centre-line strain and that there is no strain outside the contact. The real situation, however, is different.

4.14.2. Transverse tangential forces

Transverse frictional forces and moments are operating when the plane of the tyre is slightly skewed to the plane of the road. This is usually called sideslip. Similar conditions arise in the response to spin when turning a corner. The usual approach to these problems is the same as that for solid bodies. The analysis starts with the contact being divided into a stick region at the front edge of the contact patch and a slip region at the rear edge. The slip region tends to spread forward with the increase in sideslip or spin. Figure 4.50 shows one-dimensional motion describing the resistance of the tyre to lateral displacement. This displacement, k, of the equatorial line of the tyre results in its lateral deformation. The displacement, k is divided into displacement of the carcass, k_c, and the displacement of tread, k_t. The carcass is assumed to carry a uniform tension R resulting from the internal pressure. This tension acts against lateral deflection. The lateral deflection is also constrained by the walls acting as a spring of stiffness G per unit length. The tread also acts as an elastic foundation. Surface traction, $g(x)$, acting in the region $-c \leqslant x \leqslant c$ deforms the tyre. The equilibrium equation is of the form

$$G_c k_c - R \frac{\partial^2 k_c}{\partial x^2} = g(x) = G_t k_t, \tag{4.178}$$

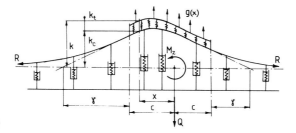

Figure 4.50

where G_t is called the tread stiffness. The velocity of lateral slip in the contact zone (rigid ground, $k_g = 0$ and one-dimensional motion) is described by

$$\frac{\dot{s}}{V} = \kappa + \psi\left(\frac{x}{c}\right) + \frac{\mathrm{d}k}{\mathrm{d}x}. \tag{4.179}$$

It should also be remembered that in a stick regime, $\dot{s} = 0$. It seems that the propositions to assume a rigid carcass and allow only for the deformation of the tread are not realistic. A more practical model is to neglect the tread deflection and only consider carcass deformation, i.e. $k = k_c$. With this assumption, eqn (4.178) becomes

$$k - \gamma^2 \frac{\mathrm{d}^2 k}{\mathrm{d}x^2} = g(x)/G_c, \tag{4.180}$$

where $\gamma = (R/G_c)^{\frac{1}{2}}$ is the relaxation length. Assuming further that $\dot{s} = 0$ in the entire contact zone, the displacement within the contact zone for a case of slideslip is given by

$$k = k_1 - \kappa x, \tag{4.181}$$

where k is the displacement at the entry to the contact zone. Outside the contact zone $g(x) = 0$, therefore eqn (4.180) yields

in front of the contact

$$k = k_1 \exp[(c + x)/\gamma]; \tag{4.182}$$

at the rear of the contact

$$k = k_2 \exp[(c - x)/\gamma]. \tag{4.183}$$

At the leading edge, the displacement gradient is continuous and therefore $k = -\gamma\kappa$. Figure 4.51 shows the equatorial line in a deflected state. In the contact zone

$$g'(x) = G_c k = -G_c\kappa(\gamma + c + x) \tag{4.184}$$

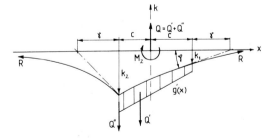

Figure 4.51

which corresponds to a force $Q' = -2G_c c(\gamma + c)$.

At the rear of the contact zone, there is a discontinuity in $\mathrm{d}k/\mathrm{d}x$ which gives rise to an infinite traction $q''(c)$ corresponding to a force

$Q'' = -2G_c\kappa\gamma(\gamma+c)$. The total cornering force is thus

$$Q = Q' + Q'' = -2G_c\kappa c^2\left(\frac{\gamma}{c}+1\right)^2. \tag{4.185}$$

Self-aligning torque can be found by taking moments about O

$$M = \int_{-c}^{c} g(x)x\,\mathrm{d}x = -2G_c\kappa c^3\left(\tfrac{1}{3}+\frac{\gamma}{c}+\frac{\gamma^2}{c^2}\right). \tag{4.186}$$

The infinite traction at the rear of the contact zone produces slip, so that the deformed shape $k(x)$ has no discontinuity in gradient and conforms to the condition $g(x)=fp(x)$ within the slip region.

4.14.3. Functions of the tyre in vehicle application

The most widely known application of pneumatic tyres is the vehicle application. In this application the pneumatic tyre fulfils six basic functions:
  (i) allows for the motion of a vehicle with a minimum frictional force;
  (ii) distributes vehicle weight over a substantial area of contact between the tyre and the road surface;
  (iii) secures the vehicle against shock loading;
  (iv) participates in the transmission of torque from the engine to the road surface;
  (v) allows, due to adhesion, for the generation of braking torque, driving and steering of the vehicle;
  (iv) provides stability of the vehicle.
When in a rolling mode the resistance to motion comes from two sources:
  – internal friction resulting from the continuous flexing of tread and walls;
  – external friction due to micro-movement within the contact area between the wheel and the road.
The tyre, from a design point of view represents a complex problem, the solution of which requires a compromise. For example, increased ride comfort means greater shock absorption, but also increased power consumption in transmitting engine torque.

4.14.4. Design features of the tyre surface

Another example of a compromise involving the tyre is the incorporation of a tread pattern into the running band. Under ideal conditions (no rain or dust deposits on a road surface) the coefficient of sliding friction of about 5 would be attained with a perfectly smooth tread since the adhesion contribution to friction is maximized by a large available contact area. The existence of a thin film of water would, in this case, easily suppress the adhesion and produce very dangerous driving conditions characterized by a coefficient of friction as low as 0.1 or less. Therefore the tread pattern is provided on the surface of the tyre to eliminate such a drastic reduction in the coefficient of friction. However, there is a price to pay, because at the

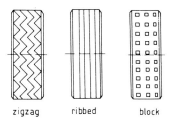

zigzag   ribbed   block

Figure 4.52

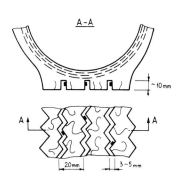

Figure 4.53

same time, the coefficient of adhesion component of friction under dry conditions is seriously reduced to a value less than unity as the area of contact is reduced by the grooves. Nevertheless, the overall effective coefficient of friction under wet conditions is considerably increased; a value of $f = 0.4$ for locked wheel skidding on a wet road is typical. The main role of the grooving on the surface of the tyre is to drain excess water from the tyre footprint in order to increase the adhesion component of friction. Thus, an adequate tread pattern offers a compromise between the higher and lower friction coefficients that would be obtained with a smooth tyre under dry and wet conditions respectively. The usual requirement of the designer is to ensure that the grooving or channeling in the tread pattern is capable of expelling the water from the tyre footprint during the time which is available at high rolling speed. Figure 4.52 shows three basic tread patterns which are used today, called zigzag, ribbed and block.

According to experimental findings the differences in performance of each type are not very significant. Apart from the basic function of bulk water removal, the tyre tread must also allow for a localized tread movement or wiping, to help with the squeezing-out of a thin water film on the road surface. This can be achieved by providing the tread with spies or cuts leading into grooves. There are a number of important design features which a modern tread should have:

(i) channels or grooves. The volume of grooving is almost constant for all tread types. On average, the grooves are approximately 3 mm wide and 10 mm deep;

(ii) spies or micro-cuts leading into the channels or feeder channels. Their main function is to allow for tread micro-movement which is characteristic of the rolling process. Usually, spies do not contribute to the removal of water from the footprint directly. Figure 4.53 shows the arrangement of channels and spies typical for the zigzag pattern;

(iii) transverse slots or feeder channels. Their size is less than the main channels which they serve. The transverse slots are not continuous but end abruptly within the tread. Their main role is to displace bulk water from the tyre footprint. They also permit the macro-movement of the tread during the wiping action.

4.14.5.   The mechanism of rolling and sliding

Both rolling and sliding can be experienced by a pneumatic tyre. Pure sliding is rather rare except in case of a locked wheel combined with flooding due to heavy rainfall. Then, the same tread elements are subjected to the frictional force and as a result of that, the wear of the tread is uneven along the tyre circumference. Severe braking but without the wheel being locked produces wear in a uniform manner along the tyre circumference since the contact zone is continually being entered by different tread elements. The extent of wear under such conditions is less, because the mean velocity of slip of the tread relative to the road surface is much lower.

During the rolling of the tyre, four fundamental elements of the process

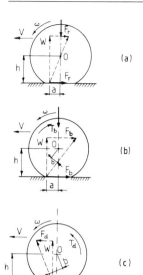

Figure 4.54

can be distinguished; free rolling, braking, accelerating, cornering or any combination of them. Figure 4.54 shows the loads acting on the tyre during (a) a free rolling, (b) a braked rolling and (c) a driven rolling. In all cases, longitudinal tractive forces are produced in the contact zone, giving rise to net forces F_r, F_b, F_d acting on the tread and the reaction force W acting at a small distance, a, ahead of the contact centre. In the case of free rolling there is no net moment about the wheel centre, and, therefore, the resultant force $(W^2 + F_r^2)^{\frac{1}{2}}$ passes through O as shown in Fig. 4.54. When the brake is on, the rolling resistance force, F_r, increases considerably and is equal to the braking force value, F_b, and the resultant force $(W^2 + F_b^2)^{\frac{1}{2}}$ is acting on a moment arm, b, about the wheel centre O. In this way, the moment equal to $b(W^2 + F_b^2)^{\frac{1}{2}}$ is produced opposing the braking torque, T_b (Fig. 4.54, case (b)). Similar reasoning is applicable to the case of driving but now the net longitudinal force, F_d, acts in the direction of motion and the moment $b(W^2 + F_d^2)^{\frac{1}{2}}$ opposes the driving torque T_d.

For steady-state conditions, the following moments about O can be taken for each of the three characteristic rolling modes

$$F_r = W(a/h), \tag{4.187a}$$

$$F_b = T_b/h + W(a/h), \tag{4.187b}$$

$$F_d = T_d/h - W(a/h), \tag{4.187c}$$

where h is the height between the axle and the ground. It can be seen from these equations that F_b and F_d are influenced by the load effect due to the eccentricity of the road surface reaction force. Now, taking into account the fact that the wheel is subjected to a load transfer effect in braking or accelerating, the normal load W is further modified by the bracketed term in the following equations

$$F_b = \frac{T_b}{h} + W\frac{a}{h}\left[1 \pm \frac{\dot{V}}{g}\left(\frac{h_g}{L}\right)4\right], \tag{4.188a}$$

$$F_d = \frac{T_d}{h} - W\frac{a}{h}\left[1 \pm \frac{\dot{V}}{g}\left(\frac{h_g}{L}\right)4\right], \tag{4.188b}$$

where h_g is the height of the centre of gravity of the vehicle above the road surface, L is the wheelbase and $\dot{V}$ the acceleration or deceleration of the vehicle. In these equations, the assumption is that each wheel of the vehicle carries an equal load W when at rest, and that the centre of gravity is at the centre of the wheelbase L. The first sign within brackets in eqns (4.188) refers to the front wheels and the second sign applies to the rear wheels.

It is important to know how the area of contact for a rolling tyre is behaving under the conditions of braking, driving and cornering. There is virtually no slip within the forward part of the contact zone, while an appreciable slip takes place towards the rear of the contact (Fig. 4.55). This is true in each case of the rolling conditions. Figure 4.56 gives details of the slip velocity distribution for a braked, driven and cornering tyre in the rolling mode. In Fig. 4.56, it is assumed that the wheel is stationary and the

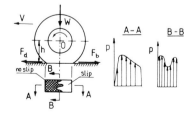

Figure 4.55

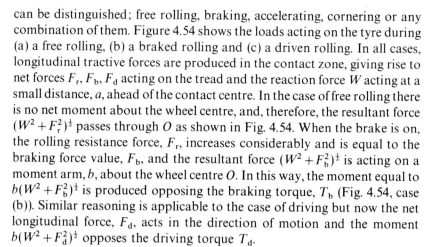

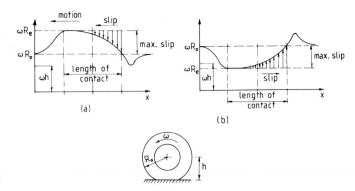

Figure 4.56

road moves with a velocity $V = \omega R_0$, where ω denotes the angular velocity of the wheel and R_0 is the effective rolling radius. During a braked rolling period, the band velocity of the tyre increases to the road velocity ωR_0 at the entrance to the contact zone and is steady until approximately one-half of the contact length has been traversed (Fig. 4.56, case (a)). From that moment onwards the tyre band velocity decreases in a non-linear way towards the rear of the contact. As a result of that a variable longitudinal slip velocity is produced in the forward direction. The slip velocity increases with the speed of the vehicle and plays a particularly significant role in promoting skidding on a wet road surface. A similar slip velocity pattern is established in the rear of the contact but this time in a backward direction (Fig. 4.56, case (b)).

4.14.6. Tyre performance on a wet road surface

Figure 4.57 shows, in a schematic way, a pneumatic tyre in contact with a wet surface of the road. The contact area length is divided into three regions. It is convenient to assume that the centre of the rolling tyre is stationary and the road moves with velocity V. Approximately, it can be said there is no relative motion between the tyre and the road within the front part of the contact zone when the former traverses the contact length. Due to the geometrical configuration, a finite wedge angle can be distinguished between the tyre and the water surface just ahead of the contact zone (Fig. 4.57), and, under conditions of heavy flooding, a hydrodynamic upward thrust P_h is generated as a result of the change in momentum of the water within the converging gap. The magnitude of this upward lift increases in proportion to the forward velocity of the tyre relative to the road surface.

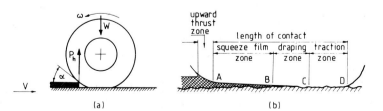

Figure 4.57

The tread elements must force their way through the water film in order to establish physical contact with the road surface asperities (Fig. 4.57, case (b)). Throughout the entire contact length the normal load on the tread elements is due to the inflation pressure of the tyre. In the region BC of the contact length, a draping of the tread about the highest asperities on the road surface takes place. The extent and rate of penetration of the tread by the road surface asperities is mainly determined by the properties of the rubber, such as hardness, hysteresis losses and resilience. The process of draping is over when an equilibrium in vertical direction is established, point C in Fig. 4.57.

The clear inference is that under wet conditions the real contact between the tyre and the road surface is taking place in the region CD (Fig. 4.57). It is then quite obvious that by minimizing the length AB, by a suitable choice of tread pattern, the length CD used for traction developing is increased, provided that the region BC remains unaffected and velocity V is unchanged. The increase in rolling velocity invariably causes growth of the squeeze-film region AB, to such an extent that it occupies almost the whole length of the contact zone AD. This leads to very low traction forces. The speed at which this happens is referred to as the viscous hydroplaning limit and is mainly defined by the ability of the front part of the contact zone to squeeze the water film out. At this critical speed the hydrodynamic pressure developed within the contact zone is quite large but is not sufficient to support the normal load, W, on the wheel. There is a second, much higher speed, at which the hydrodynamic pressure is equal to the load on the wheel and is called the dynamic hydroplaning limit. The dynamic hydroplaning limit is reached only in a few practical situations, for instance, during the landing of an aeroplane. More commonplace is the viscous hydroplaning limit which represents a critical rolling velocity for all road vehicles when the region AB takes a significant part of the contact zone AD.

During braking and driving periods the characteristic feature of the rear part of the contact zone is an increase in the velocity of relative slip between the tyre and the road surface. The separation between the tyre and the road surface increases with the slip velocity and the contact is disrupted first in the rearmost part of the contact zone as the forward velocity increases. Further increase in speed results in the rapid growth of separation between the tyre and the road surface. Simultaneously, the front part of the contact zone is being diminished by a backward moving squeeze-film separation. The situation existing in the contact zone prior to the viscous hydroplaning limit is shown in Fig. 4.58.

The rare case (for road vehicles) of the dynamic hydroplaning limit is shown in Fig. 4.58, case (a). It is not difficult to show that, according to hydrodynamic theory, twice the speed is required under sliding compared with rolling to attain the dynamic hydroplaning when $P_h = W$. This is because both surfaces defining the converging gap attempt to drag the water into it when rolling, whereas during sliding usually only one of the surfaces, namely the road surface, is acting in this way.

Figure 4.59 shows, in a schematic way, the behaviour of tyres during

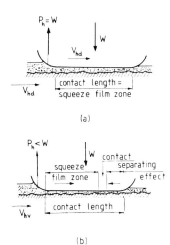

(a)

(b)

Figure 4.58

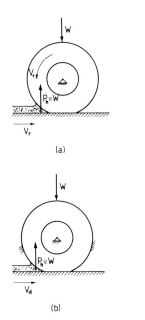

(a)

(b)

Figure 4.59

rolling and sliding under the dynamic hydroplaning conditions. When, at a certain speed, V_r, viscous hydroplaning conditions are reached, the interaction between the tyre and the ground rapidly decreases to a low value which is just sufficient to balance the load reaction eccentricity torque but not to rotate the tyre. This is characteristic for locked wheel sliding at a speed, V_r, which is significantly lower than the velocity, V_d, at which the dynamic hydroplaning is established.

4.14.7. The development of tyres with improved performance

As stated earlier, automobile tyres are very complex structures and many advances have been made in the fabrication, the type of ply, the material of the cords, the nature of the rubber and the tread pattern. As a result, modern tyres have achieved greater fuel economy and longer life than those available say some 30 years ago.

Tribology has made a significant contribution to the development of a tyre that is more skid resistant and at the same time achieves a further reduction in fuel consumption. The friction between the tyre and the road surface consists of two main parts. The first and major component arises from atomic forces across the interface. The bonds formed between the tyre and the road surface have to be broken for sliding to occur. Although the interfacial forces are not particularly strong, rubber has a relatively small elastic modulus so that the area of contact is large. This is illustrated in Fig. 4.60 which shows the sliding of rubber over a single model asperity on the road surface. As a result, the frictional force is relatively high; for a tyre on a clean, dry, fine-textured road surface f is about unity. This gives very good grip and provides good braking power, stability on cornering and a general sense of safety. If, however, the road is wet or greasy and the water film cannot be wiped away by the tread pattern quickly enough, intimate contact between the road surface and the tyre may be prevented and then the adhesion component of the friction may become too low and skidding may result. At this stage another component of the friction becomes apparent and it arises as shown in Fig. 4.61. As the rubber slides over the asperity the rubber is deformed elastically over the region BC and work is done on the rubber. The rubber over the region AB is recovering and urging the rubber forward. If the energy which emerges over the region AB were exactly equal to the energy expanded over the region BC, no energy would be lost and in the absence of adhesion there would be zero resistance to skidding. However, in the deformation cycle between BC and AB, energy is lost by interfacial friction or hysteresis in the rubber. The greater the hysteresis the greater the energy loss and the greater the force required to move the rubber over the asperity, i.e. the greater the resistance to skidding.

Similar losses occur for unlubricated surfaces but they are swamped by the much larger adhesion component of the friction. It is only when the adhesion vanishes as in wet or greasy conditions that the deformation component of friction becomes important. A coefficient of friction, $f = 0.2$ to 0.3 from this mechanism is probably not very large, it is much better,

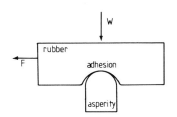

Figure 4.60

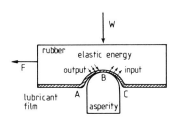

Figure 4.61

however, than $f=0$. There is, however, the problem of energy-loss in the rolling of the tyre. As the tyre rolls over the road the rubber is cyclically loaded and unloaded in the contact zone. The energy lost in rolling thus also depends on the hysteresis losses in the rubber. Indeed the rolling resistance will increase more or less in the same way as the skid resistance, i.e. a skid resistant tyre will consume more fuel during rolling. At the early stages of a research programme aimed to improve the performance of a tyre it was suggested that this could be overcome by making the main structure of the tyre of a low-loss rubber to give low-energy consumption in rolling, and then moulding on a thin surface layer of a high-loss rubber for the tread. A much neater solution to the problem was provided by R. Bond of Dunlop. He observed that during braking or during skidding the deformation of the rubber by surface asperities involves rather high-frequency loading-unloading cycles and considerable local heat generation. On the other hand, the asperity and bulk deformation of the tyre as it rolls over the road is a relatively low-frequency process and there is little bulk heating. He argued that it should thus be possible to produce a rubber which has high losses at high frequencies and elevated temperatures and low losses at low frequencies and modest temperatures. A new polymer with a unique microstructure that secures the required properties was developed and thus a new type of tyre was produced that provides both a better grip and a lower fuel consumption.

4.15. Tribodesign aspects of mechanical seals

The primary function of a seal is to limit the loss of lubricant or the process fluid from systems and to prevent contamination of a system by the operating environment. Seals are among the mechanical components for which wear is a prevailing failure mode. However, in the case of contact seals, wear during initial operation can be essential in achieving the optimum mating of surfaces and, therefore, control of leakage. With continued operation, after break-in, wear is usually in the mild regime and the wear rates are quite uniform; thus wear life may be predicted from typical operating data. In Fig. 4.62 a face seal configuration is shown. Solid contact takes place between two annular flat surfaces where one element of the primary sealing interface rotates with a shaft and the other is stationary. This contact gives rise to a series of phenomena, such as wear, friction and frictional heating. Similar problems occur with shaft riding or circumferential seals, both with carbon and other materials for rings and for elastomeric

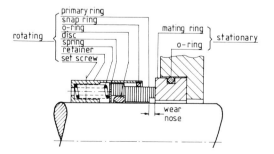

Figure 4.62

lip seals. Lubrication of the sealing interface varies from nil to full hydrodynamic and wear can vary accordingly. Wear of abradable shroud materials is utilized to achieve minimum operating clearance for labyrinth seals and other gas-path components like turbine or compressor blade tips to achieve minimum leakage. The functions of seals are also of great importance to the operation of all other lubricated mechanical components. Wear in seals can occur by a variety of mechanisms. A cause of wear in many types of mechanical systems is contamination by abrasive particles that enter the system through the seals. Design features in seals that exclude external contamination from mechanical systems may be of vital importance. Seals are also important to energy conservation design in all types of machines. The most effective leakage control for contact seals is achieved with a minimum leakage gap and when both sliding faces, moving and stationary, are flat and parallel. This condition is perhaps never achieved. That is probably fortunate, since a modest degree of waviness or nodel distortion can give rise to fluid film lubrication that would not be anticipated with the idealized geometry. With distortion, wear of either internal or external edges can cause the nose piece to form a leakage gap that can be convergent or divergent. Changes in the leakage gap geometry have significant effects on the mechanics of leakage, on the pressure balance, and on the susceptibility of lubrication failure and destructive wear.

One of the wear mechanisms which occur in seals is adhesive wear. With adhesive wear, the size of the wear particles increases with face loading. An anomaly of sealing is that as the closing forces on the sealing faces are increased to reduce the leakage gap, the real effect can be larger wear particles that establish and increase the gap height and thereby increase leakage. Also, greater closing force can introduce surface protuberances or nodes from local frictional heating, termed thermoelastic instability, that may determine the leakage gap height. The leakage flow through a sealing gap obeys the usual fluid mechanics concepts for flow. In addition, there are likely boundary layer interactions with the surfaces in an immediate proximity. In this chapter, design considerations to control the wear and to optimize wear reducing fluid lubrication will be discussed. For guidance on the selection of a proper seal for a particular application, the reader is referred to ESDU–80012 and ESDU–83031.

4.15.1. Operation fundamentals

The most important mechanism for sealing fluids between solid bodies is that of surface tension. By using various concentrations of a surface-active agent in the water phase, it is easy to demonstrate that the rate of leakage in an oil-water system depends on the oil-water interfacial tension. The usual formula to calculate the pressure due to capillarity is

$$\Delta p = \gamma \left[\frac{1}{R_1} + \frac{1}{R_2} \right], \tag{4.189}$$

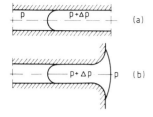

Figure 4.63

where γ is the surface tension and R_1 and R_2 are the radii of the meniscus in mutually perpendicular planes. In the case of parallel plane surfaces R can be taken as infinity and R_1 as approximately $h/2$, where h is the separation of the surfaces. Assuming a surface tension of $0.02\,\mathrm{N\,m^{-1}}$, the thickness of the fluid film is about 5×10^{-7} m and the pressure difference resisted by the seal can amount to 8×10^4 Pa. Thus, in the situation depicted in Fig. 4.63, where the fluid wets the surface, a pressure of 8×10^4 Pa acting from right to left can be resisted. However, in the absence of this pressure the fluid would continue to be drawn into the cavity with the interface advancing to the right. It has been shown experimentally, that when the meniscus reaches the end of the constricted passage it begins to turn itself inside out as indicated in Fig. 4.63.

Owing to the contamination of engineering surfaces, the contact angles of oil against synthetic rubber and steel under industrial conditions are found to be high, so that the sealed oil does not spread along the steel shaft. In addition to the equilibrium meniscus effect, any local variation of the surface tension of a liquid induces a driving force to a fluid. This is known as the Marangoni effect and its implications for the action of seals have been investigated. When a temperature gradient exists on a solid surface, a droplet of liquid laid on that surface will spread out more rapidly towards the lower-temperature side. Even when conditions are generally isothermal, differential evaporation of the constituents of a multicomponent liquid may produce local variations in surface tension which markedly affect spreading behaviour. The constituents of mineral oils having higher molecular weights will tend to spread more rapidly by reason of their greater surface tensions. This process promotes segregation of the constituents of blended oils, thus depriving the high-temperature side (where lubrication is more difficult) of the more effective components. Thus, the Marangoni effect can account for differences in the sealing behaviour of apparently similar oils. Generally, those with a narrow-ranged molecular weight distribution are easier to seal than are blended oils characterized by wide-ranged distributions.

4.15.2. Utilization of surface tension

The bearings of watches and fine instruments are lubricated by droplets of fine oil which are kept in place by a surface tension mechanism known as epilaming. The surface of the metal surrounding the joint is treated with a surface active substance such as a fatty acid which prevents the lubricant applied to the bearing from spreading.

4.15.3. Utilization of viscosity

If pressure is applied to a seal over and above that required to overcome the surface tension, an estimate of the volume of leakage may be made using the

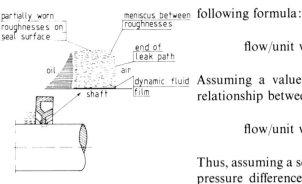

Figure 4.64

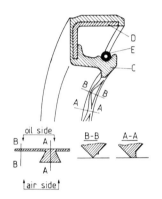

Figure 4.65

following formula:

$$\text{flow/unit width} = -\frac{1}{12\mu}h^3\frac{\mathrm{d}p}{\mathrm{d}x}. \qquad (4.190)$$

Assuming a value of 0.1 Pa s for viscosity, μ, and $h = 5 \times 10^{-7}$ m, the relationship between the flow and the pressure gradient becomes

$$\text{flow/unit width} = -4.17 \times 10^{-21}\frac{\mathrm{d}p}{\mathrm{d}x}. \qquad (4.191)$$

Thus, assuming a seal face of size 1 cm measured in the direction of flow, a pressure difference of 20 MPa, and $\mathrm{d}p/\mathrm{d}x = 2 \times 10^9$, the flow would be 4.34×10^{-12} m^3 s^{-1}. This would hardly keep pace with evaporation and it may be accepted that viscous resistance to flow, whilst it can never prevent leakage, may reduce it to a negligible quantity. In practice, surfaces will not be flat and parallel as assumed in the foregoing treatment, and in fact there will be a more complicated flow path as depicted schematically in Fig. 4.64. Some substances, such as lubricating greases may possess yield values which will prevent leakage until a certain pressure is exceeded and some microscopic geometrical feature of the surfaces may cause an inward pumping action to counteract the effect of applied pressure. Under favourable circumstances hydrodynamic pressure may be generated to oppose the flow due to the applied pressure.

A seal as shown in Fig. 4.65 employs the Rayleigh step principle to cause oil to flow inwards so as to achieve a balance. The configuration of the lip, as shown in sections AA and BB is such that the action of the shaft in inducing a flow of oil in the circumferential direction is used to generate hydrostatic pressure which limits flow in the axial direction. The component denoted by C is made of compliant material, ring D is of rigid metal, and E is a circumferential helical spring which applies a uniform radial pressure to the lip of the seal.

4.15.4. Utilization of hydrodynamic action

A number of seal designs can be devised where the moving parts do not come into contact, leakage being prevented by the hydrodynamic action. A commonly used form is the helical seal shown schematically in Fig. 4.66. The important dimensions are the clearance c, the helix angle α and the proportions of the groove. The pressure generated under laminar conditions is given by

$$\Delta p = \frac{6\mu UL \tan\alpha\gamma(1-\gamma)(\beta^3-1)(\beta-1)}{c^2\beta^3(1-\tan^2\alpha)+\tan^2\alpha(\beta^3-1)(1-\gamma)}, \qquad (4.192)$$

where $\beta = (h+c)/c$ or $h/c+1$, $\gamma = b/(a+b)$ and L is the effective length of the screwed portion.

At high Reynolds numbers (Re $\geqslant$ 600–1000), turbulent flow conditions lead to a more effective sealing action. Typical values for the design

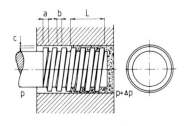

Figure 4.66

parameters of helical seals are as follows: $\alpha = 10°$–$20°$; $\beta = 4$–6; $\gamma = 0.5$–0.8. Taking mean values, eqn (4.192) becomes

$$\Delta p = 1.5\mu \frac{UL}{c^2}. \tag{4.193}$$

Clearance is usually between 2.5 and 5×10^{-5} m. Face seals may also act on the viscoseal principle. Then, spiral grooves are incorporated into the diametral plane. Such grooves are often incorporated into the contacting faces of seals made of elastomers in order to induce a self-pumping action.

4.15.5.   Labyrinth seals

Labyrinth seals are based on positive, finite mechanical clearances which are sufficiently large to preclude the possibility of contact between the parts in relative motion. They may be used either in the radial or axial flow configurations and are effective by reason of the generation of eddies within the cavities. The spacing of the barriers between the cavities is usually about twenty times the radial clearance. The most critical aspect of labyrinth seal design is the provision for the thermal expansion of the equipment being sealed. The adverse effects of inadvertent contact may be minimized by the use of a relatively soft material, for example carbon, for one of the components. Instances of failure of the barrier elements by fatigue are usually due to aeroelastic instability which could be avoided by suitable design. There are computer programmes available to design a labyrinth seal.

4.15.6.   Wear in mechanical seals

The sealing elements (the primary ring and the mating ring), of a nominally contact type seal, usually operate in uni-directional sliding. Reciprocating motion and various modes of oscillatory motion are common. Most often those elements are in an impregnated carbon-graphite nose-piece sliding against a harder material, such as ceramic, tungsten carbide or silicon carbide as listed in Table 4.1. These materials are usually selected to be chemically compatible with the lubricant or process fluid, as well as the operating environment and the conditions of operation. All these factors can contribute to the seal wear mechanisms that must be mitigated to achieve wear control.

Adhesive wear is the dominant type of wear in a well-designed seal. Even when there is a hydrodynamic lubricating film at the interface, solid contacts occur during startup, shutdown, and operating perturbations; the carbon-graphite nose-piece is usually considered the primary wearing part and the mating surface wears to a lesser extent. Details of the adhesive wear process, as such, were discussed in Chapter 2. In the special case of seals the face loads are sufficiently low so that the mild adhesive wear process occurs. The process is dominated by transfer films. The *PV* (product of specific contact pressure and sliding velocity) criterion used in the design of seals is

Table 4.1. *Coefficient of wear in order of magnitude for seal face materials*

Sliding material		Wear coefficient K
rotating	stationary	
carbon-graphite (resin filled)	cast iron	10^{-6}
carbon-graphite (resin filled)	ceramic (85% Al_2O_3)	10^{-7}
carbo-graphite (babbitt filled)	ceramic (85% Al_2O_3)	10^{-7}
carbon-graphite (bronze filled)	tungsten carbide (6% cobalt)	10^{-8}
tungsten carbide (6% cobalt)	tungsten carbide	10^{-8}
silicon carbide (converted carbon)	silicon carbide (converted carbon)	10^{-9}

an expression of the limit of mild adhesive wear. Table 4.2 gives the *PV* limitations for frequently used seal face materials.

Physical and chemical bonds can cause adhesion between surfaces; thus transfer films are formed that are basic to friction and wear processes. With relative motion, shear occurs in the direction of sliding along the weakest shear plane in the surface region. With carbon-graphite materials, graphite usually contributes to that weak shear plane. Inherent in this process is the development of a transfer film of carbon on the mating surface. The surface of that transfer film can be expected to be graphitic and highly oriented with the basal plane essentially parallel to the direction of shear. The example of transfer films of graphite is analogous to the development of highly orientated films with solid polymeric materials, especially PTFE. The surface chemistry of the base material as well as its surface topography influences the formation of transfer films.

The surfaces of sealing interfaces are usually very smooth. The lack of roughness is a fortuitous result of a manufacturing process aimed at providing physical conformance of the mating surfaces to minimize the potential gap for leakage flow. It is very clear that for many seal applications a matte type surface texture of the type obtained by lapping, hard or fine abrasive blasting and ion bombardment provides a good physical base for achieving the mechanical adherence of a transfer film.

Abrasive wear is a condition of wear in seals, that frequently limits the life of the seals. Many abrasive wear problems for seals result from the operating environment. For example, road dust or sand enters the sealing gap and the particles may move freely to abrade both interfaced surfaces by a lapping action; that is the surfaces are subjected to three-body abrasive wear. Alternatively, the particles become partially embedded in one of the surfaces and can then act as a cutting tool, shearing metal from the mating surface by a two-body wear mechanism. Abrading particles can also come

Table 4.2. *The frequently used seal face materials and their PV limits*

	Sliding material		*PV* limit	Comments
	rotating	stationary	Pa × ms^{-1}	
unlike materials	carbon graphite	ceramic (85% Al_2O_3)	3.5×10^6	poor thermal shock resistance but quite good corrosion resistance
		ceramic (99% Al_2O_3)	3.5×10^6	better corrosion resistance than 85% Al_2O_3
		tungsten carbide (6% Co)	17.5×10^6	with bronze filled carbon-graphite, the *PV* is up to 3.5×10^6 Pa ms^{-1}
		silicon carbide converted carbon	17.5×10^6	good wear resistance
		tungsten carbide (6% Ni)	17.5×10^6	nickel binder for better corrosion resistance
		silicon carbide (solid)	17.5×10^6	better corrosion resistance than tungsten carbide but poorer thermal shock strength
like materials	carbon-graphite		1.75×10^6	low *PV* but very good against face blistering
	ceramic		0.35×10^6	good service on sealing paint pigments
	silicon carbide converted carbon		17.5×10^6	excellent abrasion resistance more economical than solid silicon carbide
	silicon carbide (solid)		17.5×10^6	excellent abrasion resistance good corrosion resistance and moderate thermal shock strength
	boron carbide		—	for extreme corrosion resistance, expensive

from within the sealed system. These wear particles can come from the mechanical components, products of corrosion like rust scale, machining burrs or casting sand from the production processes. The sealed material may also be a slurry of abrasives or the process fluid may degrade to form hard solid particles. It should be noted that one of the functions of a seal is to keep external abrasives from mechanical systems. Frequently, seals will have external wipers or closures to limit the entrance of particles into the seal cavity and are most effective at high shaft speeds. In fluid systems, centrifugal separators can provide seal purge fluid relatively free from abrasives.

Corrosive wear or chemical wear is common in industrial seals exposed to a variety of process fluids or other products that are chemically active. Sliding surfaces have high transient flash temperatures from frictional heating that has been demonstrated to promote chemical reactivity. The high flash temperatures of the asperities characteristic of sliding friction, initiate reactions that are further accelerated by increasing contact pressure. Also, the ambient temperature level is important since rates of chemical reaction approximately double with each 10 °C temperature increase.

The chemistry of the process fluid or environment is very important in the selection of seal materials. Consideration must be given to both the normal corrosion reactions and the possibility of corrosive wear. Some surface reaction is essential to many useful lubrication processes in forming films that inhibit adhesive wear. However, excessive active chemical reactions are the basis for corrosive or chemical wear. It is important to remember, however, that air is perhaps the most influential chemical agent in the lubrication process and normal passive films on metals and the adsorbates on many materials, are a basic key to surface phenomena, critical to lubrication and wear.

Pitting or fatigue wear and blistering are commonly described phenomena in the wear of seal materials that can be, but are not necessarily, related. Carbon has interatomic bonding energies so high that grain growth or migration of crystal defects is virtually impossible to obtain. Accordingly, one would expect manufactured carbon and graphite elements to have excellent fatigue endurance. Pitting is usually associated with fatigue but may have other causes on sealing interfaces. For example, oxidative erosion on carbons can cause a pitted appearance. Cavitation erosion in fluid systems can produce a similar appearance. Carbon blistering may produce surface voids on larger parts. Usually blistering is attributed to the subsurface porosity being filled with a sealed liquid and subsequently vaporized by frictional heating. The vapour pressure thus created lifts surface particles to form blisters. Thermal stress cracks in the surface may be the origin for blisters with the liquids filling such cracks. In addition, the hydraulic wedge hypothesis suggested for other mechanical components might also be operative in seals. In that case, the surface loading forces may deform and close the entrance to surface cracks, also causing bulk deformation of adjacent solid material so as to create a hydraulic pressure that further propagates the liquid-filled void or crack. The blister phenomena is of primary concern with carbon seal materials, but no single approach to the problem has provided an adequate solution.

Impact wear occurs when seals chatter under conditions of dynamic instability with one seal element moving normal to the seal interface. Sometimes, very high vibration frequencies and acceleration forces might develop. Rocking or precessing of the nose-piece relative to the wear plate occurs and impact of the nose-piece edges is extremely destructive. This type of phenomena occurs in undamped seals with low face pressures and may be excited by friction or fluid behaviour, such as a phase change, as well as by misalignment forces.

Fretting usually occurs on the secondary sealing surfaces as the primary sealing interface moves axially to accommodate thermal growth, vibrations and transient displacements including wear. Fretting of the piston ring secondary seal in a gas seal can significantly increase the total seal leakage. Some seal manufacturers report that 50 to 70 per cent of the leakage is past the secondary seal and specific tests show that a fretted installation may leak more rapidly. Fretting is initiated by adhesion and those conditions that reduce adhesion usually mitigate fretting.

4.15.7. Parameters affecting wear

Three separate tests are usually performed to establish the performance and acceptability of seal face materials. Of these the most popular is the PV test, which gives a measure for adhesive wear, considered to be the dominant type of wear in mechanical seals. Abrasive wear testing establishes a relative ranking of materials by ordering the results to a reference standard material after operation in a fixed abrasive environment. A typical abrasive environment is a mixture of water and earth. The operating temperature has a significant influence upon wear. The hot water test evaluates the behaviour of the face materials at temperatures above the atmospheric boiling point of the liquid. The materials are tested in hot water at 149 °C and the rate of wear measured. None of the above mentioned tests are standardized throughout the industry. Each seal supplier has established its own criteria. The PV test is, at the present time, the only one having a reasonable mathematical foundation that lends itself to quantitative analysis.

The foundation for the test can be expressed mathematically as follows:

$$PV = [\Delta p(b - \xi) + F_s]V_f, \tag{4.194}$$

where PV is the pressure $\times$ velocity, Δp is the differential pressure to be sealed, b is the seal balance, ξ is the pressure gradient factor, F_s is the mechanical spring pressure and V is the mean face velocity.

All implicit values of eqn (4.194), with the exception of the pressure gradient factor, ξ, can be established with reasonable accuracy. Seal balance, b, is further defined as the mathematical ratio of the hydraulic closing area to the hydraulic opening area. The pressure gradient factor, ξ, requires some guessing since an independent equation to assess it has not yet been developed. For water it is usually assumed to be 0.5 and for liquids such as light hydrocarbons, less than 0.5 and for lubricating oils, greater than 0.5. The product of the actual face pressure, P, and the mean velocity, V, at the seal faces enters the frictional power equation as follows:

$$N_f = (PV)fA, \tag{4.195}$$

where N_f is the frictional power, PV is the pressure $\times$ velocity, f is the coefficient of friction and A is the seal face apparent area of contact.

Therefore, PV can be defined as the frictional power per unit area. Coefficients of friction, at $PV = 3.5 \times 10^6$ Pa m s^{-1}, for frequently used seal materials are given in Table 4.3. They were obtained with water as the lubricant. The values could be from 25 to 50 per cent higher with oil due to the additional viscous drag. At lower PV levels they are somewhat less, but not significantly so; around 10 to 20 per cent on the average. The coefficient of friction can be further reduced by about one-third of the values given in

Table 4.3. *Coefficient of friction for various face materials at*
$PV = 3.5 \times 10^6 \ Pa\,m/s$

	Sliding	material	Coefficient of friction
	rotating	stationary	
unlike materials	carbon-graphite (resin filled)	cast iron ceramic tungsten carbide	0.07
		silicon carbide	0.02
		silicon carbide (converted carbon)	0.015
like materials	silicon carbide	tungsten carbide	0.02
	silicon carbide converted carbon		0.05
	silicon carbide		0.02
	tungsten carbide		0.08

Table 4.3 by introducing lubrication grooves or hydropads on the circular flat face of one of the sealing rings. In most cases a slight increase in leakage is usually experienced. As there is no standardized PV test that is used universally throughout the industry, individual test procedures will differ.

4.15.8. Analytical models of wear

Each wear process is unique, but there are a few basic measurements that allow the consideration of wear as a fundamental process. These are the amount of volumetric wear, W, the material hardness, H, the applied load, L, and the sliding distance, d. These relationships are expressed as the wear coefficient, K

$$K = WH/(Ld). \qquad (4.196)$$

By making a few simple algebraic changes to this basic relationship it can be modified to enable the use of PV data from seal tests. With sliding distance, d, being expressed as velocity $\times$ time, that is $d = Vt$, load L as the familiar pressure relationship of load over area, $P = L/A$, and linear wear, h, as volumetric wear over contact area, $h = W/A$, the wear coefficient becomes

$$K = hH/t(PV) \qquad (4.197a)$$

or

$$K = (\text{linear wear/time}) \times (\text{hardness}/PV). \qquad (4.197b)$$

Expressing each of the factors in the appropriate dimensional units will yield a dimensionless wear coefficient, K. Since several hardness scales are used in the industry, Brinell hardness or its equivalent value, should be used for calculating K. At the present time the seal industry has not utilized the wear coefficient, but as is readily seen it can be obtained, without further testing and can be established from existing PV data, or immediately be part of the PV evaluation itself, without the necessity of running an additional separate test.

4.15.9. Parameters defining performance limits

The operating parameters for a seal face material combination are established by a series of PV tests. A minimum of four tests, usually of 100 hours each, are performed and the wear rate at each level is measured. The PV value and the wear rate are recorded and used to define the operating PV for a uniform wear rate corresponding to a typical life span of about two years. Contrary to most other industrial applications that allow us to specify the most desirable lubricant to suppress the wear process of rubbing materials, seal face materials are required to seal a great variety of fluids and these become the lubricant for the sliding ring pairs in most cases. Water, known to be a poor lubricant, is used for the PV tests and for most practical applications reliable guidelines are achieved by using it.

4.15.10. Material aspects of seal design

In the majority of practical applications about twelve materials are used, although hundreds of seal face materials exist and have been tested. Carbon has good wear characteristics and corrosion resistance and is therefore used in over 90 per cent of industrial applications. Again, over hundreds of grades are available, but by a process of careful screening and testing, only the best grades are selected for actual usage. Resin-filled carbons are the most popular. Resin impregnation renders them impervious and often the resin that fills the voids enhances the wear resistance. Of the metal-filled carbons, the bronze or copper–lead grades are excellent for high-pressure service. The metal filler gives the carbon more resistance to distortion by virtue of its higher elastic modulus. Babbitt-filled carbons are quite popular for water-based services, because the babbitt provides good bearing and wear characteristics at moderate temperatures. However, the development of excellent resin-impregnated grades over recent years is gradually replacing the babbitt-filled carbons. Counterface materials that slide against the carbon can be as simple as cast-iron and ceramic or as sophisticated as the carbides. The PV capability can be enhanced by a factor of 5 by simply changing the counterface material from ceramic to carbide. For frequently used seal face materials, the typical physical properties are given in Table 4.4.

Table 4.4. *Physical properties of frequently used seal face materials*

Properties	Material							
		ceramic		carbide			carbon	
	cast iron	85% (Al$_2$O$_3$)	99% (Al$_2$O$_3$)	tungsten 6% Co	silicon Si-C	boron B$_4$C	resin	bronze
Modulus of elasticity ($\times 10^3$ MPa)	90–110	221	345	621	331–393	290–448	17.2–27.6	20–30
Tensile strength (MPa)	448–827	138	269	850	142	155	31–62	52–62
Coefficient of thermal expansion ($\times 10^{-6}$ cm/cm/K)	11.88	7.02	7.74	4.55	3.38	3.1–5.79	4 14–6.12	4.32–5.58
Thermal conductivity (Watts/m°K)	39.7–50.2	14.7	25.1	70.9–83.0	70.9–103.8	27.7	6.57–20.8	13.84–14.7
Density (kg/m^3)	7169–7418	3405	3792	16 331	2879	2408	1771–1910	2297–2685
Hardness	217–269	87	87	92	86–88	2800	80–105	70–92
	Brinell	Rockwell A				Knoop	Shore	Shore

4.15.11.   Lubrication of seals

The initial assumptions used in analyses of narrow seal face lubrication are based on the one-dimensional incompressible Reynolds equation

$$\frac{\partial}{\partial r}\left(rh^3\frac{\partial p}{\partial r}\right)=6\mu\omega r\frac{\partial h}{\partial\Theta}, \tag{4.198}$$

where r is the radial coordinate, h is the film thickness, p is the pressure, μ is the viscosity, ω is the angular velocity and Θ is the angular coordinate.

Fluid film models for seals do not allow for the dynamic misalignment and other motions that are characteristic of all seal faces; in real seal applications there are important deviations from the concepts of constant face loads and uniform circumferential and radial film thicknesses. Also, the interface geometry is markedly influenced by the manufacturing processes, deformations and the interface wear processes, as well as by the original design considerations for film formation. The properties and states of the fluids in the seals vary, so that solid particles, corrosive reactions, cavitation phenomena and theology changes may be critical to the formation of a lubricating film. Also, it has been observed that the size of the wear particles and the surface roughness can determine the leakage gap and thereby establish the film thickness. Circumferential waviness in seal faces may result from planned or unplanned features of the manufacturing processes, from the geometry of the structure supporting the nose-piece or the primary ring, from the mechanical linkage, i.e. drive pins, restraining radial motion in the seal assembly and perhaps from several other factors. These fluid film-forming features seem to occur because of random processes that cause inclined slider geometry on both macro and micro bases. Micro-geometry of the surface may be determined by random wear processes in service. It is reasonable, however, to anticipate that desired macro-geometry waviness can be designed into a sealing interface by either modifying one or both of the sealing interface surfaces or their supporting structures.

Hydrodynamic effects of misalignment in seal faces have been analytically investigated and shown to provide axial forces and pressures in excess of those predicted for perfectly aligned faces. Misalignment of machines, however, cannot usually be anticipated in the design of seals for general industrial use. Misalignment can be designed into either the mating ring, the primary ring or the assembly supporting the primary seal ring. Using a floating primary seal ring nose-piece, misalignment can be conveniently achieved. However, with a rotating seal body (including the seal ring) the misalignment would be incorporated into the mounting of the mating ring. Hydrostatic film formation features have been achieved in several commercial face seals (in several instances with a converging gap) by a radial step configuration, and by assorted types of pads and grooves. These are essentially so-called tuned seals that work well under a limited range of operating conditions, but under most conditions will have greater leakage than hydrodynamically-generated lubricating films at the sealing interfaces.

Coning of the rotating interface element occurs as a result of wear or by thermal pressure or mechanical forces. Depending on the type of pressurization (that is internal or external) coning may enhance the hydrostatic effects or give instability with a diverging leakage flow path. The thermoelastically generated nodes can determine the leakage gap in seals so that greater axial pressures on the sealing interface may increase leakage flow. With moving points of contact and subsequent cooling, the worn nodes become recesses and a progressive alteration of the seal interface geometry occurs. There does not seem to be a predictable method of using the features described above to achieve lubricant film formation. The effects can be minimized by the proper selection of interface materials.

Recently reported investigations have mostly concentrated on isolated modes of seal face lubrication. The fact that many modes may be functioning and interacting in the operation of seals has not been questioned, but simplifying assumptions are essential in achieving tractable analyses. To utilize those research studies in a design for service requires that the modes identified be considered with respect to interactions and designed into a seal configuration that can have industrial applications. Analytical appraisal of dynamic behaviour like that associated with angular misalignment can provide a significant step towards integration. Experimental determinations will be required to document the interactions in seal face lubrication and supplement further analytical design.

References to Chapter 4

1. C. E. Wilson and W. Michels. *Mechanism – Design Oriented Kinematics.* Chicago, Ill.: American Technical Society, 1969.
2. *Belt Conveyors for Bulk Materials.* Conveyor Equipment Manufacturers Association. Boston, Mass.: Cahners Publishing Co., 1966.
3. V. M. Faires. *Design of Machine Elements.* New York: The Macmillan Company, 1965.
4. J. Gagne. Torque capacity and design of cone and disc clutches. *Mach. Des.*, **24** **(12)** (1953), 182.
5. P. Black. *Mechanics of Machines.* Elmsford, New York: Pergamon Press, 1967.
6. H. S. Rothbart. *Mechanical Design and Systems Handbook.* New York: McGraw-Hill, 1964.
7. J. N. Goodier. The distribution of load on the thread of screws. *J. Appl. Mech., Trans. ASME*, **62** (1940), 000.
8. E. T. Jagger. The role of seals and packings in the exclusion of contaminants. *Proc. Instn Mech. Engrs*, **182** (3A) (1967), 434.
9. C. M. White and D. F. Denny. *The Sealing Mechanism of Flexible Packings.* London: His Majesty's Stationary Office, 1947.

5 Sliding-element bearings

Sliding-element bearings, as distinguished from the rolling-element bearings to be discussed in Chapter 7, are usually classified as plain journal or sleeve, thrust, spherical, pivot or shoe-type thrust bearings. Another method of classification is to designate the bearing according to the type of lubrication used. A hydrodynamically-lubricated bearing is one that uses a fluid lubricant (liquid or gas) to separate the moving surfaces. If the fluid film gets thinner and is no longer able to separate the moving surfaces, partial metal–metal contact can occur; this type of lubrication is referred to as mixed lubrication. When the lubricating film gets even thinner and the two contacting surfaces are separated by a film of a few angstroms thick the bulk properties of the lubricant are not any longer important and its physico-chemical characteristic comes into prominence. This type of lubrication is usually called boundary lubrication. Boundary lubrication is usually not planned by the designer. It depends on such factors as surface finish, wear-in, and surface chemical reactions. Low-speed bearings, heavily-loaded bearings, misaligned bearings and improperly lubricated bearings are usually more prone to operate under mixed or boundary lubrication. Boundary lubrication presents yet another problem to the designer: it cannot be analysed by mathematical methods but must be dealt with on the basis of experimental data. A completely separate class of sliding element bearings constitute bearings operating without any external lubrication. They are called self-lubricating or dry bearings.

In this chapter mainly hydrodynamically-lubricated bearings are examined and discussed. The problem of bearing type selection for a particular application is covered by ESDU–65007 and ESDU–67033. Calculation methods for steadily loaded bearings are presented in ESDU–84031 and ESDU–82029. The design and operation of self-lubricating bearings are also briefly covered in this chapter. However, the reader is referred to ESDU–87007 where there is more information on this particular type of bearing.

5.1. Derivation of the Reynolds equation

It is well known from fluid mechanics that a necessary condition for pressure to develop in a thin film of fluid is that the gradient and slope of the velocity profile must vary across the thickness of the film (see Chapter 2 for details). Three methods for establishing a variable slope are commonly used:

(i) fluid from a pump is directed to a space at the centre of the bearing, developing pressure and forcing fluid to flow outward through the narrow space between the parallel surfaces. This is called a hydrostatic lubrication or an externally pressurized lubrication;

(ii) one surface rapidly moves normal to the other, with viscous resistance to the displacement of the oil. This is a squeeze-film lubrication;

(iii) by positioning one surface so that it is slightly inclined to the other, then by relative sliding motion of the surfaces, lubricant is dragged into the converging space between them. It is a wedge-film lubrication and the type generally meant when the word hydrodynamic lubrication is used.

Positioning of the surfaces usually occurs automatically when the load is applied if the surfaces are free of certain constraints. Under dynamic loads the action of a bearing may be a combination of the foregoing and hence general equations are going to be derived and used to illustrate the preceding three methods.

Let a thin film exist between the two moving bearing surfaces 1 and 2, the former flat and lying in the X-Z plane, the latter curved and inclined, as illustrated in Fig. 5.1. Component velocities u, v and w exist in directions X, Y and Z, respectively. At any instant, two points having the same x, z coordinates and separated by a distance h will have absolute velocities which give the following set of boundary conditions

$$y = 0, \quad u = U_1, \quad v = V_1, \quad w = W_1,$$
$$y = h, \quad u = U_2, \quad v = V_2, \quad w = W_2. \tag{5.1}$$

The pressure gradients, $\partial p/\partial x$ and $\partial p/\partial z$ in the X and Z directions are independent of y in a thin film, and $\partial p/\partial y = 0$.

Recalling the fundamental relationship between pressure and velocity as would be discussed in a fluid mechanics course

$$\frac{\partial p}{\partial x} = \mu \frac{\partial^2 u}{\partial y^2}$$

and integrating it with respect to y gives

$$\frac{\partial u}{\partial y} = \frac{1}{\mu} \frac{\partial p}{\partial x} y + C_1 \quad \text{and} \quad u = \frac{1}{\mu} \frac{\partial p}{\partial x} \frac{y^2}{2} + C_1 y + C_2$$

and from the conditions of eqn (5.1)

$$C_2 = U_1 \quad \text{and} \quad C_1 = -\frac{1}{\mu} \frac{\partial p}{\partial x} \frac{h}{2} + \frac{U_2 - U_1}{h}.$$

Thus

$$u = -\frac{1}{2\mu} \frac{\partial p}{\partial x} (hy - y^2) + \left[U_1 + \frac{y}{h}(U_2 - U_1) \right]. \tag{5.2a}$$

Similarly

$$w = -\frac{1}{2\mu} \frac{\partial p}{\partial z} (hy - y^2) + \left[W_1 + \frac{y}{h}(W_2 - W_1) \right]. \tag{5.2b}$$

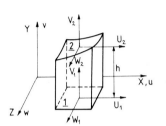

Figure 5.1

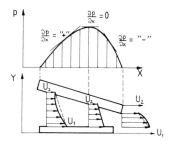

Figure 5.2

Each equation shows that a velocity profile consists of a linear portion, the second term to the right of the equals sign, and a parabolic portion which is subtracted or added depending upon the sign of the first term. For velocity u the second term is represented in Fig. 5.2 by a straight line drawn between U_1 and U_2. Since $-(hy-y^2)/2\mu$ is always negative, the sign of the first term is the opposite of the sign of $\partial p/\partial x$ or $\partial p/\partial z$, which are the slopes of the pressure versus the position curves. Notice the correspondence between the positive, zero and negative slopes of the pressure curve, shown in Fig. 5.2, and the concave (subtracted), straight and convex (added) profiles of the velocity curves also shown in Fig. 5.2.

The flow q_x normal to and through a section of area $h\,dz$ is estimated next, as illustrated in Fig. 5.3. By substitution for u eqn (5.2a), integration and application of limits

$$q_x = \int_0^h u(dy\,dz) = -\frac{h^3}{12\mu}\frac{\partial p}{\partial x}dz + \frac{U_1+U_2}{2}h\,dz. \tag{5.3a}$$

Similarly, through area $h\,dx$

$$q_z = \int_0^h w(dy\,dx) = -\frac{h^3}{12\mu}\frac{\partial p}{\partial z}dx + \frac{W_1+W_2}{2}h\,dx. \tag{5.3b}$$

Figure 5.3

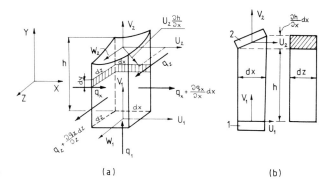

(a)   (b)

Note that these flows are through areas of elemental width. Second integrations $\int q_x$ and $\int q_z$ must be made to obtain the total flows Q_x and Q_z through a bearing slot.

Case (a) in Fig. 5.3 represents an elemental geometric space within the fluid, at any instant extending between the bearing surfaces but remaining motionless. Through its boundaries oil is flowing. A positive velocity V_1 of the lower bearing surface pushes oil inwards through the lower boundary of the space and gives a flow q_1 in the same sense as the inward flows q_x and q_z. Surface velocities U_1 and W_1 do not cause flow through the lower boundary, since the surface is flat and in the X-Z plane. Hence $q_1 = V_1\,dx\,dz$. Because the top bearing surface is inclined, its positive velocity V_2 causes outward flow $V_2\,dx\,dz$. Furthermore, positive velocities U_2 and W_2 together with the positive surface slopes $\partial h/\partial x$ and $\partial h/\partial z$ cause inward flow. In Fig. 5.3, case (a), there is shown a velocity component

$U_2(\partial h/\partial x)$ normal to the top area, that may be taken as $dx\,dz$ because of its very small inclination in bearings. In Fig. 5.3, case (b), flow at velocity U_2 is shown through the projected area $(\partial h/\partial x)\,dx\,dz$, which is shaded. Either analysis gives the same product of velocity and area. Hence the total flows q_1 inwards through the lower boundary of the geometric space and q_2 outwards through the upper boundary area, are respectively

$$q_1 = V_1\,dx\,dz, \tag{5.4}$$

$$q_2 = V_2\,dx\,dz - U_2\frac{\partial h}{\partial x}dx\,dz - W_2\frac{\partial h}{\partial z}dz\,dx.$$

Continuity with an incompressible fluid requires that the total inward flow across the boundaries equals the total outward flow, or

$$q_x + q_z + q_1 = \left(q_x + \frac{\partial q_x}{\partial x}dx\right) + \left(q_z + \frac{\partial q_z}{\partial z}dz\right) + q_2. \tag{5.5}$$

For the case of a compressible fluid (gas bearings), mass flows instead of volume flows wound be equated. A relationship between density and pressure must be introduced. With substitution from eqns (5.3) and (5.4) into eqn (5.5), selective differentiation, and elimination of the product $dx\,dz$, the result is

$$V_1 = -\frac{1}{12}\frac{\partial}{\partial x}\left(\frac{h^3}{\mu}\frac{\partial p}{\partial x}\right) + \frac{U_1+U_2}{2}\frac{\partial h}{\partial x} + \frac{h}{2}\frac{\partial}{\partial x}(U_1+U_2)$$

$$-\frac{1}{12}\frac{\partial}{\partial z}\left(\frac{h^3}{\mu}\frac{\partial p}{\partial z}\right) + \frac{W_1+W_2}{2}\frac{\partial h}{\partial z} + \frac{h}{2}\frac{\partial}{\partial z}(W_1+W_2) \tag{5.6}$$

$$+V_2 - U_2\frac{\partial h}{\partial x} - W_2\frac{\partial h}{\partial z}.$$

With rearrangement

$$\frac{1}{6}\left[\frac{\partial}{\partial x}\left(\frac{h^3}{\mu}\frac{\partial p}{\partial x}\right) + \frac{\partial}{\partial z}\left(\frac{h^3}{\mu}\frac{\partial p}{\partial z}\right)\right] = (U_1-U_2)\frac{\partial h}{\partial x} - 2(V_1-V_2)$$

$$+ (W_1-W_2)\frac{\partial h}{\partial z} + h\frac{\partial}{\partial x}(U_1+U_2) + h\frac{\partial}{\partial z}(W_1+W_2). \tag{5.7}$$

The last two terms are nearly always zero since there is rarely a change in the surface velocities U and W, which represents the stretch-film case. The stretch-film case can occur when there is a lubricating film separating a wire from the die through which it is being drawn. Reduction in the diameter of the wire gives an increase in its surface velocity during its passage through the die.

This basic equation of hydrodynamic lubrication was developed for a less general case in 1886 by Osborne Reynolds. As usual, the eqn (5.7) and its reduced forms in any coordinate system shall be referred to as the Reynolds equation. Equation (5.7) transformed into the cylindrical coordinates is

$$\frac{1}{6}\left[\frac{1}{r}\frac{\partial}{\partial r}\left(r\frac{h^3}{\mu}\frac{\partial p}{\partial r}\right)+\frac{1}{r^2}\frac{\partial}{\partial\Theta}\left(\frac{h^3}{\mu}\frac{\partial p}{\partial\Theta}\right)\right]=(R_1-R_2)\frac{\partial h}{\partial r}$$

$$-2(V_1-V_2)+(T_1-T_2)\frac{1}{r}\frac{\partial h}{\partial\Theta}+\frac{h}{r}\left\{\frac{\partial}{\partial r}\left[r(R_1+R_2)\right]+\frac{\partial}{\partial\Theta}(T_1+T_2)\right\}, \quad (5.8)$$

where the velocities of the two surfaces are R_1 and R_2 in the radial direction, T_1 and T_2 in the tangential direction, and V_1 and V_2 in the axial direction across the film. For most bearings many of the terms may be dropped, and particularly those which imply a stretching of the surfaces.

5.2. Hydrostatic bearings

Figure 5.4 shows the principle of a hydrostatic bearing action. Lubricant from a constant displacement pump is forced into a central recess and then flows outwards between the bearing surfaces, developing pressure and separation and returning to a sump for recirculation. The surfaces may be cylindrical, spherical or flat with circular or rectangular boundaries. If the surfaces are flat they are usually guided so that the film thickness h is uniform, giving zero values to $\partial h/\partial x$ and $\partial h/\partial r$, $\partial h/\partial\Theta$ in the Reynolds' equations. These appear in, and cancel out, the terms containing the surface velocities, an indication that the latter do not contribute to the development of pressure. Hence, with u considered constant, Reynolds' equation, eqn (5.7), is reduced to

$$\frac{\partial^2 p}{\partial x^2}+\frac{\partial^2 p}{\partial z^2}=0. \quad (5.9)$$

If the pad is circular as shown in Fig. 5.4 and the flow is radial, then $\partial p/\partial\Theta=0$ from symmetry, and eqn (5.8) is reduced to

$$\frac{\partial}{\partial r}\left(r\frac{\partial p}{\partial r}\right)=0. \quad (5.10)$$

Equation (5.10) is readily integrated, and together with the boundary conditions of Fig. 5.4, namely $p=0$ at $r=D/2$ and $p=p_0$ at $r=d/2$, the result is

$$p=p_0\frac{\ln(D/2r)}{\ln(D/d)}. \quad (5.11)$$

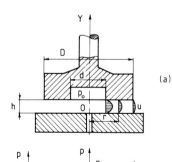

The variation of pressure over the entire circle is illustrated in Fig. 5.4, case (b), from which an integral expression for the total load P may be written and the following expression obtained for p_0

$$p_0=\frac{8P}{\pi}\frac{\ln(D/d)}{(D^2-d^2)}. \quad (5.12)$$

An equation for the radial flow velocity u_r may be obtained by substituting r for x and $U_1=U_2=0$ in eqn (4.2a)

$$u_r=-\frac{1}{2\mu}\frac{\partial p}{\partial r}(hy-y^2).$$

Figure 5.4

Substitution for $\partial p/\partial r$, obtained by the differentiation of eqn (5.11), gives

$$u_r = \frac{p_0(hy-y^2)}{2\mu r \ln(D/d)} \frac{4P(hy-y^2)}{\pi\mu r(D^2-d^2)} \qquad (5.13)$$

the latter term is obtained by substitution for p_0 from eqn (5.12). The total flow through a cylindrical section of total height h, radius r and length $2\pi r$, is

$$Q = 2\pi r \int_0^h u_r dy = \frac{\pi p_0 h^3}{6\mu \ln(D/d)} \frac{4Ph^3}{3\mu(D^2-d^2)}. \qquad (5.14)$$

This is the minimum oil delivery required from the pump for a desired film of thickness h. Let V be the average velocity of the flow in the line, A its cross-sectional area and η the mechanical efficiency of the pump. Then the power required from the pump is

$$\text{power required} = \frac{(p_0A)V}{\eta} = \frac{p_0(AV)}{\eta} = \frac{p_0Q}{\eta}. \qquad (5.15)$$

If the circular pad, shown in Fig. 5.4, is rotated with speed n about its axis, the tangential fluid velocity w_t may be found from eqn (5.2b) by substituting $W_1 = 0$, $W_2 = 2\pi rn$ and for $\partial p/\partial z$ the quantity $\partial p/\partial(r\Theta) = (1/r)\partial p/\partial\Theta$. But since $h = \text{const}$, $\partial p/\partial\Theta = 0$. Thus

$$w_t = \frac{y}{h}(2\pi rn) \qquad \text{and} \qquad \tau = \mu\frac{\partial w_t}{\partial y} = \frac{2\pi\mu rn}{h}.$$

The torque required for rotation is

$$T = \int r\, dF = \int r\tau\, dA = \frac{4\pi^2\mu n}{h} \int_{d/2}^{D/2} r^3\, dr$$

therefore

$$T = \frac{\pi^2\mu n}{16h}(D^4 - d^4). \qquad (5.16)$$

If, over a portion of the pad, the flow path in one direction X is short compared with that in the other direction Z, as shown in Fig. 5.5, the flow velocity w and the pressure gradient $\partial p/\partial z$ will be relatively small and eqn (5.9) may be approximated by $\partial^2 p/\partial x^2 = 0$, i.e. parallel flow is assumed for a distance b through each slot of approximate width l. Integration of the differential equation, together with the use of the limits $p = p_0$ at $x = 0$ and $p = 0$ at $x = b$ gives the pressure distribution. Integration $\int_{-\frac{1}{2}}^{\frac{1}{2}} q_x$ from eqn (5.3) gives the flow Q across one area bl. The slot equations for one area are

$$p = p_0\left(1 - \frac{x}{b}\right) \qquad \text{and} \qquad Q = \frac{p_0lh^3}{12\mu b}. \qquad (5.17)$$

The force or torque required to move a hydrostatic bearing at slow speed is extremely small, less than in ball- or roller-bearings. Also, there is no

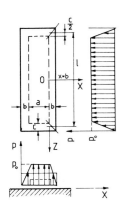

Figure 5.5

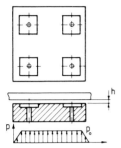

Figure 5.6

difference between static friction and kinematic friction. Here, a coefficient of friction for rotating bearings, is defined as the tangential moving force at the mean radius of the active area divided by the applied normal load.

Hydrostatic bearings are used for reciprocating platens, for rotating telescopes, thrust bearings on shafts and in test rigs to apply axial loads to a member which must be free of any restriction to turning. Journal bearings support a rotating shaft by a different method, but the larger bearings will have a built-in hydrostatic lift for the shaft before it is rotated, to avoid initial metal–metal contact.

To give stability to the pad, three or four recesses should be located near the edges or in the corners, as shown in Fig. 5.6. However, if one pump is freely connected to the recesses, passage of all the fluid through one recess may occur, tipping the pad and giving no flow or lift at an opposite recess. Orifices must be used in each line from the pump to restrict the flow to a value well below the displacement of the pump, or a separate pump can be used to feed each recess.

Air or inert gases are used to lift and hydrostatically or hydrodynamically support relatively light loads through flat, conical, spherical and cylindrical surfaces. Unlike oils, air is nearly always present, it does not contaminate a product being processed by the machine, its viscosity increases with temperature, and its use is not limited by oxidation at elevated temperatures. Its viscosity is much lower, giving markedly less resistance to motion at very high speeds. Air and gases are compressible, but the equations derived for incompressible fluids may be used with minor error if pressure differences are of the order of 35 to 70×10^3 Pa.

Numerical example

Design an externally-pressurized bearing for the end of a shaft to carry 4536 N thrust at 1740 r.p.m., with a minimum film thickness of 0.05 mm using SAE20 oil at 60 °C, pumped against a pressure of 3.5 MPa. The overall dimensions should be kept low because of space restrictions. Assume that the mechanical efficiency of the pump is 90 per cent.

Solution

The choice of a recess diameter, d, is a compromise between pad size and pump size. Since a small outside diameter is specified, a relatively large ratio of recess to outside diameter may be tried, giving a 3.5 MPa uniform pressure over a large interior area. Let $d/D = 0.6$. From eqn (5.12)

$$p_0 = \frac{8P}{\pi} \frac{\ln(D/d)}{D_2 - d^2}; \qquad 3.5 \times 10^6 = \frac{8(4536)\ln 1.67}{\pi D^2 (1 - 0.6^2)}$$

$$D = 5.12 \times 10^{-2}\,\text{m}; \qquad d = 3.10 \times 10^{-2}\,\text{m}.$$

From a viscosity-temperature diagram, the viscosity of SAE20 oil at 60 °C is 0.023 Pa s, and from eqn (5.14), the pump must deliver at least

$$Q = \frac{\pi p_0 h^3}{6\mu \ln(D/d)} = \frac{\pi(3.5 \times 10^6)(0.05 \times 10^{-3})^3}{6(0.023)\ln 1.67} = 0.19 \times 10^{-4}\,\text{m}^3\,\text{s}^{-1}.$$

This is a high rate for a small bearing. It can be decreased by using a smaller recess or lower pressure, together with a larger outside diameter, or by using a more viscous oil. Also smoother and squarer surfaces will reduce the film thickness requirement, which is halved to 0.025 mm and would decrease Q to one-eighth.

The input power to the pump is, by eqn (5.15)

$$\text{power required} = \frac{p_0 Q}{\eta} = \frac{3.5 \times 10^6 (0.19 \times 10^{-4})}{0.9} = 0.74 \times 10^2 \text{ Watt.}$$

The rotational speed is $n' = 1740/60 = 29$ r.p.s., and the torque required to rotate the bearing is, by eqn (5.16)

$$T = \frac{\pi^2 \mu n'}{16h}(D^4 - d^4) = \frac{\pi^2 (0.023)(29)}{16(0.05 \times 10^{-3})}(0.0512^4 - 0.031^4)$$

$$T = 4.8 \times 10^{-2} \text{ Nm.}$$

The power required to rotate the bearing is

$$\text{power required} = T\omega = T\frac{\pi n}{30} = 4.8 \times 10^{-2} \frac{\pi 1740}{30} = 8.74 \text{ Watt.}$$

The mean radius of the section of the pad where the film shear is high is

$$r_f = (5.12 \times 10^{-2} + 3.10 \times 10^{-2})/4 = 2.055 \times 10^{-2} \text{ m.}$$

At this radius, we may imagine a tangential, concentrated friction force,

$$F = T/r_f = \frac{4.8 \times 10^{-2}}{2.055 \times 10^{-2}} = 2.33 \text{ N.}$$

The coefficient of friction is the tangential force divided by the normal force, or

$$f = F/P = 2.33/4536 = 0.00051.$$

If lubrication were indifferently provided, with no recess, and the coefficient of friction f were 0.05 at a radius $r_f = 17$ mm, the power requirement would be

$$\text{power required} = \frac{frP\pi n}{30} = \frac{(0.05)(0.017)(4536)(3.14)(1740)}{30} = 702.2 \text{ Watt.}$$

This is eight times the power lost altogether at the bearing and pump in the externally pressurized bearing.

5.3. Squeeze-film lubrication bearings

Bearings which are subjected to dynamic loads experience constantly changing thicknesses of the oil film. Also, as a result of fluctuating loads, the lubricant is alternately squeezed out and drawn back into the bearing. Together with the oil supplied through correctly located grooves, a parabolic velocity profile with changing slope is obtained. This is illustrated in Fig. 5.7. The load-carrying ability, in such cases, is developed without the sliding motion of the film surfaces. The higher the velocity, the greater is the

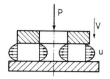

Figure 5.7

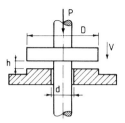

Figure 5.8

force developed. The squeeze effect may occur on surfaces of all shapes, including shapes that are flat and cylindrical. For an easy example, the case of a flat circular bearing ring and shaft collar is chosen and the relationship between the applied force, velocity of approach, film thickness and time is determined. The case being analysed is shown in Fig. 5.8.

In the Reynolds equation, all surface velocities except V_2 will be zero, and by symmetry $\partial p / \partial \Theta = 0$. With the upper surface approaching at a velocity V, $V_2 = dh/dt = -V$. Thickness h is independent of r and Θ but a function of time t. Equation (5.8) becomes

$$\frac{1}{6}\left[\frac{1}{r}\frac{\partial}{\partial r}\left(r\frac{h^3}{\mu}\frac{\partial p}{\partial r}\right)\right] = \frac{h^3}{6\mu}\left[\frac{1}{r}\frac{\partial}{\partial r}\left(r\frac{\partial p}{\partial r}\right)\right] = -2V.$$

Thus $(\partial/\partial r)(r\partial p/\partial r) = -12\mu V r/h^3$, and by integrating twice with respect to r

$$r\frac{\partial p}{\partial r} = -\frac{6\mu V}{h^3}r^2 + C_1 \quad \text{and} \quad \frac{\partial p}{\partial r} = -\frac{6\mu V}{h^3}r + \frac{C_1}{r},$$

whence

$$p = -\frac{3\mu V}{h^3}r^2 + C_1 \ln r + C_2.$$

The boundary conditions are $p=0$ at $r=D/2$ and $r=d/2$. Substitution, and simultaneous solution for C_1 and C_2, and resubstitution of these values gives for the pressure

$$p = \frac{3\mu V}{4h^3}\left[D^2 - 4r^2 - (D^2 - d^2)\frac{\ln(D/2r)}{\ln(D/d)}\right]. \tag{5.18}$$

The total force developed at a given velocity and a given film thickness is found by integration over the surface of the force on an elemental ring, or

$$P = \int_{d/2}^{D/2} p(2\pi r \, dr) = \frac{3\pi\mu V}{32h^3}\left[D^4 - d^4 - \frac{(D^2 - d^2)}{\ln(D/d)}\right]. \tag{5.19}$$

If the force is known as a function of time, the time for a given change in the film thickness may be found from eqn (5.19) by the substitution of $-dh/dt$ for V, the separation of variables, and integration between corresponding limits t', t'' and h', h'', thus

$$\int_{t'}^{t''} P \, dt = -\frac{3\pi\mu}{32}\left[D^4 - d^4 - \frac{(D^2 - d^2)}{\ln(D/d)}\right]\int_{h'}^{h''} \frac{dh}{h^3}.$$

If P is a constant of value W, such as obtained by a weight

$$t'' - t' = \frac{3\pi\mu}{64W}\left[D^4 - d^4 - \frac{(D^2 - d^2)}{\ln(D/d)}\right]\left[\left(\frac{1}{h''}\right)^2 - \left(\frac{1}{h'}\right)^2\right]. \tag{5.20}$$

The boundary condition for a solid circular plate at $r=0$ is different, namely, $\partial p/\partial r = 0$. Use of this, beginning with the equation preceding eqn

(5.18), gives

$$p = \frac{3\mu V}{4h^3}(D^2 - 4r^2) \qquad W = \frac{3\pi\mu VD^4}{32h^3}$$

and

$$t'' - t' = \frac{3\pi\mu D^4}{64W}\left[\left(\frac{1}{h''}\right)^2 - \left(\frac{1}{h'}\right)^2\right]. \tag{5.21}$$

Flat plates of other shapes are not solved so readily. If the length is much greater than the width, it may be treated as a case of unidirectional flow, as was done for the hydrostatic bearing. For a square plate with sides of length D (see Fig. 5.9) the average pressure can be taken as 4/3 of that on a circular plate of diameter D, to allow for the increased length of path of the corners. Thus

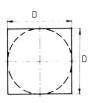

Figure 5.9

$$p_{sr} = \frac{4}{3}\left(\frac{P}{\pi D^2/4}\right) = \frac{4}{3}\left(\frac{3\pi\mu VD^4}{32h^3(\pi D^2/4)}\right) = \frac{\mu VD^2}{2h^3},$$

$$p = p_{sr}D^2 = \frac{\mu VD^4}{2h^3}. \tag{5.22}$$

The action of the fluctuating loads on cylindrical bearing films is more difficult to analyse. Squeeze-film action is important in cushioning and maintaining a film in linkage bearings such as those joining the connecting rods and pistons in a reciprocating engine. Here, the small oscillatory motion does not persist long enough in one direction to develop a hydrodynamic film.

5.4. Thrust bearings

Thrust bearing action depends on the existence of a converging gap between a specially shaped or tilted pad and a supporting flat surface of a collar. The relative sliding motion will force oil between the interacting surfaces and develop a load-supporting pressure. In Fig. 5.10, the surface velocities are U_1 and U_2. With constant viscosity μ and with $\partial h/\partial z = 0$, Reynolds' equation, eqn (5.7), becomes

$$\frac{\partial}{\partial x}\left(h^3\frac{\partial p}{\partial x}\right) + \frac{\partial}{\partial z}\left(h^3\frac{\partial p}{\partial z}\right) = 6\mu(U_1 - U_2)\frac{\partial h}{\partial x}. \tag{5.23}$$

This complete equation has not been solved analytically, but numerical analysis and digital computers may be used for solving particular cases. It is

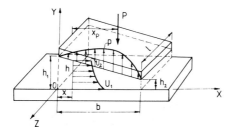

Figure 5.10

the usual practice to assume no side leakage, i.e. a bearing of infinite dimension *l* such that velocity *w* and $\partial p/\partial z$ are zero. Equation (5.23) is then simplified to

$$\frac{\partial}{\partial x}\left(h^3 \frac{\partial p}{\partial x}\right) = 6\mu(U_1 - U_2)\frac{\partial h}{\partial x}. \tag{5.24}$$

Integrating once

$$\frac{\partial p}{\partial x} = 6\mu(U_1 - U_2)\left(\frac{1}{h^2} + \frac{C_1}{h^3}\right).$$

For the bearing of Fig. 5.10 with a film thickness at the entrance of h_1 and at the exit of h_2 (shown greatly exaggerated), let the inclination be $\alpha = (h_1 - h_2)/b$. Then $h = h_1 - = h_2 + \alpha(b - x)$ and $\partial h/\partial x = -\alpha$. Hence

$$p = 6\mu(U_1 - U_2)\int \left[(h_1 - \alpha x)^{-2} + C_1(h_1 - \alpha x)^{-3}\right]\mathrm{d}x + C_2$$

$$= \frac{6\mu(U_1 - U_2)}{2\alpha(h_1 - \alpha x)^2}[2h_1 - 2\alpha x + C_1] + C_2.$$

The boundary conditions $p = 0$ at $x = 0$ and $x = b$ are utilized to obtain

$$p = \frac{6\mu(U_1 - U_2)\alpha x(b - x)}{(2h_1 - \alpha b)(h_1 - \alpha x)^2} = \frac{6\mu(U_1 - U_2)\alpha x(b - x)}{(2h_2 + \alpha b)[h_2 + \alpha(b - x)]^2}, \tag{5.25}$$

where the latter is in terms of the minimum film thickness h_2. The total load *P* is found by integration over the surface.

Machining or mounting the pads within the tolerances required for the very small angle α is difficult to attain and thus the pads are usually pivoted. The relationship between pivot distance, x_p, and the other variables may be found by taking moments about one edge of the pad. Since side leakage does occur, correction factors for the derived quantities have been determined experimentally and are available, for instance, in ESDU–82029. The theory for flat pads indicates that the maximum load capacity is attained by locating the pivot at $x_p = 0.578b$, but there is no capacity if the motion is reversed. For bearings with reversals, a natural location is the central one, $x_p = 0.5b$, but the flat pad theory indicates zero capacity for this location. However, bearings with central pivots and supposedly flat surfaces have been operating successfully for years.

5.4.1. Flat pivot

The simplest form of thrust bearing is the flat pivot or collar. In such cases the separating film of lubricant is of uniform thickness everywhere and the pressure at any given radius is constant, i.e. the pressure gradient is only possible in a radial direction. If the oil is introduced at the inner edges of the bearing surfaces it will flow in a spiral path towards the outer circumference as the shaft rotates. It is clear, however, that maintenance of the film will depend entirely upon the oil pressure at the inlet, and that this pressure will largely govern the carrying capacity of the bearing. For the purposes of design calculation it may be assumed that the film is in a state of simple torsional shear.

Numerical example

A vertical shaft of 75 mm diameter rests in a footstep bearing and makes 750 r.p.m. If the end of the shaft and the surface of the bearing are both perfectly flat and are assumed to be separated by a film of oil 0.025 mm thick, find the torque required to rotate the shaft, and the power absorbed in overcoming the viscous resistance of the oil film. The coefficient of viscosity of the oil is $\mu = 40 \times 10^{-3}$ Pa s.

Solution

Let

h = the film thickness,
r_1 = the shaft radius,
ω = the angular velocity of the shaft = $\pi n/30 = 25\pi$ rad n s^{-1},

then

$$\frac{\omega r}{h} = \text{rate of shear at radius } r$$

and

$$\frac{\mu \omega r}{h} = \text{shear stress at radius } r.$$

Considering an elementary ring of radius r and width dr,

$$\text{tangential drag on the element} = 2\pi r \, dr \frac{\mu \omega r}{h}$$

and so

$$\text{frictional torque} = \frac{2\pi\mu\omega}{h} \int_0^{r_1} r^3 \, dr$$

$$T = \frac{2\pi\mu\omega}{h} \frac{r_1^4}{4} \tag{5.26}$$

or

$\mu = 40 \times 10^{-3}$ Pa s
$h = 0.025 \times 10^{-3}$ m
$\omega = 25\pi = 78.54$ s^{-1}
$r_1 = 75 \times 10^{-3}$ m

thus,

$$T = \frac{2(3.1416)(78.54)}{0.025 \times 10^{-3}} \frac{(75 \times 10^{-3})^4}{4} (40 \times 10^{-3}) = 6.32 \, \text{Nm}$$

and frictional horsepower $= T\omega = 6.32 \times 78.54 = 496.4 \, \text{W} \approx 0.5 \, \text{kW}$.

Referring to the footstep bearing discussed in the above example, if

$A =$ the bearing area $= \pi r_1^2$,
$N =$ the speed of rotation in rev s^{-1},
$p =$ the bearing pressure per unit area, assumed uniform,
$f =$ the friction coefficient,

then, regarding the bearing as a flat pivot

frictional torque $T = \frac{2}{3} f p A r_1$.   (5.27)

Equating this value of T to that given by eqn (5.26)

$$\frac{2}{3} f p A r_1 = \frac{\mu A \omega r_1^2}{2h}$$

i.e.

$$f = \frac{3}{4} \frac{\mu \omega}{p} \frac{r_1}{h}$$

or

$$f = \frac{3}{2} \pi \frac{N \mu}{p} \frac{r_1}{h}.$$ (5.28)

5.4.2. The effect of the pressure gradient in the direction of motion

In the early, simple types of thrust bearing, difficulty was experienced in maintaining the film thickness. By the introduction of a pressure gradient in the direction of motion, i.e. circumferentially in a pivot or collar-type bearing, a much higher maximum pressure is attained between the surfaces, and the load that can be carried is greatly increased.

Michell (in Australia and Kingsbury in the USA, working independently) was the first to give a complete solution for the flow of a lubricant between inclined plane surfaces. He designed a novel thrust bearing based on his theoretical work. The results are important and may be illustrated by considering the simple slider bearing in which the film thickness varies in a linear manner in the direction of motion (Fig. 5.11). Here the slider moves with uniform velocity V and is separated from the bearing or pad by the lubricant, flow being maintained by the motion of the slider. The inlet and outlet ends are assumed filled with the lubricant at zero gauge pressure.

Let $B =$ the breadth of the bearing in the direction of motion and consider the unit length in a direction perpendicular to the velocity V. Leakage is

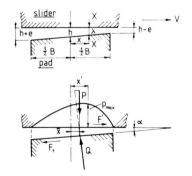

Figure 5.11

neglected, i.e. the pad is assumed to be infinitely long. Let

h = the mean thickness of the film,
$h + e$ = the thickness at inlet,
$h - e$ = the thickness at outlet,
λ = the thickness at a section X–X, at a distance x from the centre of the breadth, so that

$$\lambda = h - \frac{2ex}{B}. \tag{5.29}$$

Adopting the same procedure as that used in fluid mechanics

$$\text{flow across } X\text{–}X = \tfrac{1}{2}V\lambda - \frac{1}{\mu}\frac{dp}{dx}\frac{\lambda^3}{12}.$$

Suppose x' is the value of x at which maximum pressure occurs, i.e. where $dp/dx = 0$, then, for continuity of flow

$$\tfrac{1}{2}V\left(h - \frac{2ex'}{B}\right) = \tfrac{1}{2}V\left(h - \frac{2ex}{B}\right) - \frac{1}{\mu}\frac{dp}{dx}\frac{\lambda^3}{12}$$

so that

$$\frac{dp}{dx} = \frac{\mu Ve}{B}\frac{12}{\lambda^3}(x' - x). \tag{5.30}$$

Similarly shear stress

$$q = \frac{\mu V}{\lambda} + \tfrac{1}{2}\lambda\frac{dp}{dx}$$

so that

$$q = \frac{\mu V}{\lambda}\left[1 + \frac{6e}{B\lambda}(x' - x)\right]. \tag{5.31}$$

Integrating eqn (5.30), the pressure p at the section X–X is given by

$$\frac{pB}{12\mu Ve} = \int \frac{x'}{\lambda^3}dx - \int \frac{x}{\lambda^3}dx + k,$$

where x' and k are regarded as constants. As $d\lambda/dx = -2e/B$, this becomes

$$\frac{pB}{12\mu Ve} = \frac{Bx'}{4e}\frac{1}{\lambda^2} + \frac{B}{4e^2}\frac{1}{\lambda} - \frac{B^2h}{8e^2}\frac{1}{\lambda^2} + k$$

$$= \frac{B}{4e}\left(x' - \frac{Bh}{2e}\right)\frac{1}{\lambda^2} + \frac{B^2}{4e^2}\frac{1}{\lambda} + k. \tag{5.32}$$

The constants x' and k are determined from the condition that $p = 0$ when $x = \pm\tfrac{1}{2}B$, i.e. when $\lambda = h - e$ and $h + e$ respectively. Hence

$$k = -\frac{B^2}{8e^2h} \quad \text{and} \quad x' = \frac{Be}{2h}$$

and the pressure equation becomes

$$\frac{p}{3\mu VB} = -\frac{h^2 - e^2}{2eh}\frac{1}{\lambda^2} + \frac{1}{e\lambda} - \frac{1}{2eh}. \tag{5.33}$$

For the maximum value of p write $x = x' = Be/(2h)$, i.e. $\lambda = (h^2 - e^2)/h$. Equation (5.33) then becomes

$$p_m = \frac{\mu VB}{h^2}\frac{3a}{2(1 - a^2)}, \tag{5.34}$$

where $a = e/h$ denotes the attitude of the bearing or pad surface.

5.4.3.   Equilibrium conditions

Referring to Fig. 5.11, P is the load on the slider (per unit length measured perpendicular to the direction of motion) and F' is the pulling force equal and opposite to the tangential drag F. Similarly Q and F_r are the reaction forces on the oil film due to the bearing, so that the system is in equilibrium (a necessary condition is that the pad has sufficient freedom to adjust its slope so that equilibrium conditions are satisfied) under the action of the four forces, P, Q, F' and F_r. Again, P and F' are equal and opposite to the resultant effects of the oil film on the slider, so that

$$P = \int_{-B/2}^{B/2} p \, dx, \tag{5.35}$$

$$F' = \int_{-B/2}^{B/2} q \, dx. \tag{5.36}$$

For the former, eqn (5.33) gives

$$\frac{P}{3\mu VB} = -\frac{h^2 - e^2}{2eh}\frac{B}{2e}\left(\frac{1}{\lambda}\right)\Big|_{-B/2}^{B/2} - \frac{1}{e}\frac{B}{2e}(\ln\lambda)\Big|_{-B/2}^{B/2} - \frac{B}{2eh}$$

and writing $a = e/h$ this reduces to

$$P = \frac{\mu VB^2}{h^2}\frac{3}{2a^2}\left(\ln\frac{1 + a}{1 - a} - 2a\right). \tag{5.37}$$

Similarly, for the tangential pulling force, eqns (5.31) and (5.36) give

$$\frac{F'}{\mu V} = \int\frac{dx}{\lambda} + \frac{3e^2}{h}\int\frac{dx}{\lambda^2} + 3\int\frac{dx}{\lambda} - 3h\int\frac{dx}{\lambda^2}$$

and integrating between the limits $\pm B/2$, this reduces to

$$F' = \frac{\mu VB}{h}\frac{2}{a}\left(\ln\frac{1 + a}{1 - a} - \frac{3a}{2}\right). \tag{5.38}$$

5.4.4.   The coefficient of friction and critical slope

If f is the virtual coefficient of friction for the slider we may write

$$F' = fP$$

so that

$$\frac{fB}{h} = \frac{4a}{3} \frac{\ln\dfrac{1+a}{1-a} - \dfrac{3a}{2}}{\ln\dfrac{1+a}{1-a} - 2a} \tag{5.39}$$

Referring again to the equilibrium conditions, suppose α to be the angle in radians between the slider and the bearing pad surface, then for equilibrium we must have

$$P = Q \cos\alpha - F_r \sin\alpha,$$
$$F' = Q \sin\alpha + F_r \cos\alpha.$$

Since α is very small we may write $\sin\alpha \approx \alpha$ and $\cos\alpha = 1$. Further, F_r is very small compared with Q, and so

$$P = Q \qquad \text{and} \qquad F' = Q\alpha + F_r \qquad \text{(approximately)},$$

i.e.

$$F_r = F' - P\alpha \qquad \text{(approximately).} \tag{5.40}$$

A critical value of α occurs when $F_r = 0$, i.e.

$$\alpha = \frac{F'}{P} = f = \tan\phi = \phi, \tag{5.41}$$

where ϕ is the angle of friction for the slider. When $\alpha > \phi$, F_r becomes negative. This is caused by a reversal in the direction of flow of the oil film adjacent to the surface of the pad. The critical value of a is given by eqn (5.39). Thus

$$\alpha = f = \frac{2e}{B},$$

therefore

$$\frac{2eB}{Bh} = \frac{4a}{3} \frac{\ln\dfrac{1+a}{1-a} - \dfrac{3a}{2}}{\ln\dfrac{1+a}{1-a} - 2a}$$

and so

$$\ln\frac{1+a}{1-a} = 3a \qquad \text{or} \qquad a = \frac{e}{h} = 0.86.$$

5.5. Journal bearings

5.5.1. Geometrical configuration and pressure generation

In a simple plain journal bearing, the position of the journal is directly related to the external load. When the bearing is sufficiently supplied with oil and the external load is zero, the journal will rotate concentrically within the bearing. However, as the load is increased the journal moves to an

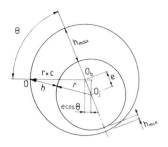

Figure 5.12

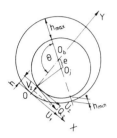

Figure 5.13

increasingly eccentric position, thus forming a wedge-shaped oil film where load-supporting pressure is generated. The eccentricity e is measured from the bearing centre O_b to the shaft centre O_j, as shown in Fig. 5.12. The maximum possible eccentricity equals the radial clearance c, or half the initial difference in diameters, c_d, and it is of the order of one-thousandth of the diameter. It will be convenient to use an eccentricity ratio, defined as $\varepsilon = e/c$. Then $\varepsilon = 0$ at no load, and ε has a maximum value of 1.0 if the shaft should touch the bearing under extremely large loads.

The film thickness h varies between $h_{max} = c(1 + \varepsilon)$ and $h_{min} = c(1 - \varepsilon)$. A sufficiently accurate expression for the intermediate values is obtained from the geometry shown in Fig. 5.12. In this figure the journal radius is r, the bearing radius is $r + c$, and is measured counterclockwise from the position of h_{max}. Distance $OO_j \approx OO_b + e \cos \Theta$, or $h + r = (r + c) + e \cos \Theta$, whence

$$h = c + e \cos \Theta = c(1 + \varepsilon \cos \Theta). \tag{5.42}$$

The rectilinear coordinate form of Reynolds' equation, eqn (5.7), is convenient for use here. If the origin of coordinates is taken at any position O on the surface of the bearing, the X axis is a tangent, and the Z axis is parallel to the axis of rotation. Sometimes the bearing rotates, and then its surface velocity is U_1 along the X axis. The surface velocities are shown in Fig. 5.13. The surface of the shaft has a velocity Q_2 making with the X axis an angle whose tangent is $\partial h/\partial x$ and whose cosine is approximately 1.0. Hence components $U_2 = Q$ and $V_2 = U_2(\partial h/\partial x)$. With substitution of these terms, Reynolds' equation becomes

$$\frac{1}{6}\left[\frac{\partial}{\partial x}\left(\frac{h^3}{\mu}\frac{\partial p}{\partial x}\right) + \frac{\partial}{\partial z}\left(\frac{h^3}{\mu}\frac{\partial p}{\partial z}\right)\right] = (U_1 - U_2)\frac{\partial h}{\partial x} + 2V_2$$

$$= (U_1 + U_2)\frac{\partial h}{\partial x} = U\frac{\partial h}{\partial x}, \tag{5.43}$$

where $U = U_1 + U_2$. The same result is obtained if the origin of coordinates is taken on the journal surface with X tangent to it. Reynolds assumed an infinite length for the bearing, making $\partial p/\partial z = 0$ and endwise flow $w = 0$. Together with μ constant, this simplifies eqn (5.43) to

$$\frac{\partial}{\partial x}\left(h^3\frac{\partial p}{\partial x}\right) = 6\mu U\frac{\partial h}{\partial x}. \tag{5.44}$$

Reynolds obtained a solution in series, which was published in 1886. In 1904 Sommerfeld found a suitable substitution that enabled him to make an integration to obtain a solution in a closed form. The result was

$$p = \frac{\mu U r}{c^2}\left[\frac{6\varepsilon(\sin \Theta)(2 + \varepsilon \cos \Theta)}{(2 + \varepsilon^2)(1 + \varepsilon \cos \Theta)^2}\right]. \tag{5.45}$$

This result has been widely used, together with experimentally determined end-leakage factors, to correct for finite bearing lengths. It will be referred to as the Sommerfeld solution or the long-bearing solution. Modern bearings are generally shorter than those used many years ago. The length-to-diameter ratio is often less than 1.0. This makes the flow in the Z direction and the end leakage a much larger portion of the whole. Michell in 1929 and Cardullo in 1930 proposed that the $\partial p/\partial z$ term of eqn (5.43) be retained and the $\partial p/\partial x$ term be dropped. Ocvirk in 1952, by neglecting the parabolic, pressure-induced flow portion of the U velocity, obtained the Reynolds equation in the same form as proposed by Michell and Cardullo, but with greater justification. This form is

$$\frac{\partial}{\partial z}\left(h^3\frac{\partial p}{\partial z}\right)=6\mu U\frac{\partial h}{\partial x}. \tag{5.46}$$

Unlike eqn (5.44), eqn (5.46) is easily integrated, and it leads to the load number, a non-dimensional group of parameters, including length, which is useful in design and in plotting experimental results. It will be used here in the remaining derivations and discussion of the principles involved. It is known as the Ocvirk solution or the short-bearing approximation. If there is no misalignment of the shaft and bearing, then h and $\partial h/\partial x$ are independent of z and eqn (5.46) may be integrated twice to give

$$p=\frac{3\mu U}{h^3}\frac{\partial h}{\partial x}z^2+\frac{C_1}{h^3}z+C_2.$$

From the boundary conditions $\partial p/\partial z=0$ at $z=0$ and $p=0$ at $z=\pm\frac{1}{2}$. This is shown in Fig. 5.14. Thus

$$p=-\frac{3\mu U}{h^3}\left(\frac{l^2}{4}-z^2\right)\frac{\partial h}{\partial x}. \tag{5.47}$$

The slope $\partial h/\partial x=\partial h/\partial(r\Theta)=(1/r)\partial h/\partial\Theta$ and from eqn (5.42), $\partial h/\partial x=-(c\varepsilon\sin\Theta)/r$.

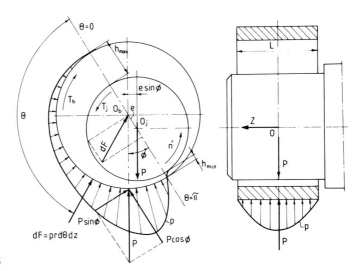

Figure 5.14

Substitution into eqn (5.47) gives

$$p = \frac{\mu U}{rc^2} \left(\frac{l^2}{4} - z^2 \right) \frac{3\varepsilon \sin \Theta}{(1 + \varepsilon \cos \Theta)^3}. \tag{5.48}$$

This equation indicates that pressures will be distributed radially and axially somewhat as shown in Fig. 5.14; the axial distribution being parabolic. The peak pressure occurs in the central plane $z = 0$ at an angle

$$\Theta_m = \cos^{-1} \left(\frac{1 - \sqrt{1 + 24\varepsilon^2}}{4\varepsilon} \right) \tag{5.49}$$

and the value of p_{max} may be found by substituting Θ_m into eqn (5.48).

5.5.2.   Mechanism of load transmission

Figure 5.14 shows the forces resulting from the hydrodynamic pressures developed within a bearing and acting on the oil film treated as a free body. These pressures are normal to the film surface along the bearing, and the elemental forces $dF = pr\, d\Theta\, dz$ can all be translated to the bearing centre O_b and combined into a resultant force. Retranslated, the resultant P shown acting on the film must be a radial force passing through O_b. Similarly, the resultant force of the pressures exerted by the journal upon the film must pass through the journal centre O_j. These two forces must be equal, and they must be in the opposite directions and parallel. In the diverging half of the film, beginning at the $\Theta = \pi$ position, a negative (below atmospheric) pressure tends to develop, adding to the supporting force. This can never be very much, and it is usually neglected. The journal exerts a shearing torque T_j upon the entire film in the direction of journal rotation, and a stationary bearing resists with an opposite torque T_b. However, they are not equal. A summation of moments on the film, say about O_j, gives $T_j = T_b + Pe \sin \phi$ where ϕ, the attitude angle, is the smaller of the two angles between the line of force and the line of centres. If the bearing instead of the journal rotates, and the bearing rotates counterclockwise, the direction of T_b and T_j reverses, and $T_b = T_j + Pe \sin \phi$.

Hence, the relationship between torques may be stated more generally as

$$T_r = T_s + Pe \sin \phi, \tag{5.50}$$

where T_r is the torque from the rotating member and T_s is the torque from the stationary member.

Load P and angle ϕ may be expressed in terms of the eccentricity ratio ε by taking summations along and normal to the line $O_b O_j$, substituting for p from eqn (5.48) and integrating with respect to Θ and z. Thus

$$P \cos \phi = 2 \int_0^l \int_0^\pi (pr\, d\Theta\, dz) \cos \Theta = \frac{\mu U l^3}{c^2} \frac{\varepsilon^2}{(1 - \varepsilon^2)^2}$$

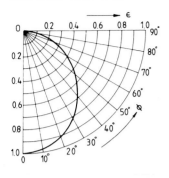

Figure 5.15

and

$$P \sin \phi = 2 \int_0^l \int_0^\pi (pr\, \mathrm{d}\Theta\, \mathrm{d}z) \sin \Theta = \frac{\mu U l^3}{c^2} \frac{\pi \varepsilon}{4(1-\varepsilon^2)^{\frac{3}{2}}},$$

whence

$$P = \sqrt{(P \cos \phi)^2 + (P \sin \phi)^2} = \frac{\mu U l^3}{c^2} \frac{\varepsilon[\pi^2(1-\varepsilon^2)+16\varepsilon^2]^{\frac{1}{2}}}{4(1-\varepsilon^2)^2} \qquad (5.51)$$

and

$$\phi = \tan^{-1}\left(\frac{P \sin \phi}{P \cos \phi}\right) = \tan^{-1}\frac{\pi\sqrt{1-\varepsilon^2}}{4\varepsilon}. \qquad (5.52)$$

With an increasing load, ε will vary from 0 to 1.0, and the angle ϕ will vary from 90° to 0°. Correspondingly, the position of minimum film thickness, h_{min}, and the beginning of the diverging half, will lie from 90° to 0° beyond the point where the lines of force P intersect the converging film. The path of the journal centre O_j as the load and eccentricity are increased is plotted in Fig. 5.15.

The fraction containing the many eccentricity terms of eqn (5.51) is equal to $Pc^2/(\mu U l^3)$, and although it is not obvious, the eccentricity, like the fraction, increases non-linearly with increases in P and c and with decreases in μ, l, U and the rotational speed n'. It is important to know the direction of the eccentricity, so that parting lines and the holes or grooves that supply lubricant from external sources may be placed in the region of the diverging film, or where the entrance resistance is low. The centre O_b is not always fixed, e.g. at an idler pulley, the shaft may be clamped, fixing O_j, and the pulley with the bearing moves to an eccentric position. A rule for determining the configuration is to draw the fixed circle, then to sketch the movable member in the circle, such that the wedge or converging film lies between the two force vectors P acting upon it. The wedge must point in the direction of the surface velocity of the rotating member. This configuration should then be checked by sketching in the vectors of force and torque in the directions in which they act on the film. If the free body satisfies eqn (5.50) the configuration is correct.

Oil holes or axial grooves should be placed so that they feed oil into the diverging film or into the region just beyond where the pressure is low. This should occur whether the load is low or high, hence, the hole should be at least in the quadrant 90°–180°, and not infrequently, in the quadrant 135°–225° beyond where the load P is applied to the film. The 180° position is usually used for the hole or groove since it is good for either direction of rotation, and it is often a top position and accessible. The shearing force $\mathrm{d}F$ on an element of surface $(r\, \mathrm{d}\Theta)\, \mathrm{d}z$ is $(r\, \mathrm{d}\Theta)\, \mathrm{d}z\mu(\partial u/\partial y)_{y=H}$, where either zero or h must be substituted for H. The torque is $r\, \mathrm{d}F$. If it is assumed that the entire space between the journal and bearing is filled with the lubricant, integration must be made from zero to 2π, thus

$$T = \mu r^2 \int_{-l/2}^{+l/2} \int_0^{2\pi} \left(\frac{\partial u}{\partial y}\right)_{y=H} d\Theta \, dz = \mu r^2 l \int_0^{2\pi} \left(\frac{\partial u}{\partial y}\right)_{y=H} d\Theta. \quad (5.53)$$

The short bearing approximation assumes a linear velocity profile such that $(\partial u/\partial y)_{y=0} = (\partial u/\partial y)_{y=H}$. Use of this approximation in eqn (5.53) will give but one torque, contrary to the equilibrium condition of eqn (5.50). However, the result has been found to be not too different from the experimentally determined values of the stationary member torque T_s. Hence we use eqn (5.53), with h from eqn (5.42), integrating and substituting $c = c_d/2, r = d/2$ and $U_1 - U_2 = \pi d(n_2' - n_1')$ where n_2 and n_1 are the rotational velocities in r.p.s; the results are

$$T = \frac{\mu r^2 l (U_2 - U_1)}{c} \int_0^{2\pi} \frac{d\Theta}{(1 + \varepsilon \cos \Theta)}$$

$$= \frac{\mu d^2 l (U_2 - U_1)}{c_d} \frac{\pi}{(1 - \varepsilon^2)^{\frac{1}{2}}}$$

$$= \frac{\mu d^3 l (n_2' - n_1')}{c_d} \frac{\pi^2}{(1 - \varepsilon^2)^{\frac{1}{2}}}. \quad (5.54)$$

Dimensionless torque ratios are obtained by dividing T_s or T_r by the no-load torque T_0 given by the formula

$$T_0 = (d/2)F = \frac{\pi^2 \mu d^3 l n'}{c_d}$$

and first setting $n' = n_2' - n_1'$. Thus

$$T_s/T_0 = \frac{1}{(1 - \varepsilon^2)^{\frac{1}{2}}}. \quad (5.55)$$

5.5.3. Thermo-flow considerations

The amount of oil flowing out at the end of a journal bearing, i.e. the oil loss at plane $z = \frac{l}{2}$ or $z = -\frac{l}{2}$ may be determined by integration of eqn (5.3b) over the pressure region of the annular exit area, substituting $r \, d\Theta$ for dx. Thus, since $W_1 = W_2 = 0$

$$Q = \int_{\Theta_1}^{\Theta_2} q_z = -\frac{r}{12\mu} \int_{\Theta_1}^{\Theta_2} h^3 \frac{\partial p}{\partial z} d\Theta. \quad (5.56)$$

To determine the flow Q_H out of the two ends of the converging area or the hydrodynamic film, $\partial p/\partial z$ is obtained from eqn (5.48), h from eqn (5.42), and $Q_H = 2Q$ from eqn (5.56). The limits of integration may be $\Theta_1 = 0$ and $\Theta_2 = \pi$, or the extent may be less in a partial bearing. However, Q_H is more easily found from the fluid rejected in circumferential flow. With the linear velocity profiles of the short bearing approximation, shown in Fig. 5.16, and

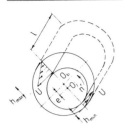

Figure 5.16

with eqn (5.42), the flow is seen to be

$$Q_H = \tfrac{1}{2}Uh_{max}l - \tfrac{1}{2}Uh_{min}l = \tfrac{1}{2}Ul[(c+e)-(c-e)]$$

or

$$Q_H = Ule = \frac{Ul\varepsilon c_d}{2} = \frac{\pi d(n_2' - n_1')l\varepsilon c_d}{2}, \qquad (5.57)$$

where c_d is the diametral clearance. Although it is not directly evident from this simple result, the flow is an increasing function of the load and a decreasing function of viscosity, indicated by the eccentricity term ε. At the ends of the diverging space in the bearing, negative pressure may draw in some of the oil previously forced out. However, if a pump supplies oil and distribution grooves keep the space filled and under pressure, there is an outward flow. This occurs through a cylindrical slot of varying thickness, which is a function of the eccentricity. The flow is not caused by journal or bearing motion, and it is designated film flow Q_f. It is readily determined whether a central source of uniform pressure p_0 may be assumed, as from a pump-fed partial annular groove. Instead of starting with eqn (5.56), an elemental flow q_z from one end may be obtained from the flat slot, eqn (5.17), by writing $r\,d\Theta$ for b and $(l-a)/2$ for l, where the new l is the bearing length and a is the width of the annular groove. Then $Q_f = 2\displaystyle\int_{\Theta_1}^{\Theta_2} q_z$, where Θ_1 and Θ_2 define the appropriate angular positions, such as π and 2π respectively. Additional flow may occur through the short slots which close the ends of an axial groove or through a small triangular slot formed by chamfering the plane surfaces at the joint in a split bearing.

Oil flow and torque are closely related to bearing and film temperature and, thereby, to oil viscosity, which in turn affects the torque. Oil temperature may be predicted by establishing a heat balance between the heat generated and the heat rejected. Heat H_g is generated by the shearing action on the oil, heat H_o is carried away in oil flowing out of the ends of the bearing, and by radiation and convection, heat H_b is dissipated from the bearing housing and attached parts, and heat H_s from the rotating shaft. In equation form

$$H_g = H_o + H_b + H_s. \qquad (5.58)$$

The heat generation rate H_g is the work done by the rotating member per unit time (power loss). Thus, if torque T_r is in Nm and n' in r.p.s., the heat generation rate is

$$H_g = 2\pi T_r n' \qquad \text{[Watt]}. \qquad (5.59)$$

Now, T_r and T_0 vary as d^3 and n'. Therefore, H_g varies as d^3 and approximately as $(n')^2$. Hence a large diameter and high speed bearings generally require a large amount of cooling, which may be obtained by a liberal flow of oil through the space between the bearing and the journal. Flowing out of the ends of the bearing, the oil is caught and returned to a sump, where it is cooled and filtered before being returned. The equation for the heat removed by the oil per unit of time is

$$H_o = c_p \gamma Q (t_o - t_i), \tag{5.60}$$

where c_p is the specific heat of the oil, and γ is the specific weight of the oil. The flow Q in eqn (5.60) may consist of the hydrodynamic flow Q_H, eqn (5.57), film flow Q_f, and chamfer flow Q_c as previously discussed, or any others which may exist. The heat lost by radiation and convection may often be neglected in well-flushed bearings.

The outlet temperature t_o represents an average film temperature that may be used to determine oil viscosity for bearing calculations, at least in large bearings with oil grooves that promote mixing. The average film temperature is limited to 70 °C or 80 °C in most industrial applications, although it may be higher in internal combustion engines. Higher temperatures occur beyond the place of minimum film thickness and maximum shear. They may be estimated by an equation based on experimental results. The maximum temperatures are usually limited by the softening temperature of the bearing material or permissible lubricant temperature.

In self-contained bearings, those lubricated internally as by drip, waste packing, oil-ring feed or oil bath (immersion of journal), dissipation of heat occurs only by radiation and convection from the bearing housing, connected members and the shaft. Experimental studies have been directed towards obtaining overall dissipation coefficients K for still air and for moving air. These dissipation coefficients are used in an equation of the form

$$H_b = K A (t_b - t_a),$$

where A is some housing or bearing surface area or projected area, t_b is the temperature of its surface, and t_a is the ambient temperature.

5.5.4. Design for load bearing capacity

It is convenient to convert eqn (5.51) into a non-dimensional form. One substitution is a commonly used measure of the intensity of bearing loading, the unit load or nominal contact pressure, p, which is the load divided by the projected bearing area $(l \times d)$, thus

$$p = \frac{p}{ld}, \tag{5.61}$$

where l is the bearing length, d is the nominal bearing diameter, and p has the same units as pressure.

The surface velocity sum, $U = U_1 + U_2$, is replaced by

$$\pi d (n_1' + n_2') = \pi \, d n',$$

where $n' = n_1' + n_2'$ is the sum of the rotational velocities. Also, c may be expressed in terms of the more commonly reported diametral clearance, c_d, where $c = c_d/2$. Let the non-dimensional fraction containing ε in eqn (5.51) be represented by E. Transposition of eqn (5.51), followed by substitution gives

$$E = \frac{Pc^2}{\mu U l^3} = \frac{(pld)(c_d/2)^2 d}{\mu(\pi dn')l^3 d} = \frac{1}{4\pi}\frac{p}{\mu n'}\left(\frac{d}{l}\right)^2\left(\frac{c_d}{d}\right)^2 = \frac{1}{4\pi}N_1,$$

where N_1, called the load number or the Ocvirk number, is the product of the three nondimensional ratios. Then, from the preceding equation and the definition of E by eqn (5.51)

$$N = \frac{p}{\mu n'}\left(\frac{d}{l}\right)^2\left(\frac{c_d}{d}\right)^2 = 4\pi E = \frac{\pi\varepsilon[\pi^2(1-\varepsilon^2)+16\varepsilon^2]^{\frac{1}{2}}}{(1-\varepsilon^2)^2}$$

$$= \frac{\pi^2\varepsilon(1+0.62\varepsilon^2)^{\frac{1}{2}}}{(1-\varepsilon^2)^2}. \tag{5.62}$$

The load number determines the dimensions and parameters over which the designer has a choice. Equation (5.62) indicates that all combinations of unit load p or P/ld, viscosity μ, speed sum n', length–diameter ratio, l/d and clearance ratio c_d/d that give the same value of N_1 will give the same eccentricity ratio. Eccentricity ratio is a measure of the proximity to failure of the oil film since its minimum thickness is $h_{min} = c(1-\varepsilon)$. Hence the load number is a valuable design parameter.

5.5.5. Unconventional cases of loading

Reynolds equation, eqn (5.43), which describes the process of pressure generation, contains the sum of surface velocities of the oil film, $U = U_1 + U_2$. In the derivation of the load number, N_1, the substitution $U = \pi dn'$ was made, where $n' = n'_1 + n'_2$, the sum of the rotational velocities of the two surfaces. The load number was found to be inversely proportional to n' and hence to $n'_1 + n'_2$. It is important to note that velocities U_1 and U_2 were taken relative to the line of action of load P, which was considered fixed. If the load rotates, the same physical relationships occur between the film surfaces and the load, provided that the surface velocities are measured relative to the line of action of the load. Thus all previous equations apply, if the rotational velocities n'_1 and n'_2 are measured relative to the rotational velocity n' of the load. Several examples are given in Fig. 5.17. Member 1 is fixed and represents the support of the machine. At least one other member is driven at a rotational speed ω relative to it. The load source and the two fluid films are not necessarily at the same axial position along the central member. Thus case (e) in Fig. 5.17, for instance, may illustrate a type of jet engine which has a central shaft with bladed rotors turning at one speed in the bearings supporting it, and a hollow spool with additional rotors turning at a different speed and supported on the first shaft by another set of bearings. The sum (n') a multiple of ω, is a measure of the load capacity, which equals ω in a standard bearing arrangement, case (a). If $n' = 2\omega$, as for the inner film of case (d), then for the same load number

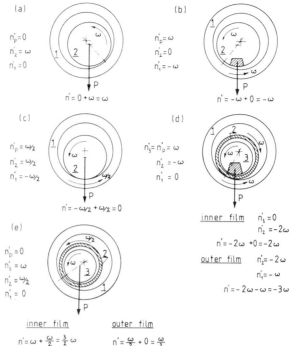

Figure 5.17

and eccentricity, the load capacities p and P are doubled. The zero capacity of the bearing in case (c) represents a typical situation for the crankpin bearings of four-stroke-cycle engines. The same is true in the case of the bushing of an idler gear and the shaft that supports it, if they turn with opposite but equal magnitude velocities relative to a non-rotating load on the gear. The analyses discussed give some ideas on relative capacities that can be attained and indicate the care that must be taken in determining n' for substitution in the load number equation. However, it should be noted that the load numbers and actual film capacities are not a function of n' alone.

The diameters d and lengths l of the two films may be different, giving different values to $p = P/ld$ and to $(d/l)^2$ in the load number, but they may be adjusted to give the same load number. Also, a load rotating with the shaft, case (b), appears to give the bearing the same capacity as the bearing illustrated by case (a). However, unless oil can be fed through the shaft to a hole opposite the load, it will probably be necessary to feed oil by a central annular groove in the bearing so that oil is always fed to a space at low pressure. With pressure dropping to the oil-feed value at the groove in the converging half, the bearing is essentially divided into two bearings of approximately half the l/d ratio. Since d/l is squared in the load-number equation, each half of the bearing has one-fourth and the whole one-half the capacity of the bearing in case (a).

Another way to deal with the problem of the rotating load vector is shown in Fig. 5.18. Let ω_1 and ω_2 be the angular velocities of the shaft or the

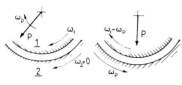

Figure 5.18

bearing. Consider the load to rotate at a uniform angular velocity ω_p. When r is the radius of the shaft

$$U_1 = \omega_1 r \quad \text{and} \quad U_2 = \omega_2 r.$$

The case of a rotating load on a stationary bearing can be equated to that of a fixed load on a complete system which is rotated as a whole at velocity $-\omega_p$. Thus, the shaft velocity becomes $\omega_1 - \omega_p$, the load vector is moving with speed $\omega_p - \omega_p = 0$ and the bearing velocity is $0 - \omega_p = -\omega_p$. Then

$$\frac{U_1 + U_2}{r} = \omega_1 - \omega_p - \omega_p$$

$$= \omega_1 - 2\omega_p.$$

The problem can be expressed in terms of a general equation

$$\frac{\text{load capacity for rotating load}}{\text{load capacity for steady load}} = \frac{U_1 + U_2}{U_1}$$

$$= 1 - 2\frac{\omega_p}{\omega_1} = 1 - 2R,$$

where $R = \text{ratio} = (\text{angular velocity of load})/(\text{angular velocity of shaft})$. When $R = \frac{1}{2}$ the load capacity is indicated as falling to zero, i.e. when the load is rotating at half the speed of the shaft.

Experimental results show that under these circumstances, bearings operate at a dangerously high value of eccentricity, any lubricating film which may be present is attributed solely to secondary effect. Where the load operates at the speed of the shaft (a very common situation when machinery is out of balance), load-carrying capacity is the same as that for a steady load. As the frequency of a load increases so does the load-carrying capacity. Sometimes a hydrodynamic film exists between a non-rotating outer shell of a bearing and its housing. An out-of-balance load might, for example, be applied to the inner housing so that, although there was no relative lateral motion of the surfaces of the bearing outer shell and its housing, a rotating load would be applied thereto. Thus both ω_1 and ω_2 are zero so that the effective speed U becomes $2\omega_p r$. Thus a pressure film of twice the intensity of the case where the load is rotated with the shaft would be generated.

5.5.6. Numerical example

In a certain shaking device, an off-centre weight provides a centrifugal force of 26,000 N, rotating at 3600 r.p.m. This force is midway between the ends of the shaft, and it is shared equally by two bearings. Self-alignment of the bushing is provided by a spherical seat, plus loosely fitting splines to prevent rotation of the bushing about the axis of the shaft. The bearing is shown in Fig. 5.19. Oil of 10.3 mPa s viscosity will be provided for lubrication of the interior surfaces at I and the exterior surfaces at E. The

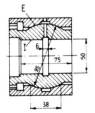

Figure 5.19

diametral clearance ratio is 0.0015 at both places, and the central annular groove at I has a width of 6 mm. Determine the load numbers and minimum film thickness at I and E.

Solution

(i) Surface I. Relative to the load, the velocity of the bushing surface is $n_1' = -3600/60 = -60$ r.p.s. and that of the shaft is $n_2' = 0$. Hence, $n' = n_1' + n_2' = -60 + 0 = -60$ r.p.s. Each bearing, carrying $26\,000/2 = 13\,000$ N, is divided by an oil groove into two effective lengths of $(75-6)/2 = 34.5$ mm, so $l/d = 34.5/50 = 0.69$ and $P = 13\,000/2 = 6500$ N.

The specific load $p = 6500/(34.5)(50) = 3.768$ N/mm^2 $(3.768 \times 10^6$ Pa$)$, and with the oil viscosity, $\mu = 10.3 \times 10^{-3}$ Pa s, the load number is

$$N_1 = \frac{p}{\mu n'}\left(\frac{d}{l}\right)\left(\frac{c_d}{d}\right)^2 = \frac{3.768 \times 10^6}{(10.3 \times 10^{-3})(60)}\left(\frac{1}{69}\right)^2 (0.0015)^2 = 28.8.$$

From the diagram of the eccentricity ratio and minimum film thickness ratio versus load number, Fig. 5.20, $h_{min}/c = 0.19$, and as $c = c_d/2 = d(c_d/d)/2 = 50(0.0015)/2 = 0.0375$ mm, then $h_{min} = (0.19)(0.0375) = 0.0071$ mm.

(ii) Surface E. As the spherical surfaces are narrow, they will be approximated by a cylindrical bearing of average diameter 92 mm, whence $l/d = 38/92 = 0.413$. The specific load becomes $p = (3.72 \times 10^6$ Pa$)$. Both stationary surfaces have a velocity of -60 r.p.s. relative to the rotating load, and $n' = n_1' + n_2' = -60 - 60 = -120$ r.p.s.

$$N = \frac{3.72 \times 10^6}{(10.3 \times 10^{-3})(120)}\left(\frac{1}{0.413}\right)^2 = 39.7.$$

As $c = (92)(0.0015)/2 = 0.069$ mm, then from the diagram (Fig. 5.20)

$$h_{min} = (0.14)(0.069) = 0.0097 \text{ mm}).$$

The film is developed and maintained because the rotating load causes a rotating eccentricity, i.e. the centre of the bushing describes a small circle of

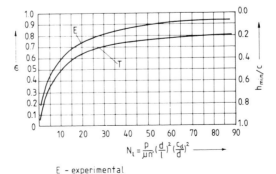

E - experimental

T - short bearing theory

Figure 5.20

radius *e* about the centre of the spherical cavity. The wedge shape formed by the film of oil rotates with the load, always pointing in the direction opposite to that of the motion of the load, and in effect, supporting it. Although the two surfaces of the oil film have no absolute tangential motion, they have a tangential motion relative to the load. Because of a complete film of oil, extremely small oscillations of alignment can occur with negligible friction or binding.

5.5.7. Short bearing theory – CAD approach

The fact that journal bearings have been so widely used in the absence of sophisticated design procedures, generally with complete success, can be attributed to the fact that they represent a stable self-adjusting fluid and thermal control system as shown in Fig. 5.21. This is attributed to two major sets of variables, one of which includes those variables which are powerfully dependent on an eccentricity ratio such as the rate of lubricant flow, friction and load-carrying capacity, whilst the other includes those factors which depend on temperature, such as viscosity.

The narrow-bearing theory or approximation arises from the difficulty of solving the Reynolds equation in two dimensions. The pressure induced component of flow in the longitudinal direction is neglected, and additionally it is assumed that the pressure in the oil film is positive throughout the converging portion of the clearance volume and zero throughout the diverging portion.

In the procedure outlined here, it is assumed that a designer's first preference will be for a standard bearing having a length-to-diameter ratio of 0.5 and a clearance ratio of 0.001 (i.e. $c/r = 0.001$). Assuming further that the load, speed and shaft diameter are determined by the designer, then to complete the design, all that is necessary is to select the operating viscosity so that the bearing will operate at an eccentricity ratio of 0.707. This value of eccentricity ratio is optimal from the temperature rise point of view. To select the viscosity, the following equation can be used

$$\mu = \frac{W}{VD} 0.8 \times 10^{-6}, \tag{5.63}$$

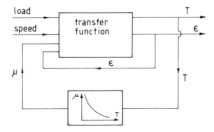

Figure 5.21

where W is the load on the bearing, V is the linear speed and D is the shaft diameter. Alternatively

$$\mu = \frac{p}{\omega} 0.8 \times 10^{-6}, \tag{5.64}$$

where ω is the angular velocity of the shaft and $p = W/LD$ is the nominal contact pressure on the projected area of the bearing. This will be satisfactory, subject to the bearing material being capable of withstanding the applied load and to the temperature of the system being kept within acceptable limits.

In the case of a white-metal bearing lining, a permissible load on the projected area can be assumed to be $8 \times 10^6\,\text{N/m}^2$. Then

$$p = \frac{W}{LD} = \frac{2W}{D^2} \leqslant 8\,\text{MN/m}^2$$

for $L/D = 1/2$, or

$$\frac{W}{D^2} < 4\,\text{MN/m}^2. \tag{5.65}$$

A reasonable temperature limitations for white metal is $120\,^\circ\text{C}$, so that

$$T_m = T_i + 1.13p(10^{-5}) < 120\,^\circ\text{C}, \tag{5.66}$$

where T_m is the maximum temperature and T_i is the inlet temperature of the oil.

Maximum temperature, T_m, having been obtained, an oil of a viscosity equal to or above μ for this temperature should be selected by reference to Fig. 5.22, which shows a normal viscosity–temperature plot. If however the selected oil has a viscosity greater than μ at the temperature T_m further adjustment will be necessary. Moreover, it is unlikely that a bearing will be required to operate at a constant single speed under an unvarying load throughout the whole of its life. In practice a machine must run up to speed from zero, the load may vary over a wide range, and, because bearing

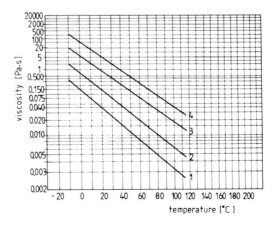

Figure 5.22

performance is determined by the combination of both factors, some method is required to predict the temperature and film parameters at other than the basic design point.

A strong relationship between temperature rise and eccentricity is quite obvious and the short bearing theory can be used to establish it. Then, knowing the eccentricity, the actual operating temperature can be predicted. If the result of eqn (5.64) does not relate precisely to a conveniently available oil then an oil having a higher viscosity at the estimated temperature must be selected. This, however, will cause the bearing to operate at a non-optimum eccentricity ratio, the temperature rise will change, and with it the viscosity. Some process of iteration is again necessary and the suggested procedure is illustrated in Fig. 5.23.

The method outlined above is best illustrated by a practical example. It is assumed that a shaft 0.25 m in diameter and rotating at 42 rad s^{-1} is required to support a load of 38 000 N. A clearance ratio of 10^{-3} and L/D ratio of 1/2 can be assumed. Then from eqn (5.64)

$$\mu = 0.023 \text{ Pa s.}$$

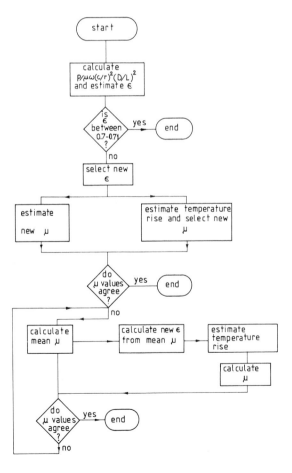

Figure 5.23

Equation (5.65) gives $p = 1.22 \times 10^6 \, \text{N/m}^2$, which is a safe value for white metal. Assuming an inlet temperature of 40 °C, eqn (5.66) yields

$$T_m = 40 + 13.6 = 53.6 \, °\text{C}.$$

As can be seen from the reference to Fig. 5.22, oil 2 meets this condition to a close approximation and the solution is complete. In a practical case, however, it may be necessary to use oil 3 at some other point in the system of which the bearing is a part and, to avoid the necessity for two oils in one machine, this oil may also be used in the bearing. Because the viscosity will be greater, the bearing will operate at a lower eccentricity and a higher temperature than when lubricated by oil 2. The exact values of eccentricity and temperature will depend on the viscosity-temperature characteristics of oil 3 and can be determined by the iterative process shown in Fig. 5.23.

Assuming a trial value of eccentricity of 0.5, the corresponding value of $(p/\mu\omega)(c/r)^2(D/L)^2$ is 1.55 from which the viscosity can be estimated at 0.075 Pa s. This value of μ produces a temperature rise of 53°C, so the operating temperature is $40 + 53 = 93 \, °\text{C}$. From Fig. 5.22 this gives a viscosity of 0.02 Pa s. The estimates of viscosity are not in agreement and therefore the assumption of 0.5 for eccentricity ratio is insufficiently accurate. A better approximation is obtained by taking the mean of the two estimates of viscosity. Thus, a new value for μ is 0.0475 Pa s and the corresponding eccentricity is 0.6 which in turn determines the temperature rise of 30 °C. The temperature rise of 30 °C, taken in conjunction with the assumption of 40 °C for the inlet temperature, gives an effective operating temperature of 70 °C. Reference to Fig. 5.22 gives the viscosity of oil 3 at this temperature as about 0.048 Pa s which is in good agreement with the assumed mean. It will be sufficient for most purposes, therefore, to accept that the result of using oil 3 in the bearing will be to reduce the eccentricity ratio to 0.6 and to increase the operating temperature to 70 °C.

If agreement within acceptable limits had not been achieved at this stage, further iteration would be carried out until the desired degree of accuracy is attained. It is clear therefore that the method presented is very convenient when a computer is used to speed-up the iteration process.

5.6 Journal bearings for specialized applications

Hydrodynamically lubricated journal bearings are frequently used in rotating machines like compressors, turbines, pumps, electric motors and electric generators. Usually these machines are operated at high speeds and therefore a plain journal bearing is not an appropriate type of bearing to cope with problems such as oil whirl. There is, therefore, a need for other types of bearing geometries. Some of them are created by cutting axial grooves in the bearing in order to provide a different oil flow pattern across the lubricated surface. Other types have various patterns of variable clearance so as to create pad film thicknesses which have more strongly converging and diverging regions. Various other geometries have evolved as well, such as the tilting pad bearings which allow each pad to pivot about some point and thus come to its own equilibrium position. This usually results in a strong converging film region for each pad.

Many of the bearings with unconventional geometry have been developed principally to combat one or another of the causes of vibration in high-speed machinery. It should be noted, however, that the range of bearing properties due to the different geometric effects is so large that one must be relatively careful to choose the bearing with the proper characteristics for the particular causes of vibration for a given machine. In other words, there is no one bearing which will satisfy all requirements.

5.6.1. Journal bearings with fixed non-preloaded pads

The bearings shown in Fig. 5.24 are, to a certain extent, similar to the plain journal bearing. Partial arc bearings are a part of a circular arc, where a centrally loaded 150° partial arc bearing is shown in the figure. If the shaft has radius R, the pad is manufactured with radius $R + c$. An axial groove bearing, also shown in the figure, has axial grooves machined in an otherwise circular bearing. The floating bush bearing has a ring which rotates with some fraction of the shaft angular velocity. All of these bearings are called non-preloaded bearings because the pad surfaces are located on a circle with radius $R + c$.

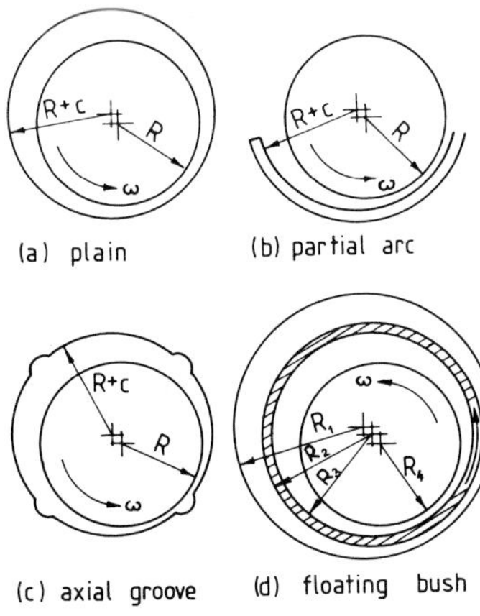

Partial arc bearings are only used in relatively low-speed applications. They reduce power loss by not having the upper pad but allow large vertical vibrations. Plain journal and axial groove bearings are rarely perfectly circular in shape. Except in very few cases, such as large nuclear water pump bearing which are made of carbon, these are crushed in order to make the bearing slightly non-circular. It has been found that over many years of practical usage of such bearings, that inserting a shim or some other means of decreasing the clearance slightly in the vertical direction, makes the machine run much better.

Cylindrical plain journal bearings are subject to a phenomenon known as oil whirl, which occurs at half of the operating speed of the bearing. Thus, it is called half-frequency whirl. Axial groove bearings have a number of axial grooves cut in the surface which provide for a better oil supply and also suppress whirl to a relatively small degree. Floating bush bearings reduce the power loss as compared to an equivalent plain journal bearing but are also subject to oil whirl. All of these bearings have the major advantage of being low in cost and easy to make.

(a) plain   (b) partial arc

(c) axial groove   (d) floating bush

Figure 5.24

5.6.2. Journal bearings with fixed preloaded pads

Figure 5.25 shows four bearings which are rather different from the conventional cylindrical bearings. The essence of the difference consists in that the centres of curvature of each of the pads are not at the same point. Each pad is moved in towards the centre of the bearing, a fraction of the pad clearance, in order to make the fluid film thickness more converging and diverging than it is in the plain or axial groove journal bearings. The pad centre of curvature is indicated by a cross. Generally these bearings give good suppression of instabilities in the system but can be subject to

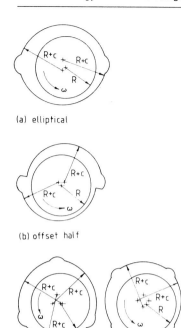

(a) elliptical

(b) offset half

(c) three lobe   (d) four lobe

Figure 5.25

subsynchronous vibration at high speeds. Accurate manufacture of these bearings is not always easy to obtain. A key parameter used in describing these bearings is the fraction of converging pad to full pad length. Ratio α is called the offset factor and is given by

$$\alpha = \text{converging pad length/pad arc length.}$$

An elliptical bearing, as shown in Fig. 5.25, indicates that the two pad centres of curvature are moved along the y-axis. This creates a pad which has each film thickness and which is one-half converging and one-half diverging (if the shaft were centred) corresponding to an offset factor $\alpha = 0.5$. Another offset half-bearing shown in Fig. 5.25 consists of a two-axial groove bearing which is split by moving the top half horizontally. This results in low vertical stiffness. Basically it is no more difficult to make than the axial groove bearing. Generally, the vibration characteristics of this bearing are such as to avoid the previously mentioned oil whirl which can drive the machine unstable. The offset half-bearing has a purely converging pad with pad arc length 160° and the point opposite the centre of curvature at 180°. Both the three-lobe and four-lobe bearings shown in Fig. 5.25 have an offset factor of $\alpha = 0.5$.

The fraction of pad clearance which the pads are moved inwards is called the preload factor, m. Let the bearing clearance at the pad minimum film thickness (with the shaft centred) be denoted by c_b. Figure 5.26 shows that the largest shaft which can be placed in the bearing has radius $R + c_b$. Then the preload factor is given by the ratio

$$m = (c - c_b)/c. \tag{5.67}$$

A preload factor of zero corresponds to having all of the pad centres of curvature coincide at the centre of the bearing, while a preload factor of 1.0 corresponds to having all of the pads touching the shaft. Figure 5.26 illustrates both of these cases where the shaft radius and pad clearance are held constant.

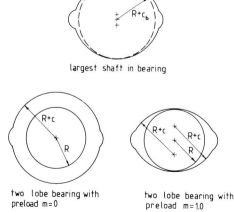

largest shaft in bearing

Figure 5.26   two lobe bearing with   two lobe bearing with
preload $m = 0$ preload $m = 1.0$

(a) dam bearing

(b) multiple dam

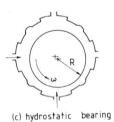

(c) hydrostatic bearing

Figure 5.27

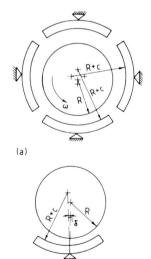

(a)

(b)

Figure 5.28

5.6.3. Journal bearings with special geometric features

Figure 5.27 shows a pressure dam bearing which is composed of a plain journal, or a two-axial-groove bearing in which a dam is cut in the top pad. If the dam height is c_d, the radius of the bearing in the dam region is $R + c + c_d$. As the fluid rotates into the dam region, a large hydrodynamic pressure is developed on top of the shaft. The resulting hydrodynamic force adds to the static load on the bearing making the shaft appear to weigh much more than it actually does. This has the effect of making the bearing appear much more heavily loaded and thus more stable. Pressure dam bearings are extremely popular with machines used in the petrochemical industry and are often used for replacement bearings in this industry. It is relatively easy to convert one of the axial groove or elliptical bearing types over to a pressure dam bearing simply by milling out a dam. With proper design of the dam, these bearings can reduce vibration problems in a wide range of machines. Generally, one must have some idea of the magnitude and direction of the bearing load to properly design the dam.

Some manufacturers of rotating machinery have tried to design a single bearing which can be used for all (or almost all) of their machines in a relatively routine fashion. An example is the multiple axial groove or multilobe bearing shown in Fig. 5.27. Hydrostatic bearings, also shown in Fig. 5.27, are composed of a set of pockets surrounding the shaft through which a high pressure supply of lubricant comes. Clearly, the use of hydrostatic bearings require an external supply of high pressure lubricant which may or may not be available on a particular machine. The bearings also tend to be relatively stiff when compared with other hydrodynamic bearings. Because of their high stiffness they are normally used in high precision rotors such as grinding machines or nuclear water pumps.

5.6.4. Journal bearings with movable pads

This widely used type of bearing is called the tilting pad bearing because each of the pads, which normally vary from three up to seven, is free to tilt about a pivot point. The tilting pad bearing is shown in Fig. 5.28. Each pad is pivoted at a point behind the pad which means that there cannot be any moment acting on the pad. The pad tilts such that its centre of curvature moves to create a strongly converging pad film. The pivot point is set from one-half the length of the pad to nearly all the way at the trailing edge of the pad. The fraction of the distance from the leading edge of the pad pivot point divided by the distance from the pad leading edge to the trailing edge is called the offset factor, similar to the offset factor for multilobe bearings. Offset factors vary from 0.5 to 1.0. An offset factor less than 0.5 creates a significant fraction of diverging wedge which is undesirable. If there is any possibility that the bearing will rotate in the direction opposite to the design direction, an offset of 0.5 should be used. An offset of 0.5 also avoids the problem of the pad being installed backwards, which has been known to occur from time to time.

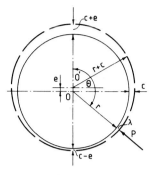

Figure 5.29

Another important consideration for tilting pad bearings is the radial location of the pad pivot point. It may be moved so that the pad centres-of-curvature do not coincide at a point at the centre of the bearing. This is a preload factor essentially the same as described for elliptical, three-, and four-lobe bearings. A preload factor of less than zero (the pad centre-of-curvature between the pad and bearing centre) creates a pad which will tend to dig the leading edge into the shaft. This is sometimes called pad lock-up. Lock-up can be prevented by placing a small bevel on the pad leading edge, which produces a small converging wedge effect, but negative preloads should be avoided.

Tilting pad bearings are very widely used to stabilize machines which have subsynchronous vibration. Because the pads are free to follow the shaft, the forces produced in the bearing are not capable of driving the shaft in an unstable mode. Their disadvantages include high cost, high horse-power loss and installation problems. Tilting pad journal bearings have been widely adapted, particularly in cases where they are not readily accessible and maintenance of alignment is important.

Referring to Fig. 5.29, it is assumed that when the journal is under load the film thickness becomes slightly reduced on those pads towards which the load is directed, and correspondingly increased on the opposite side of the bearing, i.e. eccentricity e is in the line of action of the load. Suppose that the centre of breadth of each pad is located by the angle Θ measured from the position of maximum film thickness. Denoting

$\lambda =$ the mean film thickness for a pad at angle Θ,
$c =$ the radial clearance when the journal is placed centrally without load, and using the notation as for a normal journal bearing,

$$\lambda = c + e\cos\Theta. \tag{5.68}$$

If P is the normal load on the pad per unit length of journal at angle Θ

$$P = \frac{1}{z}\mu V\frac{B^2}{\lambda^2}$$

and the upward vertical component of P is

$$P\cos(\pi - \Theta) = -\frac{1}{z}\mu V\frac{B^2}{\lambda^2}\cos\Theta,$$

where z is a dimensionless constant,

$$z = \left(\frac{B}{c}\right)^2\frac{\mu V}{P} = \frac{2\varepsilon^2}{3\left[\ln\left(\dfrac{1+a}{1-a}\right) - 2a\right]}$$

and

$$\varepsilon = e/c.$$

As in the Reynolds theory we may neglect the effect of tangential drag in estimating the load carried, so that, if N is the number of pads and Q the total vertical load per unit length

$$Q = -\frac{1}{z}\mu V B^2 \sum \frac{\cos \Theta}{\lambda^2},$$

where the summation sign covers N terms.

From the table of the Reynolds integrals, discussed in Chapter 2, it follows

$$I_3 = \int_0^{2\pi} \frac{\cos \Theta \, d\Theta}{\lambda^2} = \frac{-2\pi e}{(c^2 - e^2)^{\frac{3}{2}}}$$

the mean value of

$$\frac{\cos \Theta}{\lambda^2} = \frac{-e}{(c^2 - e^2)^{\frac{3}{2}}}$$

so that

$$Q = \frac{1}{z}\mu V B^2 \frac{Ne}{(c^2 - e^2)^{\frac{3}{2}}}.$$

Following usual practice, take the effective arc of action of the pads to be 80 per cent of the complete ring, then $(NB) = 5r$; hence neglecting leakage

$$Q = \frac{25}{Nz}\mu V r^2 \frac{e}{(c^2 - e^2)^{\frac{3}{2}}}.$$

For the pads, assume $z = 1.69$ and $(fB)/\lambda = 2.36$, then

$$Q = \frac{14.8}{N}\mu V r^2 \frac{e}{(c^2 - e^2)^{\frac{3}{2}}}. \tag{5.69}$$

The tangential drag in each pad is fP, and so

$$F' = 2.36\frac{\lambda}{B}\frac{1}{z}\mu V \frac{B^2}{\lambda^2} = \frac{2.36}{z}\mu V \frac{B}{\lambda}.$$

If $\bar{F}$ is the total frictional resistance exerted on the journal per unit length

$$\bar{F} = \frac{2.36}{z}\mu V B \sum \frac{1}{\lambda}.$$

Again, from the table of integrals (see Chapter 2), the mean value of $1/\lambda$ from 0 to 2π is

$$\frac{I_1}{2\pi} = \frac{1}{\sqrt{(c^2 - e^2)}}$$

so that

$$\bar{F} = \frac{2.36}{z}\mu V B \frac{N}{\sqrt{(c^2 - e^2)}}$$

$$= \frac{11.8}{z}\frac{\mu V r}{\sqrt{(c^2 - e^2)}}.$$

Again, taking $z = 1.69$ and assuming an axial length large compared with the breadth B, so that leakage may be neglected

$$\bar{F} = 6.98 \frac{\mu V r}{\sqrt{(c^2 - e^2)}}. \tag{5.70}$$

Hence the virtual coefficient of friction for the journal is

$$f = \frac{\bar{F}}{Q} = 0.47 \frac{N}{re}(c^2 - e^2). \tag{5.71}$$

5.7.   Gas bearings

Fluid film lubrication is an exceptional mechanical process in which viscous shear stress contributes directly to the useful function of developing a load capacity. Although viscosity also causes bearing friction, the equivalent lift-to-drag ratio of a typical hydrodynamic wedge is of the order of 1000 to 1, which compares favourably to a high-performance wing. In the case of gas bearings, in contrast to the more common liquid lubricated bearings, lubricant compressibility is the distinctive feature.

Although basic concepts such as the hydrodynamic wedge are still applicable to gas bearings despite gas compressibility, many additional features of gas bearings are unique and require separate attention. The potential for large-scale industrial application of gas bearings was recognized in the late 1950s. Advocates of gas lubrication have emphasized the following advantages:
- the gaseous lubricant is chemically stable over a wide temperature range;
- atmospheric contamination is avoided by the use of gas bearings;
- the viscosity of a gas increases with temperature so that the heating effect in overloading a gas bearing tends to increase the restoring force to overcome the overload;
- a gas bearing is more suitable for high-speed operation;
- there is no fire hazard;
- use of gas bearings can reduce the thermal gradient in the rotor and enhance its mechanical integrity and strength;
- for high-speed applications, the gas bearing is inherently more noise-free than the rolling-contact bearing;
- system simplicity is enhanced by the use of self-acting gas bearings, which do not require cooling facilities.

These optimistic views must be tempered with more subtle engineering considerations before one can confidently substitute gas bearings for more conventional oil lubricated bearings in actual applications.

Intense development of gas lubrication technology was triggered by the demands of sophisticated navigation systems, by the prospects for gas-cooled nuclear reactors, by the proliferation of magnetic peripheral devices in the computer industry and by the everlasting quest for machinery and devices in aerospace applications.

Although not all the early expectations have been realized, the advantages of gas lubrication are fully established in the following areas:
(i) Machine tools. Use of gas lubrication in grinding spindles allows attainment of high speeds with minimal heat generation.

(ii) Metrology. Air bearings are used for precise linear and rotational indexing without vibration and oil contamination.

(iii) Dental drills. High-speed air-bearing dental drills are now a standard equipment in the profession.

(iv) Airborne air-cycle turbomachines. Foil-type bearings have been successfully introduced for air-cycle turbomachines on passenger aircraft. Increased reliability, leading to reduced maintenance costs, is the benefit derived from air bearings.

(v) Computer peripheral devices. Air lubrication makes possible high-packing-density magnetic memory devices, including tapes, discs and drums. Read-write heads now operate at submicrometer separation from the magnetic film with practically no risk of damage due to wear.

In the development of each of these successful applications, effective utilization of analytical design tools was crucial. This section gives only an introduction to the problems associated with gas bearing design. There is a quite sophisticated theory of gas lubrication, which forms the foundation of all analytical design tools. However, detailed presentation and discussion of this theory is beyond the scope of this text and reader is referred to the specialized books listed at the end of this chapter. It is, however, appropriate to review briefly, lessons that were learned in the past so that future designers will not be misled by too optimistic views of supporters of gas lubrication.

Most important problems identified in the past can be summarized as follows:

I. Inadvertent contact between the bearing surfaces is unavoidable. Even if the surfaces are coated with a boundary lubricant, the coefficient of friction is expected to be at least 0.3. This is more than three times as large as that between oil-lubricated metal surfaces. Thus, a gas bearing is substantially more vulnerable to wear damage than an oil-lubricated bearing. For this reason, the gas bearing surface is usually a hard material.

II. Even when a nominal separation between the bearing surfaces is maintained under normal operation, particulate debris may occasionally enter the bearing clearance and cause solid-debris-solid contact with high normal and tangential local stresses. In a conventional oil lubricated bearing, one of the surfaces is usually a soft material such as bronze or babbitt; the intruding debris become embedded in the soft surface with no damage done to the bearing. Since the wear-life requirement precludes use of a soft gas bearing surface, one has to resort to the other extreme; the bearing surface, together with its substrate, must be hard enough to pulverize the debris.

III. Gas bearings generally operate at very high sliding velocities; $50 \, \text{m s}^{-1}$ is quite common, and this is at least ten times higher than the sliding speed of a typical oil-lubricated bearing. Intense local heating results when dry contact occurs or debris is encountered. Together with the three times higher coefficient of friction, the thermal-mechanical distress in a gas bearing is potentially thirty times more severe than that in an oil-lubricated bearing under the same normal load. An even more serious situation exists for the self-acting gas bearings during the period of attaining nominal velocity. Because the viscosity of a gaseous lubricant is about 1/1000 of that of a typical oil, the speed at which there is complete separation of contacting surfaces would be 1000 times higher for the same normal load and surface topography, and the thermal-mechanical distress up to the above-mentioned speed would be 3000 times more severe.

IV. Chemical breakdown of an oil lubricant under an extreme thermal-mechanical load is, in a way, desirable. The endothermic latent heat of the chemical breakdown process serves to limit the local temperature rise and forestalls catastrophic failure of the bearing surface. Because gaseous lubricants are chemically stable, all thermal-mechanical load is converted into a severe bearing surface temperature rise, which tends to initiate irreparable material damage.

These factors combine to make gas bearings more susceptible to mechanical damage and thus preclude widespread application of gas bearings in heavy-duty equipment. The same considerations also have a dominating influence in the choice of satisfactory materials for gas bearings. Beneficial use of gas bearings must be predicted on avoidance of these limiting factors. Gas lubrication theory is generally regarded as an extension of the liquid film lubrication theory based on the Reynolds equation, which was originally derived for an incompressible lubricant. The main additional issue is the concern for an appropriate account of the density variation within the lubricating film, such that the basic principles of thermodynamics are satisfied, to a degree consistent with the approximations already invoked in momentum considerations.

In certain ways, gas bearings are more easily analysed than liquid-lubricated bearings. In a gas bearing film, the temperature may be regarded as constant, even though viscous heating necessarily causes some temperature rise above that of the bearing surfaces. Since the viscosity coefficient of most gases is dependent solely on temperature, an isoviscous approximation is satisfactory for studying gas bearings. In a liquid bearing film, the isoviscous approximation is less reliable. The gas bearing film is inherently a single-phase constituent. Irrespective of local pressure level relative to ambient pressure, the gaseous lubricating film remains a homogeneous medium. However, in a liquid bearing it has been established empirically that a homogeneous liquid state is ensured only when the local pressure is near or above atmospheric pressure. Where the pressure tends to become subambient in a self-acting liquid bearing film, a two-phase flow structure is prevalent. In fact, a completely rigorous treatment of this aspect of the liquid-lubricant film has yet to be demonstrated.

5.8. Dynamically loaded journal bearings

Journal bearings used in, for instance, reciprocating compressors and internal combustion engines are subjected to fluctuating loads. When studying the performance of such bearings, it is necessary to determine the bearing loads and the change in magnitude and direction of these loads with time. As an illustration of the problem, let us analyse a two-mass system for a single cylinder arrangement shown in Fig. 5.30. It is convenient

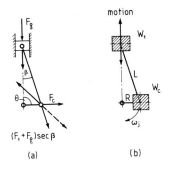

Figure 5.30

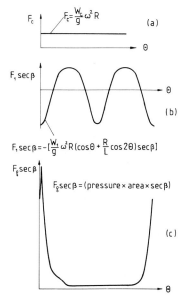

Figure 5.31

to represent time by the crank angle position and numbers indicating these positions are usually marked on the polar load diagram for constant increments of the crank angle. The distance from the polar origin to such a point represents the magnitude of the load on the bearing in vector form.

5.8.1. Connecting-rod big-end bearing

The loads on the connecting-rod big-end bearing can be attributed to three component loads due to the reciprocating inertia forces, rotating inertia forces and gas forces. The system shown in Fig. 5.30 is used to obtain the inertia forces, i.e. a reciprocating mass at the small-end and a rotating mass at the big-end of the connecting-rod. The reciprocating mass W_1 at the small-end consists of the mass of the piston, gudgeon pin and part of the connecting-rod (usually about one third of the connecting-rod is included). The remainder of the connecting-rod mass is the rotating mass W_c acting at a big-end. Both the reciprocating forces F_1 (resulting from reciprocating mass W_1) and the gas forces F_g are applied in line with the cylinder axis, but the force on the crankpin itself, due to these two forces, will be larger by a factor of $\sec \beta$ because allowance must be made for connecting-rod obliquity. Typical component loads acting on a big-end bearing are shown in Fig. 5.31. Relative to the bearing, the reaction to the rotating inertia force F_c, is constant in magnitude but has a continuously changing value of angular velocity. The reaction to the reciprocating inertia force, $F_1 \sec \beta$, will act on the bearing in line with the connecting-rod axis and will vary in magnitude and direction as shown in Fig. 5.31. The component due to cylinder pressure will force the connecting-rod down on the crankpin causing a load reaction on the rod half of the bearing.

5.8.2. Loads acting on main crankshaft bearing

These loads are partly due to force reactions from the big-end bearings and partly due to the out-of-balance of the crankshaft. The out-of-balance of the crankshaft is sometimes reduced by the use of balance weights. When considering the forces from the big-end bearing, it is necessary to orientate them to the same non-rotating datum as the main bearing. It should be noted that there is no difference in magnitude of the loads on the bearing, for a particular crank-angle position, whether the loads are plotted relative to connecting-rod, crankpin or cylinder axis, as it is the angular velocity of the load vector relative to the chosen datum axis which changes and thus produces differently shaped load diagrams.

In a multicylinder engine which has a main bearing between each big-end bearing, it is usual to consider the main bearing forces as resulting from the component forces associated with the crank system between two adjacent cylinder axes. Such a system is shown in Fig. 5.32, which is for a six-throw crankshaft. The forces F_1 and F_2 are the reactions from the adjacent big-end bearings, while C_1 and C_2 are the resultant crankshaft out-of-balance forces between adjacent cylinders.

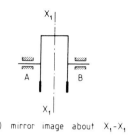

(a) mirror image about $X_1 - X_1$

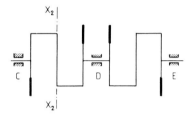

(b) non-mirror image about $X_2 - X_2$

Figure 5.33

Figure 5.32

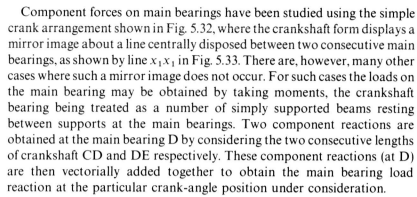

Component forces on main bearings have been studied using the simple crank arrangement shown in Fig. 5.32, where the crankshaft form displays a mirror image about a line centrally disposed between two consecutive main bearings, as shown by line $x_1 x_1$ in Fig. 5.33. There are, however, many other cases where such a mirror image does not occur. For such cases the loads on the main bearing may be obtained by taking moments, the crankshaft bearing being treated as a number of simply supported beams resting between supports at the main bearings. Two component reactions are obtained at the main bearing D by considering the two consecutive lengths of crankshaft CD and DE respectively. These component reactions (at D) are then vectorially added together to obtain the main bearing load reaction at the particular crank-angle position under consideration.

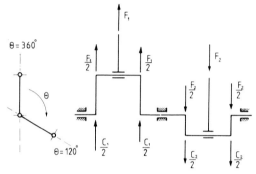

5.8.3. Minimum oil film thickness

The problem of predicting the minimum oil film thickness in a relatively simple dynamic load case which consists of rotating loads of both constant magnitude and angular velocity will now be considered. In such a case, a modified steady load theory, known also as the equivalent speed method, can be used. This method for predicting minimum oil film thickness is applicable to load diagrams where the magnitude of the load W and the angular velocity of the load vector ω_1 are constant, as shown in Fig. 5.34. It should be noted that while the angular velocity of the load vector ω_1 is constant, it is not necessary (for this method) that it be equal to the journal angular velocity ω_j and furthermore that it may rotate in the opposite direction to ω_j.

However, when the load vector does rotate at the journal speed and in the same direction (i.e. ω_1 equal to ω_j), this represents a similar case to that of the steady load. Imagine the whole system mounted on a turntable which rotates at the journal speed in the opposite direction to both journal and load line. The load and journal would then become stationary and the bearing would rotate at $-\omega_j$. A similar load-carrying system as the steady load case is then created, with one surface moving at ω_j, the other stationary and the load stationary. Thus one of the conventional steady-load-bearing capacity versus eccentricity-ratio charts may be employed. For the cases

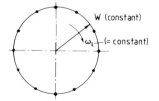

Figure 5.34

where the load rotates at speeds other than the journal speed, a correction may be made to the journal speed term to account for this.

If, however, a rigorous view is adopted about the equivalence of rotating-load and steady-load cases, then it should be made apparent that oil grooving in the bearing and/or oil feed holes in the journal will not give a true similarity. Neither will the heat distribution in the bearing be the same. For the rotating-load case all the bearing surface will be subjected to the shearing of small oil films as the load passes over it, whereas for the steady case, small films and associated high temperatures are confined to one local region of the bearing. The bearing surface, at any point, is subjected to fluctuating developed pressure in the oil film due to the rotating load although this load is of constant magnitude. Such a condition could give rise to fatigue of the bearing material. These factors, although they may be of secondary importance, illustrate that one must be aware of realities when considering a so-called equivalent system.

We return to the equivalent speed method and consider a journal bearing arrangement which has a rotating journal of constant angular velocity, ω_j, a rotating load of constant magnitude and constant angular velocity, ω_l, and a fixed or rotating bearing of constant angular velocity ω_b. The load-carrying capacity of such a system is proportional to the average angular velocity of the bearing and journal relative to the load line. This particular case was discussed in more detail in Section 5.5.5. Thus

$$\frac{\text{(load capacity of rotating load case)}}{\text{(load capacity of steady load case)}} = 1 - \frac{2\omega_l}{\omega_j}.$$

The load-carrying capacity for a steadily loaded bearing, although proportional to ω_j, also depends on the bearing length L, diameter d, radial clearance c and operating viscosity μ in the bearing. These variables together with the load W form a dimensionless load number S which is given by

$$S = \frac{W/L}{U\mu}\left(\frac{c}{r}\right)^2 = \frac{W/L}{\omega_j r\mu}\left(\frac{c}{r}\right)^2.$$

This is usually referred to as the Sommerfeld number and was derived in Chapter 2. There are a number of cases where the load-carrying capacity can be deduced from the load number. Thus: for steady load $\omega_l = 0$ the load capacity is proportional to ω_j, for counter rotation $\omega_l = -\omega_j$ the load capacity is proportional to $3\omega_j$, for rotating in phase $\omega_l = \omega_j$ the load capacity is proportional to ω_j, for a load rotating at half-speed $\omega_l = \omega_j/2$ the load capacity is zero. For the stationary bearing ($\omega_b = 0$) and a rotating journal, the oil in the clearance space can be considered as basically rotating at half-shaft-speed. A particular case when the load vector rotates at half-shaft-speed, i.e. $\omega_l = \omega_j/2$, is shown in Fig. 5.35. The combination of these two factors, that is the load rotating at the same speed as the oil, is such that there is no net drag flow relative to the load line and hence no hydrodynamic wedge action is created. The oil is then forced out due to

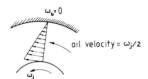

Figure 5.35

squeeze action. For the case when the value of both ω_1 and W are changing with time, the equivalent speed method, using a steady load-capacity relationship, is not applicable. There are several ways to demonstrate that this is the case. For instance if ω_1 happened to pass through a half-speed load vector condition almost instantaneously, the equivalent speed method would give zero oil film thickness at that instant. In practice, however, the oil cannot be squeezed out of the bearing instantaneously. It takes time, during which the load vector has changed and the half-speed vector no longer exists. Another point which is often overlooked is, that the position and direction of motion of the journal centre in the bearing, depend on the velocity variation of the journal centre along its path. Such variations are not taken into account in the equivalent speed method. In consequence, this method which relies on wedge action should not be used to predict oil film thickness in engine bearings where the load and ω_1 are varying. The above method is, however, useful to indicate in an approximate manner, where periods of zero load capacity due to collapse of the wedge action exist and during such periods squeeze-action theory can be applied.

It is quite clear from the method discussed previously that when the load is rotating at or near half-shaft-speed, the load capacity due to wedge action collapse and another mechanism, called squeeze action, is operational. This is shown schematically in Fig. 5.36. Consequently, during such a period, the eccentricity ratio will increase and continue to squeeze the oil out until there is a change in conditions when this squeezing period is no longer predominant. The squeeze film action has a load capacity due to radial displacement of the journal at the load line. As we have seen in the pure rotating load case, for example, the wedge action load capacity collapses if the angular velocity of the oil is zero relative to the load line. This velocity can be associated with $\bar{\omega}$ which denotes the average angular velocity between the journal and bearing relative to the load line. Thus:

(i) for a main bearing (e.g. stationary bearing)

$$\bar{\omega} = \frac{\omega_j}{2}\left[1 - \frac{2\omega_1}{\omega_j}\right];$$

(ii) for a connecting-rod bearing where the polar load diagram is relative to the engine cylinder axis

$$\bar{\omega} = \frac{\omega_j}{2}\left[1 - \frac{2\omega_1}{\omega_j} + \frac{\omega_1}{\omega_j}\right];$$

(iii) for a connecting-rod bearing where the polar load diagram is relative to the connecting-rod axis,

$$\bar{\omega} = \frac{\omega_j}{2}\left[1 - \frac{2\omega_1}{\omega_j} - \frac{\omega_b}{\omega_j}\right].$$

Since the angular velocity of the bearing has to be taken into account in a big-end connecting-rod bearing, one should not consider ω_1/ω_j equal to 0.5 as indicating collapse of the load capacity due to wedge action. Zero load

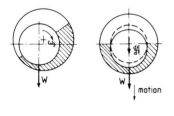

(a) wedge action   (b) squeeze action

Figure 5.36

capacity for wedge action will occur in all the above cases when $\bar{\omega}/\omega_j$ is zero. For a particular load diagram, use of this fact can be made by plotting $\bar{\omega}/\omega_j$ against the crank angle Θ and noting when this value is small compared with the load W. If this should happen for a comparatively long period during the load cycle, then a squeeze interval is predominant and squeeze action theory can be applied as an approximation for the solution of minimum oil film thickness. Typical squeeze paths in the clearance circle resulting from squeeze action are shown relative to the load line in Fig. 5.37.

The performance characteristics of the journal travelling along the central squeeze path are used in this quick method for predicting minimum oil film thickness. Other more exact methods, however, are available using performance data for offset squeeze paths and for mapping out the whole journal centre cyclic path. Designers are generally interested in on-the-spot solutions, and this quick approximate method predicting the smallest oil film thickness based on central squeeze action, will give the required trends if predominant squeeze action prevails.

Usually, a design chart shows an impulse, J_a, plotted against the ratio of minimum oil film thickness/radial clearance for central squeeze action and is based on the impulse capacity concept. Generally, impulse can be considered as $\int_{t_1}^{t_2} P \, dt$, which forms part of the dimensionless expression for J_a

$$J_a = \left(\frac{c}{r}\right)^2 \int_{t_1}^{t_2} \frac{P}{\mu} \, dt, \tag{5.72}$$

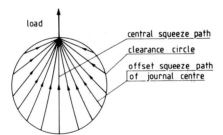

Figure 5.37

where (c/r) is the relative radial clearance, P is the specific load, μ is the absolute viscosity and t is time. All these parameters have to be expressed in consistent units.

5.9. Modern developments in journal bearing design

Thin-wall bearings, defined as lined inserts which, when assembled into a housing conform to that housing, are commonly used in modern medium-speed internal combustion engines. They are almost invariably steel-backed to take advantage of the greater thermal stability, choice of bearing surface material and homogeneity of this material. The thin-wall bearings have a thickness/diameter ratio varying from 0.05 at 40 mm diameter to 0.02 at 400 mm. However, there are still other factors which have to be considered. From the very definition of a thin-wall bearing, its form is dictated by the housing into which it fits. This implies that if the housing contains errors or irregularities, these will be reflected in the assembled bearing and, therefore, have to be contained within tight limits.

5.9.1. Bearing fit

To ensure conformity of the bearing shell to its housing, an accurate interference fit has to be provided between the two, thereby restricting manufacturing tolerances of peripheral lengths of both the housing and the bearing. The interference fit is derived from an excess peripheral length in each half bearing which has to be closely defined to enable bearings of the same part number to be directly interchangeable. On assembly, the excess peripheral length, or so called crush, creates a hoop or circumferential stress around the bearing and a radial contact pressure between the bearing back and the housing bore. This contact pressure resists relative movement between the bearing housing and the bearing back thus preventing fretting. Unfortunately there is a theoretically correct level – housings with a great flexibility require a higher contact pressure than stiffer ones. On early engines, having thin-wall bearings, a contact pressure as low as 2 MPa was usually sufficient to resist fretting, but as engine ratings increase, and housing stress analysis becomes more sophisticated, higher pressures are necessary, often reaching 8–10 MPa today. In these very high interference fit assemblies, particular care has to be taken to ensure that the joint face clamping bolts have sufficient capacity to assemble the bearing, yet with sufficient reserve to resist the dynamic separating forces from engine operation. As the contact pressure is increased for any given bearing size, the hoop stress increases to the point where the steel backing begins to yield, adjacent to the joint face, and of course, this must be avoided. Knowing the combined effect of bearing steel yield strength and the friction force for bearing assembly, a wall thickness can be determined which will avoid yield. It is worth noting that the yield strength of the bearing back, in finished form, varies considerably with the method of manufacture. For instance, a bearing which is roll formed at some stage, but not fully annealed, will have a considerably greater yield strength than that of the raw steel.

An increased contact pressure requires a greater bolt tension for fitting bearing caps to their opposite half-housings. Proper bolt preload is very important because if it is insufficient, the housing joints will separate dynamically, giving rise to a high dynamic loading and to probable fatigue cracking of the bolts.

To reduce the tendency to fretting, even though the majority of engine bearings suffer fretting to some degree, it is recommended that the housing bore surface finish should not exceed 1.6 μm c.l.a. Bearing backs are typically 0.8 μm surface finish or better and in highly loaded zones should always be supported. Cyclic variation of the hydrodynamic oil pressure on the bearing surface will attempt to make the bearing back conform to the housing and if for example there are grooves or oil holes behind the plain bearing, fretting will almost be certain. If flexure is sufficiently great, the lining material on the bearing surface can also suffer fatigue damage.

5.9.2. Grooving

Both the theoretical and the actual oil film thicknesses are influenced significantly by the extent of oil grooving in the bearing surface. The simplest form of oil groove in a bearing is a central, fully circumferential, type. Some bearings have such grooves misplaced from the centre, but this should usually be avoided, because oil will flow from this groove preferentially across the narrower of the two bearing lands. Also, a multiplicity of circumferential grooves should be avoided, because any bearing land with pressure fed oil grooves on both ends will not generate any oil flow across it and will overheat.

A central circumferential groove has the advantage that it allows the simplest method of supplying oil around the full circumference of the bearing, and also the simplest method of transferring oil from one location to another, for example in a diesel engine from the main bearing to the big-end bearing, then to the small-end bush and so to the piston cooling passages.

Hydrodynamically, oil grooves are detrimental to load-carrying capacity, and if the minimum oil film thickness is likely to be so low as to present a potential wiping problem, ungrooved or partially grooved bearings may be employed.

Regardless of bearing design, if a fine level of oil filtration is not maintained throughout the engine life, ferrous debris can contaminate the bearing surface, but often not be totally embedded. Such particles penetrate the thin oil film and rub against the crankshaft thereby causing them to work-harden to a significantly higher level of hardness than the crankshaft. This inevitably results in wear of the crankshaft surface, and with fully-grooved bearings a circumferential ridge is eventually produced; the area of crankshaft corresponding to the oil groove remaining unworn. If the wear is not excessive no real problem is created, other than eventual excessive clearance. However, with partially-grooved bearings, a similar ridge is still produced on the journal surface, caused by wear particles entrapped in the grooved region of the bearing. This differential wear of the journal surface then results in wiping, wear and even fatigue of the ungrooved region of the bearing surface, directly in line with the partial groove.

The final major disadvantage of partially-grooved bearings is cavitation erosion of the bearing surface. Partially-grooved bearings have become much more common as engine ratings have increased, and bearing loadings more severe; a combination which in recent years has given rise to a much greater incidence of bearing damage caused by cavitation erosion.

5.9.3. Clearance

High values of diametral clearance could lead to excessive damage due to cavitation erosion. Even where bearings have been designed with a reduced vertical clearance, but a high horizontal clearance, in an attempt to induce a cooling oil flow through the bearing, so-called eccentric wall suction cavitation erosion has occurred in the centres of the bearing lands in the high clearance regions. Theoretically, low level of clearance is required for good load-carrying capacity and generation of the hydrodynamic oil film, but the oil flow through the bearing is restricted, leading to increased temperature of operation, and consequently a reduced viscosity within the oil film which in turn results in a thinner film. Thus, clearance has to be a compromise of several factors. Theoretical calculations, based on a heat balance across steadily loaded bearings, point to a standardized minimum clearance of $0.00075 \times$ journal diameter which has been verified by satisfactory hydrodynamic, cavitation-free operation in a whole range of different engine types under normal, varied, service conditions. The same order of clearance has also been confirmed theoretically in computer simulation of minimum film thickness with due allowance for the variation in viscosity as a consequence of a clearance change.

5.9.4. Bearing materials

The modern medium-speed diesel engine, especially at the higher outputs at increased running speeds, uses either tin–aluminium or copper–lead lining materials in various compositions. With increasing requirements for high load carrying capacity in a multitude of operating conditions there is a general tendency for both types of lining material to be given a thin galvanically-plated overlay of soft lead–tin or lead–tin–copper. This layer derives its strength from the underlying lining, and becomes weaker with increasing thickness. The basic advantage of an overlay is that it will accommodate significant levels of built-in dirt particles, oil-borne contaminants, misalignment and distortion, together with minor manufacturing inaccuracies of all the relevant parts.

With these inherent advantages, it is usual to design bearings for the modern engine with a view to retaining the overlay, but this must be considered semisacrificial in allowing for the above mentioned defects. With copper–lead lined bearings, the overlay provides a much more important service, that of corrosion protection. If the lubricating oil contains organic acids and peroxides, for example, from leakage of sulphur-containing fuel oils or blow-by of exhaust gases, the lead phase in a copper–lead matrix can be leached out leaving an extremely weak porous copper matrix which is easily fatigued by the dynamic loads applied to the surface. Tin–aluminium is not subject to the same type of damage, and only suffers corrosion if directly contacted by water, in the absence of oil. Relative to the stronger copper–lead materials, the tin–aluminium materials are weaker, but if the fatigue limit of the composition lining and overlay are taken into consideration, the levels are the same, being dependent upon the fatigue strength of the overlay. As it is the intention to retain this overlay, this becomes the design criterion. If the overlay is worn away, for example, to accommodate misalignment then compatibility between the shaft and the lining material becomes important. Rig tests to establish the relative compatibility rates of various compositions of tin–aluminium and copper–lead verify the superiority of the tin–aluminium.

The large slow-speed direct-drive engines still principally use white-metal lined bearings, although more commonly these are thin-wall bearings, but predominantly overlay-plated to gain the benefits mentioned earlier. Tin–aluminium, however, and in some instance copper–lead are increasingly being adopted, and as with medium-speed engines, this will become more and more usual to take advantage of the higher fatigue strength of white-metal linings. It is considered that the more compatible, corrosion resistant high tin–aluminium alloys will be the more satisfactory in these engine types.

5.10. Selection and design of thrust bearings

Thrust bearings come in two distinct types, which involve rather different technical levels; first, the bearing which is mainly an end-clearance limiting or adjusting device, and second, a bearing which has to carry a heavy load. A typical example of the first type is the bearing used to locate the crankshaft of the reciprocating engine. The loading in these bearings is not usually known with any reasonable accuracy, arising as it does from shocks or tilting of the engine.

It is obviously advantageous to take advice of a bearing manufacturer regarding material and maximum loading, however, the most practical approach is usually to be guided by past experience and comparable machines, but to allow space for possible future modifications in the light of further experience. Bearings of this kind are no longer made by lining a casing with white metal. It is an almost universal practice to stamp complete rings or half-rings from steel-backed strip, faced with white metal, overlay-plated copper–lead, aluminium–tin or one of the self-lubricating or dry-running bearing composition materials. These rings can then be removed if necessary without disturbing the main shaft. In the past, thrust rings – often of solid bronze – were prevented from rotation by means of deeply countersunk screws that secured them to the housings. The current trend, however, is to clamp them into undercut recesses by tightening down the bearing cap. Figure 5.38 shows a typical section of such a device, which is both cheap and convenient. Many of these simple thrust bearing rings are lubricated by oil flowing from the end of the adjacent journal bearing and it is usual to provide a few radial grooves across the bearing face, not only to assist in spreading the oil over the thrust face but, more importantly, to minimize the restrictive effect on the oil emerging from the journal bearing.

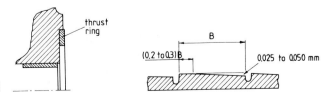

Figure 5.38

A practical point needs to be watched, since failure to appreciate it has resulted in expensive damage to quite a few main bearings. The radial width of the thrust face on the shaft must always be greater than that of the thrust ring. It is easy for a design which was originally satisfactory to become dangerous when, at some later stage, the width of the thrust ring is increased. The edge of the much harder thrust face will then bite into the soft bearing material, forming a step which severely restricts the outward flow of oil. The result is overheating and usually the complete failure of the main bearing also with possible damage to an expensive shaft. It is therefore prudent to allow ample radial width of the thrust faces in case the need is felt later, as a result of service experience, to increase the bearing area. It is obvious too, that differential thermal expansion must be estimated in order to prevent endwise tightening of the assembly. In the design of thrust bearings for more precisely defined conditions, especially where the loads are very heavy, the performance required and the disposal of the heat generated, are likely to be the controlling considerations. The plain flat thrust bearing with radial oil grooves can carry surprisingly high loads. Although the thrust face is machined flat, pressure-viscosity effects in the oil film, combined with small thermal and mechanical deflections of the pads between the oil grooves, enable an effective oil film to be built up in accordance with hydrodynamic theory.

Apart from this simple configuration there are three other types of thrust bearing. The fixed-pad type is the plain, flat, grooved thrust washer, but with the pads inclined to form ramps to promote the development of the hydrodynamic oil film. The tilting-pad bearings have pads supported on a central or offset step or pivot, or on some articulating device, to improve the load-sharing between pads. The hydrostatic bearing prevents contact and hence excessive friction and wear between the thrust collar and the bearing block by applying a static fluid pressure to one or more annular cavities in the bearing block. It is usual to supply the fluid by constant-volume pumps, so that the peripheral gaps through which it leaks to the drains vary according to the applied load, and the pressure in the cavity is thereby adjusted to balance the load.

The characteristics of these three types of thrust bearing are known mainly from experiments carried out on full scale bearings loaded with a wide range of loads. Of particular interest to the designer is that:

(i) the load capacity is enormously influenced by the slope of the ramps;
(ii) ideally the slope should be very small; in practice and with commonly used dimensions, a slope of 0.025 to 0.050 mm over the pad width gives acceptable results while remaining within attainable manufacturing standards;
(iii) with suitably designed pads, the load capacity increases rapidly with speed, even from zero. Starting or stopping under load is therefore not a serious problem;
(iv) under conditions of misalignment, the pads in the more heavily loaded arc of circumference operate with smaller clearance than those in the opposite arc and therefore develop higher hydrodynamic pressures. A

righting couple is thus set up which tends to correct the misalignment by taking up some of the available clearance in the adjacent journal bearings. Though it is obviously desirable to avoid misalignment, the bearing will automatically reduce its ill effects;

(v) the fixed pad bearing is a simple, effective and compact device capable of functioning under the severe conditions associated with stopping and starting under load. It is particularly useful where axial length has to be kept to a minimum.

5.10.1.   Tilting-pad bearing characteristics

The tilting-pad bearing is a complex arrangement because of the intricate interplay between a number of design features. The conventional form consists of a ring of pads, each supported on pivots, which may be either at the optimum point, 0.4 of the pad width from the trailing edge, or, if rotation in both directions has to be allowed for, at the centre of the pad. Better still, at the cost of some design complication the pads may be supported on some form of mechanical or hydrostatic articulation system with a view to equalizing the loads on them.

For some 50 years after the original Michell bearing was invented it was assumed that the pads tilted so as to adopt something like an ideal angle of inclination with respect to the thrust collar, and thus to induce the formation of effective hydrodynamic lubrication. During this period the limiting specific loading on the thrust bearings for steam turbines, vertical hydroelectric machines and similar plant remained around 0.021 MPa. Little or no attempt was made to improve this, so as to reduce the large size, weight, cost and power losses of these bearings. When these problems were investigated, some very interesting facts were established. First of all it was found that under typical current conditions of load and speed the pads did not tilt and their hydrodynamic action is due to thermal and mechanical distortion of the surfaces. The load shearing between pads is often extremely poor, the ratio of the highest to the lowest load being as high as 7 or 8 in a typical installation. Even with extreme care in fitting the pads to gauge room standards of accuracy it is between 2 and 4. This and the very thin oil film accounted for the failure of many such bearings in service. Experiments with alternate pairs of pads removed resulted in substantially increasing the permissible loading. For example, whereas with eight pads seizure of at least one of them occurred at an overall nominal specific load of roughly 0.07 MPa, with only two pads this figure became at least 0.28 MPa in the most favourable speed range (1000 to 1750 r.p.m.) and over 0.21 MPa at all speeds between 500 and 3000 r.p.m. Reducing the number of pads increased the chances of load sharing, proving that one or more of the pads in the full bearing were almost certainly carrying more than the overall average of 0.07 MPa specific load. The conventional thrust bearing is lubricated and cooled by pumping oil into the housing at a low point and allowing it to flow out from somewhere near the top. The whole assembly thus becomes a fluid brake resulting in heat generation and hence the need for a copious flow of oil. If only to reduce the losses, something in the nature of a dry sump lubrication system is obviously best. In order to feed oil to the individual pads in a dry sump system the oil is sprayed in high velocity jets radially outwards and obliquely onto the surface of the thrust collar between pairs of pads. This supplies an adequate film to the downstream pads while quickly draining the hot oil emerging from the upstream pads. The scrubbing action on the collar surface removes much of the heat carried here. The total oil flow required is much reduced, the pad surface temperature is lower and in a typical bearing, losses are reduced to some 30 per cent of those with flooded lubrication. Theoretically, both circumferential and radial tilting of the pads is desirable which suggests support by a hydraulic capsule under each pad. By linking the individual capsules hydraulically and keeping the fluid-filled volume as small as possible and absolutely free from air a workable articulating system can be arranged. The main design problem involved is to accommodate the relatively high pressure in the capsules. Such a system can improve the load ratio to about 2. A mechanical method of achieving this goal, the Kingsbury lever system, seems to be insufficiently sensitive and ineffective in practice, while rubber-bonded capsules have stress limitations which reduce their usefulness.

Before setting out to design a tilting-pad thrust bearing, it is useful to consider some of the more general factors which may determine its capacity limits under various conditions.

First, it is not true, particularly with a large, heavily loaded bearing, that it can accept a larger load at moderate speeds. On the contrary, in any device depending on hydrodynamic lubrication the oil film pressure decreases with the speed. Power losses plotted against specific loads at various speeds, show clearly the limiting loads at each speed. At the lowest speeds, the load is limited by the reduction of hydrodynamic pressure and therefore oil film thickness. At the highest speeds, performance is limited by the high rate of shear in the oil film, which generates high temperatures and thus reduces the oil viscosity. The best performance is thus obtained at intermediate speeds and the limit is determined by straightforward seizure and wiping of the bearing surfaces initiated by a very thin, hot oil film.

Summarizing, the tilting-pad thrust bearing has some adverse features; it is relatively complex and heavy and requires more axial space than the simpler forms. It is difficult to ensure that the heights of the pad surfaces are uniform and the internal losses are very high unless special steps are taken to provide directed lubrication and dry sump drainage. Self-aligning properties are rather similar to those of a fixed-pad bearing. In the view of the evidence that the pads do not tilt, at least with conventional design and under normal conditions, it would seem pointless to design as though they did. However, some form of flexible support for the individual pads should be helpful towards achieving reasonably good load sharing. Allowance should be made however, for the development of such a system.

The Michell type bearing is traditionally selected for marine thrust blocks and large steam turbines, but designers may well give serious consideration to the simpler fixed-pad bearing for these and similar applications. Adequately designed and made, the latter is likely to carry larger loads while occupying less space. With oil fed from an inner annular 30 per cent of those in a conventional flooded tilting-pad assembly.

The kind of application to which the tilting pad type is well suited is in a large, vertical hydroelectric machine, where each pad can be separately supported on a jack for adjustment of height. With inserted thermocouples or pressure transducers in the most heavily loaded area of each pad, the readings obtained can be used to monitor the loading on them so that appropriate adjustments can be made. Such conditions permitted one particular machine to be fitted with pads of much reduced area so that the mean specific load was increased from 0.021 MPa to about 0.07 MPa and the performance was reported as satisfactory for several years afterwards.

5.10.2. Design features of hydrostatic thrust bearings

This type of bearing depends for its operation on a balance between the pressure in one or more cavities in the thrust block and the applied load. The fluid escapes through an annular gap between the block and the thrust collar. It is the variation of these gaps with the load which automatically adjusts the hydraulic pressure over the whole area of the bearing. The load limit is set by the maximum pressure the feed pump can maintain at small delivery rates, i.e. at the highest thrust load, when the gaps are reduced towards zero. Obviously, a failure or sudden decrease of fluid supply will allow the peripheral lands of the thrust block to come into contact with the collar, and rapid scoring or seizure is to be expected. Safety, depends therefore on the reliability of the oil supply.

In one important application safety was ensured by duplicating the pump, the feed being through simple non-return valves, so that, if the flow from one pump stopped, that from the other took over. In view of the need for complete reliability, a third pump was also provided, driven by a motor drawing its power from a separate emergency supply.

The internal losses of a well-designed hydrostatic thrust bearing can be very small compared with those of any hydrodynamic bearing with pads. The only area of high shear is in the peripheral gaps and these can be made quite narrow radially and very smooth. The fluid can be an oil of low viscosity so that churning losses are negligible with adequate drainage. The heat to be removed from the central zone of the bearing is therefore significantly reduced and much of the work in supporting the thrust load is transferred outside, to the pumps, where it can much more easily be dealt with and where a breakdown can more readily be attended to.

The type of bearing with a single annular cavity has no self-aligning properties, but by substituting a series of separate cavities, each fed through a suitable restriction, it is possible to design for almost any righting moment. However, the greater this effect, the greater also will be the total losses in the system.

A final caution; the designer should not line one side or the other of the gaps with white metal, in the hope of reducing the bad effects of metal–metal contact at this point. Should this occur the damage to the white metal, being a great deal deeper than between the normal steel and cast-iron, allows the fluid pressure in the cavity to fall, thus still further reducing the safety of the bearing.

5.11. Self-lubricating bearings

Although self-lubricating bearings account for only a small fraction of the world's total bearing usage, they are now firmly established in certain applications. Two classes of self-lubricating material widely used in general engineering applications are plastics and metals containing solid lubricant fillers. Self-lubricating materials are most useful either as dry bearings or as bearings in marginally lubricated applications.

Bearings based on thermosetting resins first appeared on the market in the 1930s. The thermoplastics came later, with nylon first on the scene in bearing applications, followed by PTFE in the early 1950s and more recently by the polyacetals. The economic grounds for preferring plastic are, at best, not very strong, and it is on the grounds of performance that their main claim must be made. The most important property of the bearing plastics, particularly thermoplastics, is their compatibility with metals, and in particular with the steel against which they have to run. There is an air of contradiction about self-lubricating bearings. They must wear to fight wear; they depend on friction to overcome friction. The precise mechanisms of their operation vary and depend on the particular material being used. There are, however, similar elements in the basic mechanisms. The contact between the bearing and its mating surface generates a lubricating film. The film may be composed of solids transferred from the bearing surface; or frictional heat and capillary action may draw a lubricant out of a porous matrix. These two processes may be at work at the same time. Most self-lubricating bearings, especially polymer-based, wear rapidly during the first few hours of operation (Fig. 5.39). The process is especially noticeable in laminated bearings, where the top layer of lubricant may be considerably softer than the underlying layer of polymer and sintered bronze. On the other hand, pre-lubricated bearings show very little initial wear. In any case, the shaft makes its footprint during this initial running-in period. It wears into the bearing material until the contact is spread over a wide area sufficiently large to support the external load on the bearing. At the same time, some of the transferred bearing material may fill the microscopic asperities in the shaft surface, smoothing the shaft and reducing friction. After running-in, the wear-rate drops off sharply and continues at a constant tolerable level for the rest of the bearing's service life. Design and service life prediction procedures are based on experimental data. As the bearing reaches the end of its useful life, the wear-rate accelerates rapidly and it is then time to replace a self-lubricating bearing.

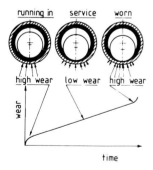

Figure 5.39

5.11.1.   Classification of self-lubricating bearings

The aim of this classification scheme is to help define some of the differences among the self-lubricating and pre-lubricated bearings. In general, four categories of self-lubricating bearings can be distinguished and within each category further subcategories can easily be identified.

A. *Homogeneous metal composites*

These are porous, sintered-metal bearings. The majority of them are iron–bronzes, but occasionally lead–bronze, lead–irons or aluminiums are used. The softer alloys resist galling longest if the lubricant film breaks down, while the aluminiums can carry nearly twice the compressive load.

A-1. Liquid-impregnated bearings are soaked in oil (often ordinary machine oil), which may make up 30 per cent of the finished bearing's volume. In operation, frictional heat and capillary action draw the oil from the pores to lubricate the shaft or thrust collar. High temperatures and high speeds can deplete the oil quickly. They are not recommended for operation in a dusty environment because the oil film can trap grit. Surface finish of the mating surfaces should be $0.2\,\mu$m or better.

A-2. Solid-impregnated bearings are typically sintered from a blend of powdered graphite and iron–bronze. They are particularly useful with shaft finishes of 0.4 to $0.8\,\mu$m.

B. *Metal-backed laminates*

Good lubricating materials usually lack structural strength. These laminates consist of a steel backbone on top of which a thin layer of lubricating material is deposited and hence they can withstand greatly increased permissible contact stresses. The lubricant layers may consist of:

B-1. Porous, sintered metal impregnated with solid lubricants. Solid lubricants like PTFE are notoriously difficult to bond to the substrate. One solution is to avoid glue completely and create a mechanical link, that is, to hot-press the lubricant into a shallow surface layer of porous bronze. This is shown schematically in Fig. 5.40. In service, the plastic's relatively high rate of thermal expansion pushes the lubricant out of the pores to coat the shaft. The sintered bronze layer also helps conduct heat away from the bearing surface. Acetal bearings may also be made in this way. Though these bearings require an initial pre-lubrication with grease, they are often listed as dry-running bearings.

B-2. Bonded fabric. A warp of lubricant fibres (usually PTFE) is laid across a woof of bondable fibres or wire, so that the lubricant threads predominate on one side and the bondable threads dominate on the other.

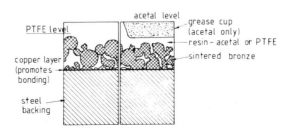

Figure 5.40

This fabric may then be bonded to a steel backing. Such bearings are limited by their adhesives. Though they are relatively insensitive to high pressures, glues may give way to environmental chemicals or high temperatures – such as those generated by high sliding speeds.

B-3. Bonded plastic-based layer. Thermoplastic tapes, thermosetting phenolic or polyamides filled with PTFE are bonded to a steel strip. The plastic layer may be moulded around wires or bondable cloth to facilitate welding or glueing to the backing.

B-4. Unbonded liners. Cylinders of moulded nylon (filled or unfilled), acetal or reinforced PTFE are easily installed and replaced in metal sleeves. They generally cannot take as much load or speed as bonded liners, though setting a reinforcing fabric into the polymer helps to improve the situation.

C.   Homogeneous non-metallic composites

C-1. Unfilled base resins are usually nylons, acetals, polyethylene (especially high density), polyamides and PTFE. Each of these has its own special advantages, though they can only support relatively low loads.

C-2. Single lubricant fillers. These are made of nylons, acetals, polyethylenes, polyimides, PTFE, phenolics and polyphenylene sulfides with lubricant fillers – MoS_2, PTFE or graphite. Additions of silicone or oil are not very popular.

C-3. Single reinforcing fillers, typically fibreglass, in proportion from 10 to 30 per cent, can increase compressive strength, cohesion and temperature resistance.

C-4. Multiple fillers. Various combinations of the materials already mentioned are used, plus bronze powder, metal oxides and sometimes carbon fibres.

C-5. Fabric-and-filler composites are usually compression moulded from phenolic resins filled with PTFE or MoS_2 onto an open-weave reinforcing fabric.

D.   Filament wound

Manufacturers can make sleeves of glass or other fibre, using techniques developed for the fabrication of pressure vessels. The sleeves are then lined in the same way as the metal sleeves.

D-1. Fibre-lined. A strand of bondable material is twisted together with a strand of lubricant polymer. The resulting thread is wound on a mandrel and encapsulated in epoxy.

D-2. Bonded fabric is of the same construction as that described in B-2.

D-3. Bonded tape is described in B-3.

5.11.2.   Design considerations

There are a number of situations where self-lubricating and pre-lubricated bearings of some kind should be considered by a designer. The need to reduce maintenance or to increase reliability is frequently encountered in practice. Inaccessible bearings on all types of equipment are ready candidates for self-lubrication, as are pieces of equipment in remote places. By replacing metal bearings with self-lubricating ones, substantial savings could be made. Similarly, self-lubrication can improve the service life of equipment bound to be neglected, as for instance, consumer appliances.

Seldom used but critical bearings are also prominent candidates; the pivots on an elevator emergency brake might remain motionless for years, but if called upon, the joint must move easily. There is no reason to prevent the designer from using a self-lubricating bearing as a hydrodynamically lubricated one. There is little point in this, however, if the bearing will be properly and continuously lubricated, but if there is a chance that the oil flow could stop, a self-lubricating bearing could prevent serious damage and the need for protracted shut-down and repair.

A significant improvement in bearing performance may often be obtained by conventional liquid lubrication, and like any well-made journal bearing, the oil lubricated self-lubricating bearing should last almost forever. There are, however, a number of subtle interfacial phenomena which are sometimes noted and some of which are deleterious to good operation. A particular type of problem arises when the fluid migrates to a significant depth into the matrix of the polymer and causes a premature failure. This problem of premature failure is especially acute at intense levels of energy dissipation within the contact area.

Another reason for selecting self-lubricating bearings is the necessity to cope with hostile environments. Self-lubricating bearings retain their load-carrying capacity at high temperatures. They can operate where rolling-element bearings fail due to fatigue, and where conventional lubricants oxidize rapidly. Furthermore, many self-lubricating polymers resist corrosion very well.

An important issue related to the operation of machines is the protection of the environment from contamination. Sliding bearings do not make as much noise as rolling-contact bearings, and the plastic liners can act as dampers absorbing some vibration energy. At the same time, many self-lubricating bearings are completely oil free, so that they cannot contaminate their surroundings with a hydrocarbon mist – a point especially important to designers of medical equipment, food processing equipment and business machines. However, it should be pointed out that some self-lubricating materials, like the various lead-filled polymers, may emit contaminants of their own. It is known that fatigue limits the service life of rolling-contact bearings, while wear constitutes the main limitation to the life of self-lubricating bearings. So it is not surprising that dry bearings should perform much better in applications that defeat rolling-element bearings. Oscillating motions of the order of a few degrees, for example, greatly accelerate needle bearing fatigue. The rolling elements do not circulate in and out of the load zone but instead, a single roller or a couple of rollers will rock in and out of the zone always under load. Under these conditions rolling elements undergo accelerated fatigue and fail quickly. Oscillating motions pose even bigger problems for hydrodynamic bearings;

there is seldom enough time for an oil film to form as the shaft starts, stops, reverses itself and starts again. Some metal–metal contact is inevitable. For that matter, any equipment that is stopped and started frequently – even if the rotation is unidirectional – will have problems with metallic contact at low operating speeds. The solution to this problem is offered by a self-lubricating bearing and actually there are a number of situations where self-lubricating bearings are run with additional external lubrication. The self-lubrication is there for start-up, shut-down, and emergencies. Lubrication also increases a bearing's load-carrying capacity. Self-lubricating bearings do not stick on start-up. For instance, a bronze bushing and babbits have a start-up coefficient of friction of around 0.3 while a PTFE based bearing is only 0.05.

Traditionally, the performance of unlubricated bearings is measured in terms of PV, the product of the bearing's unit loading and the relative sliding velocity of the bearing and the mating surface. The dimensions of $PV - (N/m^2) \times (m/s)$ are the dimensions of the energy flux. This is to be expected as the energy lost to friction and subsequently dissipated as heat should be proportional to the frictional force multiplied by the sliding distance. The energy lost per unit time, per unit area of contact, should be the product of the unit load, coefficient of friction and sliding velocity. As such, PV should give a good indication of the heat produced in a bearing. If the bearing dissipates heat at a constant rate, then PV should be the measure of the pressure and sliding velocity that the bearing can tolerate. The only complication is that the coefficient of friction is not constant but changes with speed and the contact pressure. Therefore it is justifiable to take the PV values with some reserve.

The most important problem for the designer intending to utilize a self-lubricating bearing is to estimate its service life. Unfortunately there is no universally accepted, comprehensive design and service life formulae, but instead each manufacturer of self-lubricating bearings has its own and usually different method of projecting wear life. This situation is partly justified by the fact that all calculation methods are based exclusively on experimental results. For general design purposes, ESDU Item No. 76029 – 'A guide on the design and selection of dry rubbing bearings' can be recommended. Also, there are a number of so-called 'designers' handbooks' produced by manufacturers giving detailed information on the selection and wear-life projection of self-lubricating bearings.

References to Chapter 5

1. D. F. Wilcock and E. R. Booser. *Bearing Design and Application*. New York: McGraw-Hill, 1957.
2. P. R. Trumpler. *Design of Film Bearings*. New York: The Macmillan Co., 1966.
3. F. T. Barwell. *Bearing Systems, Principles and Practice*. Oxford: Oxford University Press, 1979.
4. O. Pinkus and B. Sternlicht. *Theory of Hydrodynamic Lubrication*. New York: McGraw-Hill, 1961.
5. D. D. Fuller. *Theory and Practice of Lubrication for Engineers*. New York: Wiley, 1956.

6. G. B. DuBois and F. W. Ocvirk. The short bearing approximation for plain journal bearings. *Trans. ASME*, **77** (1955), 1173–8.

7. W. Gross. *Gas Film Lubrication*. New York: Wiley, 1962.

8. J. Campbell, P. P. Love, F. A. Martin and S. O. Rafique. Bearings for reciprocating machinery. A review of the present state of theoretical, experimental and service knowledge. *Proc. Instn. Mech. Engrs*, **182** (3A) (1967), 14–21.

6 Friction, lubrication and wear in higher kinematic pairs

6.1. Introduction

It is well known in the theory of machines that if the normals to three points of restraint of any plane figure have a common point of intersection, motion is reduced to turning about that point. For a simple turning pair in which the profile is circular, the common point of interaction is fixed relatively to either element, and continuous turning is possible. A pair of elements in which the centre of turning changes its position at the completion of an indefinitely small rotation, i.e. the new position is again the common point of intersection of the normals at three new points of restraint. For this to be possible the profiles will, in general, have differing geometric forms, and are then referred to as a higher pair of elements. Again, since the elements do not cover each other completely as in lower pairing and are assumed to be cylindrical surfaces represented by the profiles, contact will occur along a line or lines instead of over a surface. Relative motion of the elements may now be a combination of both sliding and rolling.

In higher pairing, friction may be a necessary counterpart of the closing force as in the case of two friction wheels in contact. Here the force on the wheels not only holds the cylinders in contact but must be sufficient to prevent relative sliding between the circular elements if closure is to be complete. In certain cases it is essential that force closure of higher pairs shall do more than maintain contact of the functional surfaces. For example, the ball-bearing functions as a lower pair or as an incomplete higher pair of elements, it is, however, usually regarded as being a higher pair.

This chapter is designed to provide familiarization and perspective to readers planning to pursue in more detail any of the various topics covered by the collective name of higher kinematic pairs. There are two pervading objectives:

(i) to develop an understanding of the basic concepts of concentrated contacts;
(ii) to develop a facility with the analytical techniques for predicting and assessing the behaviour of concentrated contacts which are typical for higher kinematic pairs.

The information contained in this chapter can be used to solve a number of problems common for all higher kinematic pairs. First, problems associated with contact between two nonconforming surfaces are discussed. They include the force transmitted at a point of contact, surface tractions, elastic hysteresis during rolling, rolling friction, and the lubrication of rollers. Next, film thickness under isothermal elastohydrodynamic conditions, inlet viscous heating, regimes of line contact lubrication are presented. Finally, contact problems in rolling element bearings, gears, and cam-follower systems are reviewed and equations to evaluate required minimum film thickness are discussed.

6.2. Loads acting on contact area

In this section loads acting on a contact area and the way they are transmitted from one surface to another shall be considered. The load on the contact can be resolved into a normal force P acting along the common normal and a tangential force T opposed by friction. The relationship between W and T is given by

$$T \leqslant fW, \tag{6.1}$$

where f is the coefficient of limiting friction. T can be resolved into components T_x and T_y parallel to axes x and y. In a purely sliding contact the tangential force reaches its limiting value in a direction opposed to the sliding velocity. The force transmitted at a normal point of contact has the effect of compressing solids so that they make contact over an area of finite size. As a result it becomes possible for the contact to transmit a resultant moment in addition to a force. This is schematically shown in Fig. 6.1. The components of this moment M_x and M_y are called rolling moments and oppose a rolling motion but are small enough to be neglected. The third component M_z, acting about the common normal, arises from friction within the contact area and is referred to as the spin moment. When spin accompanies rolling, the energy dissipated by the spin moment is combined with that dissipated by the rolling moments to make up the overall rolling resistance.

Free rolling is defined as a rolling motion in which spin is absent and where the tangential force T at the contact point is zero. This is the condition of the unpowered and unbraked wheels of a vehicle if the rolling resistance and the friction in the bearings are neglected. It is in marked contrast with the driving wheels or the braked wheels which transmit sizeable tangential forces at their points of contact with the road or rail.

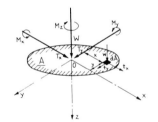

Figure 6.1

6.3. Traction in the contact zone

The forces and moments discussed above are transmitted across the contact interface by surface tractions at the interface. The normal traction (pressure) is denoted here by w and the tangential traction (due to friction) by t, shown acting on the lower surface in Fig. 6.1. For overall equilibrium

$$W = \int_A w \, dA, \tag{6.2}$$

$$T_x = \int_A t_x \, dA, \qquad T_y = \int_A t_y \, dA. \tag{6.3}$$

With contacts formed by the convex surfaces the contact area lies approximately in the x-y plane. Therefore

$$M_x = \int_A wy \, dA, \qquad M_y = - \int_A wx \, dA, \tag{6.4}$$

and

$$M_z = \int_A (t_y x - t_x y) \, dA. \tag{6.5}$$

When the bodies have closely conforming curved surfaces, as for example in a deep-groove ball-bearing, the contact area is warped appreciably out of the tangent plane and the expressions for M_x and M_y, eqn (6.4), have to be modified to include terms involving the shear tractions t_x and t_y.

6.4. Hysteresis losses

Some energy is always dissipated during a cycle of loading and unloading even within the so-called elastic limit. This is because no solid is perfectly elastic. The energy loss is usually expressed as a fraction α of the maximum elastic strain energy stored in the solid during the cycle where α is referred to as the hysteresis loss factor. For most metals, stressed within the elastic limit, the value of α is very small, less than 1 per cent, but for polymers and rubber it may be much larger.

In free rolling, the material of the bodies in contact undergoes a cycle of loading and unloading as it flows through the region of contact deformation (Fig. 6.2). The strain energy of material elements increases up to the centre-plane due to the work of compression done by the contact pressure acting on the front half of the contact area. After the centre-plane the strain energy decreases and work is done against the contact pressures at the back of the contact. Neglecting any interfacial friction the strain energy of the material arriving at the centre-plane in time dt can be found from the work done by the pressure on the leading half of the contact. For a cylindrical contact of unit width

$$dP = \omega \, dt \int_0^a p(x)x \, dx, \tag{6.6}$$

where $\omega = V/R$ is the angular velocity of the roller. Taking $p(x)$ to be given by the Hertz theory

$$\dot{P} = \frac{2}{3\pi} Wa\omega, \tag{6.7}$$

where W is the contact load. If a small fraction α of this strain energy is now assumed to be dissipated by hysteresis, the resultant moment required to maintain the motion is given by equating the net work done to the energy dissipated, then

$$M_y \omega = \alpha \dot{P} = \frac{2}{3\pi} \alpha Wa\omega$$

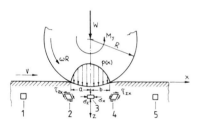

Figure 6.2

or

$$f_r = \frac{M_y}{WR} = \alpha \frac{2a}{3\pi R}, \tag{6.8}$$

where f_r is defined as the coefficient of the rolling resistance. Thus the resistance to rolling of bodies of imperfectly elastic materials can be expressed in terms of their hysteresis loss factor. This simple theory of rolling friction is due to Tabor. Using the same calculation for an elliptical contact area given the result

$$f_r = \frac{M_y}{WR} = \alpha \frac{3}{16} \frac{a}{R}, \tag{6.9}$$

where a is the half-width of the contact ellipse in the direction of rolling. For a sphere rolling on a plane, a is proportional to $(WR)^{\frac{1}{3}}$ so that the effective rolling resistance $F_r = M_y/R$ should be proportional to $W^{\frac{4}{3}}R^{-\frac{2}{3}}$. This relationship is reasonably well supported by experiments with rubber but less well with metals.

There are basically two problems with this simple theory. First, the hysteresis loss factor α is not usually a material constant. In the case of metals it increases with strain (a/R), particularly as the elastic limit of the material is approached. Second, the hysteresis loss factor in rolling cannot be identified with the loss factor in a simple tension or compression cycle. The deformation cycle in the rolling contact, illustrated in Fig. 6.2, involves rotation of the principal axes of strain between points 2, 3 and 4, with very little change in total strain energy. The hysteresis loss in such circumstances cannot be predicted from uniaxial stress data.

The same deformation cycle in the surface would be produced by a rigid sphere rolling on an inelastic deformable plane surface as by a frictionless sphere sliding along the surface. In spite of the absence of interfacial friction the sliding sphere would be opposed by a resistance to motion due to hysteresis in the deformable body. This resistance has been termed the deformation component of friction. Its value is the same as the rolling resistance F_r given by eqn (6.9).

6.5. Rolling friction

Rolling motion is quite common in higher kinematic pairs. Ideally it should not cause much power loss, but in reality energy is dissipated in various ways giving rise to rolling friction. The various sources of energy dissipation in rolling may be classified into:

(i) those which arise through micro-slip and friction at the contact interface;

(ii) those which are due to the inelastic properties of the material;

(iii) those due to the roughness of the rolling surfaces.

Free rolling has been defined as a motion in the absence of a resultant tangential force. Resistance to rolling is then manifested by a couple M_y which is demanded by the asymmetry of the pressure distribution, that is, by higher pressures on the front half of the contact than on the rear. The trailing wheels of a vehicle, however, rotate in bearings assumed to be frictionless and the rolling resistance is overcome by a tangential force T_x applied at the bearing and resisted at the contact interface. Provided that the rolling resistance is small ($T_x \ll W$) these two situations are the same within the usual approximations of small strain contact stress theory, i.e. to first order in (a/R). It is then convenient to write the rolling resistance as a non-dimensional coefficient f_r expressed in terms of the rate of energy dissipation $\dot{P}$, thus

$$f_r = \frac{M_y}{WR} = \frac{T_x}{W} = \frac{\dot{P}}{WV}. \tag{6.10}$$

The quantity $\dot{P}/V$ is the energy dissipated per unit distance travelled.

Energy dissipated due to micro-slip

Energy dissipation due to micro-slip occurs at the interface when the rolling bodies have dissimilar elastic contacts. The resistance from this cause depends upon the difference of the elastic constants expressed by the parameter β (defined by eqn (6.11)) and the coefficient of sliding friction f

$$\beta = \frac{1}{2}\left[\frac{[(1-2v_1)/G_1] - [(1-2v_2)/G_2]}{[(1-v_1)/G_1] + [(1-v_2)/G_2]}\right]. \tag{6.11}$$

The resistance to rolling reaches a maximum value of

$$f_r = \frac{M_y}{WR} \approx 15 \times 10^{-4}\beta\left(\frac{a}{R}\right) \tag{6.12}$$

when $\beta/f \approx 5$. Since, for typical combinations of materials, β rarely exceeds 0.2, the rolling resistance due to micro-slip is extremely small. It has been suggested that micro-slip will also arise if the curvatures of two bodies are different. It is quite easy to see that the difference in strain between two such surfaces will be second-order in (a/R) and hence negligible in any small strain analysis. A special case is when a ball rolls in a closely conforming groove. The maximum rolling resistance is given by

$$f_r = \frac{M_y}{WR} = 0.08f\left(\frac{a}{R}\right)^2\left(\frac{b}{a}\right)^2. \tag{6.13}$$

The shape of the contact ellipse (b/a) is a function of the conformity of the ball and the groove; where the conformity is close, as in a deep groove ball-bearing, $b \gg a$ and the rolling resistance from this cause becomes significant.

In tractive rolling, when large forces and moments are transmitted between the bodies, it is meaningless to express rolling resistance as T_x or M_y/R. Nevertheless, energy is still dissipated in micro-slip and, for comparison with free rolling, it is useful to define the effective rolling resistance coefficient $f_r = \dot{P}/VW$. This gives a measure of the loss of efficiency of a tractive drive such as a belt, a driving wheel or a continuously variable speed gear.

Energy dissipated due to plastic deformations

In the majority of cases, resistance to rolling is dominated by plastic deformation of one or both contacting bodies. In this case the energy is dissipated within the solids, at a depth corresponding to the maximum shear component of the contact stresses, rather than at the interface. With materials having poor thermal conductivity the release of energy beneath the surface can lead to high internal temperatures and failure by thermal stress. Generally metals behave differently than non-metals. The inelastic properties of metals, and to some extent hard crystalline non-metallic solids, are governed by the movement of dislocations which, at normal temperatures, is not significantly influenced either by temperature or by the rate of deformation.

The rolling friction characteristics of a material which has an elastic range of stress, followed by rate-independent plastic flow above a sharply defined yield stress, follow a typical pattern. At low loads the deformation is predominantly elastic and the rolling resistance is given by the elastic hysteresis equation (6.8). The hysteresis loss factor as found by experiment is generally of the order of a few per cent.

At high loads, when the plastic zone is no longer contained, i.e., the condition of full plasticity is reached, the rolling resistance may be estimated by the rigid-plastic theory. The onset of full plasticity cannot be precisely defined but, from the knowledge of the static indentation behaviour, where full plasticity is reached when $W/2a \approx 2.6$ and $Ea/YR \approx 100$, it follows that $GW/kR \approx 300$, where k is the yield stress in shear of the solid.

Energy dissipated due to surface roughness

It is quite obvious that resistance to the rolling of a wheel is greater on a rough surface than on a smooth one, but this aspect of the subject has received little analytical attention. The surface irregularities influence the rolling friction in two ways. First, they intensify the real contact pressure so that some local plastic deformation will occur even if the bulk stress level is within the elastic limit. If the mating surface is hard and smooth the asperities will be deformed plastically on the first traversal but their deformation will become progressively more elastic with repeated traversals. A decreasing rolling resistance with repeated rolling contact has been observed experimentally. The second way in which roughness influences resistance is through the energy expended in climbing up the irregularities. It is significant with hard rough surfaces at light loads. The centre-of-mass of the roller moves up and down in its forward motion which is therefore unsteady. Measurements of the resistance force show very large, high-frequency fluctuations. Energy is dissipated in the rapid succession of small impacts between the surface irregularities. Because the dissipation is by impact, the resistance due to this cause increases with the rolling speed.

6.6. Lubrication of cylinders

It is generally necessary to use a lubricant to ensure satisfactory operation of engineering surfaces in sliding contact. Even surfaces in nominal rolling contact, such as ball-bearings, normally experience some micro-slip, which necessitates lubrication if surface damage and wear are to be avoided. A lubricating fluid acts in two ways. First, it provides a thin adsorbed film to the solid surfaces, preventing the adhesion which would otherwise take place and reducing friction through an interfacial layer of low shear strength. This is the action known as boundary lubrication. The film is generally very thin and its behaviour is very dependent upon the physical and chemical properties of both the lubricant and the solid surfaces. The lubricant may act in a quite different way. A relatively thick coherent film is drawn in between the surfaces and sufficient pressure is developed in the film to support the normal load without solid contact. This action is known as hydrodynamic lubrication. It depends only upon the geometry of the contact and the viscous flow properties of the fluid. The way in which a load-carrying film is generated between two cylinders in rolling and sliding contact is described in this section. The theory can be applied to the lubrication of gear teeth, for example, which experience a relative motion which, as shown in Section 6.2, is instantaneously equivalent to the combined rolling and sliding contact of two cylinders.

A thin film of an incompressible lubricating fluid, viscosity μ, between two solid surfaces moving with velocities V_1 and V_2 is shown in Fig. 6.3. With thin, nearly parallel films, velocity components perpendicular to the film are negligible so that the pressure is uniform across the thickness. At a low Reynolds number, for the case of a thin film and a viscous fluid, the inertia forces are negligible. Then, for two-dimensional steady flow, equilibrium of the fluid element gives

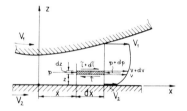

Figure 6.3

$$\frac{\partial p}{\partial x} = \frac{\partial \tau}{\partial z} = \frac{\partial}{\partial z}\left(\mu\frac{\partial v}{\partial z}\right) = \mu\frac{\partial^2 v}{\partial z^2}, \tag{6.14}$$

where v is the stream velocity. Since $\partial p/\partial x$ is independent of z, eqn (6.14) can be integrated with respect to z. Putting $v = V_2$ and V_1 at $z = 0$ and h, gives a parabolic velocity profile, as shown in Fig. 6.3, expressed by

$$v(z) = \frac{1}{2\mu}\frac{\mathrm{d}p}{\mathrm{d}x}(z^2 - hz) + (V_1 - V_2)\left(\frac{z}{h}\right) + V_2. \tag{6.15}$$

The volume flow rate Q across any section of the film is

$$Q = \int_0^h v(z)\,\mathrm{d}z = -\frac{h^3}{12\mu}\left(\frac{\mathrm{d}p}{\mathrm{d}x}\right) + (V_1 + V_2)\frac{h}{2}. \tag{6.16}$$

For continuity of flow, Q is the same for all cross-sections, i.e.

$$Q = (V_1 + V_2)\frac{h_1}{2}, \tag{6.17}$$

where h_1 is the film thickness at which the pressure gradient $\mathrm{d}p/\mathrm{d}x$ is zero.

Eliminating Q gives

$$dp/dx = 6\mu(V_1 + V_2)\left(\frac{h - h_1}{h^3}\right). \qquad (6.18)$$

This is Reynolds equation for a steady two-dimensional flow in a thin lubricating film. Given the variation in thickness of the film $h(x)$, it can be integrated to give pressure $p(x)$ developed by hydrodynamic action. For a more complete discussion of the Reynolds equation the reader is referred to the books on lubrication listed at the end of Chapter 5.

Now, eqn (6.18) will be used to find the pressure developed in a film between two rotating cylinders.

Case (i) – Rigid cylinders

The geometry of two rotating rigid cylinders in contact is schematically shown in Fig. 6.4. An ample supply of lubricant is provided on the entry side. Within the region of interest the thickness of the film can be expressed by

$$h \approx h_0 + x^2/2R, \qquad (6.19)$$

where $1/R = 1/R_1 + 1/R_2$ and h is the thickness at $x = 0$. Substituting eqn (6.19) into (6.18) gives

$$dp/dx = \frac{6\mu(V_1 + V_2)}{h_0^2}\left\{\frac{1 - \left(\frac{h_1}{h_0}\right) + \left(\frac{x^2}{2Rh_0}\right)}{\left(1 + \frac{x^2}{2Rh_0}\right)^3}\right\}. \qquad (6.20)$$

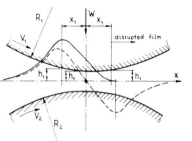

Figure 6.4

By making the substitution $\zeta = \tan^{-1}[x/(2Rh)^{\frac{1}{2}}]$ eqn (6.20) can be integrated to give an expression for the pressure distribution

$$\frac{h_0^2}{(2Rh_0)^{\frac{1}{2}}}\frac{p}{6\mu(V_1 + V_2)} = \frac{\zeta}{2} + \frac{\sin 2\zeta}{4}$$

$$- \sec^2\zeta_1\left[\frac{3\zeta}{8} + \frac{\sin 2\zeta}{4} - \frac{\sin 4\zeta}{32}\right] + A, \qquad (6.21)$$

where $\zeta_1 = \tan^{-1}[x_1/(2Rh_0)^{\frac{1}{2}}]$ and x_1 is the value of x where $h = h_1$ and $dp/dx = 0$. The values of ζ_1 and A are found from the end conditions.

At the start it is assumed that the pressure is zero at distant points at entry and exit, i.e. $p = 0$ at $x = \pm\infty$. The resulting pressure distribution is shown by the dotted line in Fig. 6.4. It is positive in the converging zone at entry and equally negative in the diverging zone at exit. The total force W supported by the film is clearly zero in this case. However this solution is unrealistic since a region of large negative pressure cannot exist in normal ambient conditions. In practice the flow at the exit breaks down into streamers separated by fingers of air penetrating from the rear. The pressure is approximately ambient in this region. The precise point of film breakdown is determined by consideration of the three-dimensional flow in the streamers and is influenced by surface tension forces. However it has been found that it can be located with reasonable accuracy by imposing the condition

$$\mathrm{d}p/\mathrm{d}x = p = 0 \tag{6.22}$$

at that point. When this condition, together with $p=0$ at $x=-\infty$ is imposed on eqn (6.21) it is found that $\xi_1 = 0.443$, whence $x_1 = 0.475(2Rh_0)^{\frac{1}{2}}$. The pressure distribution is shown by the solid line curve in Fig. 6.4. In this case the total load supported by the film is given by

$$W = \int_{-\infty}^{x_1} p(x)\,\mathrm{d}x = 2.45(V_1 + V_2)R\mu/h_0. \tag{6.23}$$

In most practical situations it is the load which is specified. Then, eqn (6.23) can be used to calculate the minimum film thickness h_0. To secure effective lubrication, h_0 must be greater than the surface irregularities. It is seen from eqn (6.23) that the load carrying capacity of the film is generated by a rolling action expressed by $(V_1 + V_2)$. If the cylinders rotate at the same peripheral velocity in opposite directions, then $(V_1 + V_2)$ is zero, and no pressure is developed in the film.

Case (ii) – Elastic cylinders

Under all engineering loads the cylinders deform elastically in the pressure zone so that the expression for the film profile becomes

$$h(x) = h_0 + \frac{x^2}{2R} + [u_{z1}(x) - u_{z1}(0)] + [u_{z2}(x) - u_{z2}(0)],$$

where u_{z1} and u_{z2} are the normal elastic displacements of the two surfaces and are given by the Hertz theory. Thus

$$h(x) = h_0 + (x^2/2R) - \frac{2}{\pi E}\int_{-\infty}^{\infty} p(s) \ln \left| \frac{x-s}{s} \right| \mathrm{d}s. \tag{6.24}$$

This equation and the Reynolds eqn (6.18) constitute a pair of simultaneous equations for the film shape $h(x)$ and the pressure $p(x)$. They can be combined into a single integral equation for $h(x)$ which can be solved numerically. The film shape obtained in that way is then substituted into the Reynolds equation to find the pressure distribution $p(x)$.

An important parameter from the point of view of the designer is the minimum film thickness $h_{\min}$. In all cases $h_{\min} \approx 0.8h_1$. The lubrication process in which elastic deformation of the solid surface plays a significant role is known as elastohydrodynamic lubrication.

Case (iii) – Variable viscosity of the lubricant

It is well known that the viscosity of most practical lubricants is very sensitive to changes in pressure and temperature. In contacts characteristic of higher kinematic pairs, the pressures tend to be high so that it is not surprising that an increase in the viscosity with pressure is also a significant factor in elastohydrodynamic lubrication. When sliding is a prevailing motion in the contact, frictional heating causes a rise in the temperature in the film which reduces the viscosity of the film. However, for reasons which will be explained later, it is possible to separate the effects of pressure and temperature.

Let us consider an isothermal film in which variation in the viscosity with pressure is given by the equation

$$\mu = \mu_0 e^{\alpha p}, \tag{6.25}$$

where μ_0 is the viscosity at ambient pressure and temperature and α is a constant pressure coefficient of viscosity. This is a reasonable description of the observed variation in the viscosity of most lubricants. Substituting this relationship into the Reynolds eqn (6.18) gives

$$e^{-\alpha p}\frac{dp}{dx} = 6\mu_0(V_1 + V_2)\left[\frac{h - h_1}{h^3}\right]. \tag{6.26}$$

This modified Reynolds equation for the hydrodynamic pressure in the field must be solved simultaneously with eqn (6.24) for the effect of elastic deformation on the film shape. The solution to this problem can be obtained numerically. There are a number of changes in the contact behaviour introduced by the pressure–viscosity effect. Over an appreciable fraction of the contact area the film is approximately parallel. This results from eqn (6.26). When the exponent αp exceeds unity, the left-hand side becomes small, hence $h - h_1$ becomes small, i.e. $h \approx h_1 = $ constant. The corresponding pressure distribution is basically that of Hertz for dry contact, but a sharp pressure peak occurs on the exit side, followed by a rapid drop in pressure and thinning of the film where the viscosity falls back to its ambient value μ_0. The characteristic features of highly loaded elastohydrodynamic contacts, that is a roughly parallel film with a constriction at the exit and a pressure distribution which approximates to Hertz but has a sharp peak near the exit, are now well established and supported by experiments. It is sufficiently accurate to assume that the minimum film thickness is about 75 per cent of the thickness in the parallel section. The important practical problem is to decide under what conditions it is permissible to neglect elastic deformation and/or variable viscosity. Some guidance in this matter can be obtained by examining the values of the two non-dimensional parameters, the viscosity parameter g_v and the elasticity parameter g_e which are presented and discussed in Chapter 2, Section 2.12.1. The mechanism of elastohydrodynamic lubrication with a pressure dependent lubricant is now clear. The pressure develops by hydrodynamic action in the entry region with a simultaneous very large increase in the viscosity. The film thickness at the end of the converging zone is limited by the necessity of maintaining a finite pressure. This requirement virtually determines the film thickness in terms of the speed, roller radii and the viscous properties of the lubricant. Increasing the load increases the elastic deformation of the rollers with only a minor influence on the film thickness. The highly viscous fluid passes through the parallel zone until the pressure and the viscosity get back to normal at the exit. This means a decrease in thickness of the film. The inlet and exit regions are effectively independent. They meet at the end of the parallel zone with a discontinuity in the slope of the surface which is associated with a sharp peak in the pressure.

6.7. Analysis of line contact lubrication

In this section line contact lubrication is presented in a way which can be directly utilized by the designer. The geometry of a typical line contact is shown in Fig. 6.5. The minimum film thickness occurs at the exit of the region and can be predicted by the formula proposed by Dowson and Higginson for isothermal conditions

$$\frac{h_{min}}{R} = 2.65 \frac{G^{0.54} V^{0.7}}{W^{0.13}},\tag{6.27}$$

where $G = \alpha E$ is the dimensionless material parameter, $V = [\mu_0 (V_1 + V_2)]/2ER$ is the dimensionless speed parameter, $W = w/ERL$ is the dimensionless load parameter, α is the pressure-viscosity coefficient based

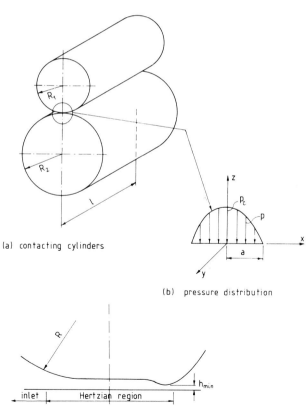

(a) contacting cylinders

(b) pressure distribution

(c) film thickness distribution

Figure 6.5

on the piezo-viscous relation $\mu = \mu_0 e^{\alpha p}$ and reflects the change of viscosity with pressure

$$1/E = \tfrac{1}{2}\left[\frac{1-v_1^2}{E_1} + \frac{1-v_2^2}{E_2}\right],$$

μ_0 is the lubricant viscosity at inlet surface temperature, V_1, V_2 are the surface velocities relative to contact region

$$R = \frac{R_1 R_2}{R_1 \pm R_2}$$

where the plus sign assumes external contact (both surfaces convex) and the minus sign denotes internal contact (the surface with the larger radius of curvature is concave), w is the total load on the contact and L is the length of the contact.

The viscosity of the lubricant at the temperature of the surface of the solid in the contact inlet region is the effective viscosity for determining the film thickness. This temperature may be considerably higher than the lubricant supply temperature and therefore the inlet viscosity may be substantially lower than anticipated, when based on the supply temperature. Usually the inlet surface temperature is an unknown quantity in design analyses. The solution to this problem is to use the lubricant system outlet temperature or an average of the inlet and the outlet temperatures to obtain an estimate of the film thickness. It should be pointed out that the predicted film thickness may be too large when the system supply temperature is used and the flow of lubricant is not sufficient to keep the parts close to the lubricant inlet temperature. Input data characterizing the lubricant, that is its viscosity, pressure-viscosity coefficient and temperature-viscosity coefficient are usually available from catalogues of lubricant manufacturers.

The best way to illustrate the practical application of eqn (6.27) is to solve a numerical problem. Two steel rollers of equal radius $R_1 = R_2 = 100$ mm and length $L = 100$ mm form an external contact. Using the following input data estimate the thickness of the lubricating film

$$R_1 = 100 \text{ mm},$$
$$R_2 = 100 \text{ mm},$$
$$w = 30.0 \text{ kN},$$
$$E_1 = E_2 = 207 \text{ GPa},$$
$$v_1 = v_2 = 0.33,$$
$$\alpha = 14.5 \,(\text{GPa})^{-1},$$
$$V_1 = V_2 = 15.0 \text{ m s}^{-1},$$
$$\mu_0 = 27.6 \text{ mPa s}.$$

Equivalent radius of the contact

$$R = \frac{R_1 R_2}{R_1 + R_2} = \frac{(100)(100)}{100 + 100} = 50 \text{ mm}.$$

Equivalent Young modulus

$$E = 2\left[\frac{1 - v_1^2}{E_1} - \frac{1 - v_2^2}{E_2}\right]^{-1} = 233\,\text{GPa}.$$

Material parameter

$$G = \alpha E = 3379.$$

Speed parameter

$$V = \frac{\mu_0(V_1 + V_2)}{2ER} = \frac{27.6 \times 10^{-3}(15 + 15)}{2 \times 233 \times 10^9 \times 50 \times 10^{-3}} = 3.55 \times 10^{-11}.$$

Load parameter

$$W = \frac{w}{ERL} = \frac{30 \times 10^3}{233 \times 10^9 \times 50 \times 10^{-3} \times 0.1} = 2.575 \times 10^{-5}.$$

Therefore

$$h_{\text{min}}/R = 2.65\frac{(3379)^{0.54}(3.55 \times 10^{-11})^{0.7}}{(2.575 \times 10^{-5})^{0.13}} = 4.579 \times 10^{-5}$$

$$\therefore h_{\text{min}} = (4.579 \times 10^{-5})(50 \times 10^{-3}) = 228.95 \times 10^{-8}\,\text{m}.$$

6.8. Heating at the inlet to the contact

The assumption in the analysis presented in the previous section, was that the lubricant properties are those at the inlet zone temperature and the system is isothermal. The inlet zone lubricant temperature can be, and frequently is, higher than the bulk lubricant temperature in the system. There are basically two mechanisms responsible for the increase in the lubricant temperature at the inlet to the contact. The first is viscous heating of the lubricant in the inlet zone and the second is the conduction of thermal energy accumulated in the bulk of the contacting solids to the inlet zone. This second mechanism is probably only important in pure sliding where the conduction can occur through the stationary solid. The heating at the inlet zone is significant only at high surface velocities and can be subjected to certain simplified analysis. Under conditions of high surface velocity or high lubricant viscosity the effect of inlet heating due to shear on effective viscosity ought to be considered. The engineering approach to this problem is to use a thermal reduction factor, T_{f}, which can be multiplied by the isothermal film thickness, h_0, to give a better estimate of the actual film thickness.

The thermal correction factor is a weak function of load and material parameters. As a first approximation, the following expression may be used to determine the thermal correction factor for line contacts

$$T_{\text{f}} = 0.857 - 0.0234T_l + 0.000168(T_l)^2. \tag{6.28}$$

In eqn (6.28) the thermal loading factor, T_l, is defined as

$$T_l = \frac{\mu_0 V^2 \delta}{\rho}, \tag{6.29}$$

where μ_0 is the lubricant viscosity at the atmospheric pressure and the inlet surface temperature, $V = (V_1 + V_2)/2$ is the average surface velocity, [m/s], δ is the lubricant viscosity-temperature coefficient, $[=(1/\mu)/\Delta\mu/\Delta T)$; $°C^{-1}]$ and ρ is the lubricant thermal conductivity.

The viscosity-temperature coefficient, δ, can be adequately estimated from a temperature viscosity chart of the lubricant. It is necessary to select two temperatures about $20°$ apart near the assumed inlet temperature and divide the viscosity difference by the average viscosity and temperature difference corresponding to the viscosity difference.

Thermal conductivity, ρ, is relatively constant for classes of lubricants based on chemical composition. For mineral oils, suitable values for these calculations are 0.12–0.15 W/m K. The lower range applies for lower viscosity either resulting from lower molecular weight or higher temperature. To illustrate the procedure outlined above the numerical example solved earlier is used. Thus, assuming that the lubricant is SAE10 mineral oil, inlet temperature is $55°C$, corresponding δ value is $0.045°C^{-1}$ and lubricant thermal conductivity is 0.12 W/m K, eqn (6.29) gives

$$T_l = \frac{(27.6 \times 10^{-3})(15)^2 \times 0.045}{0.12} = 2.32.$$

Then, using eqn (6.28)

$$T_f = 0.857 - 0.0234(2.32) + 0.000168(2.32)^2 = 0.8.$$

Finally, the thermally corrected film thickness is

$$h = T_f h_0 = (0.8)(2.28) = 1.824 \, \mu m.$$

This means a 20 per cent reduction of the film thickness as a result of the heating at the inlet zone.

6.9. Analysis of point contact lubrication

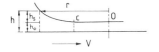

(a) geometry of entry

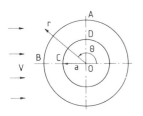

(b) geometry of contact

Figure 6.6

In contrast with the heavily loaded line contacts which have been investigated very fully, the understanding of point contact lubrication is less advanced. Any analysis of the problem naturally relies to a considerable extent on a knowledge of the local shape of the contact, which usually is not known in detail. The foundation of the theoretical solution to the problem was laid down by Grubin. He proposed that:

(i) the pressure distribution under lubricated conditions was almost Hertzian;

(ii) the shape of the entry gap was determined by the Hertz pressure alone, i.e. the fluid pressure at the entry to the contact zone had negligible effect.

The application of the Grubin approximation is quite simple in the case of line contacts. The point contact case is far more complex due to side leakage effects. Figure 6.6 shows the geometry of the point contact. The various formulae for load, peak pressure, contact radius and surface deformation can be easily found in any standard textbook on elasticity (see Chapter 3 for more details). The relationships needed here are as follows:

(i) the applied load W producing a circular contact of radius a:

$$W = \tfrac{2}{3}\pi a^2 p_{max},\tag{6.30}$$

where p_{max} is the peak pressure at the centre;

(ii) the contact radius a is found from the relationship

$$a^3 = \tfrac{3}{4}\pi \frac{RW}{E},\tag{6.31}$$

where R is the equivalent radius defined by $1/R = 1/R_1 + 1/R_2$ and E is the equivalent materials modulus

$$1/E = \frac{1}{\pi}\left[\frac{1 - v_1^2}{E_1} + \frac{1 - v_2^2}{E_2}\right],$$

where v denotes the Poisson ratio and the subscripts 1 and 2 refer to the two materials in contact;

(iii) the gap outside the main contact region at any radial distance r is (see Fig. 6.6)

$$h = h_0 + \frac{3W}{4aE}\left[\left(\left(\frac{r^2}{a^2}\right) - 1\right)^{\frac{1}{2}} - \left(2 - \frac{r^2}{a^2}\right)\cos^{-1}\frac{a}{r}\right].\tag{6.32}$$

The expression in square brackets can be approximated over the region $a \leqslant r \leqslant 2a$ by

$$3.81(r/a - 1)^{1.5}.$$

By denoting that $\varepsilon = (11.43W)/4aEh_0$, eqn (6.32) becomes

$$h = h_0\left\{1 + \varepsilon\left[\frac{r}{a} - 1\right]^{1.5}\right\}.\tag{6.33}$$

By writing a non-dimensional version of the Reynolds equation in polar coordinates, and considering the quadrant of the annulus $ADCBA$, only the final differential equation can be derived. Obviously, the equation does not have an exact analytical solution and it is usually solved numerically. This has been done by many workers and the solution in a simplified form is as follows

$$\frac{h_0}{R} = \frac{3\alpha\mu_0(V/R)}{(W/ER^2)^{\frac{1}{3}}},\tag{6.34}$$

where μ_0 is the oil viscosity at atmospheric pressure, α is the pressure-viscosity coefficient and V is the surface velocity. Equation (6.34) has been verified experimentally and it is now clear that despite an unfavourable geometric configuration, an elastohydrodynamic lubrication film exists at the nominal point contacts over a very wide range of conditions.

6.10. Cam-follower system

The schematic geometry of a cam-flat-follower nose is shown in Fig. 6.7. The film parameter λ for a cam-follower system can be calculated by the

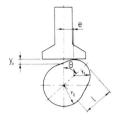

Figure 6.7

following equation

$$\lambda = \frac{1}{\sigma} 4.35 \times 10^{-3} [bSn]^{0.74} R^{0.26}, \tag{6.35}$$

where n is the cam shaft speed [r.p.m.], $S = (\mu_0 \alpha 10^{11})$ lubricant parameter and $b = |2r_1 - l|$, where l is the distance from the nose tip to the shaft axis, and r_1 is the nose radius.

$$R = \left[\frac{1}{r_1} + \frac{1}{r} \right],$$

where r is the follower radius and $\sigma = \sqrt{\sigma_1^2 + \sigma_2^2}$ is the composite roughness; where $\sigma_{1,2}$ is the root mean square (r.m.s.) surface roughness of surfaces 1 and 2 in μm. If the arithmetic average R_a (also known as c.l.a.) is available, multiply by 1.3 to convert to r.m.s.

In general, the value of λ in cam systems is well below one. In this regime, elastohydrodynamic lubrication is not very effective and we must rely heavily on surface film or boundary lubrication to protect the surfaces against scuffing and excessive wear.

References to Chapter 6

1. F. F. Ling. *Surface Mechanics*. New York: Wiley, 1973.
2. A. W. Crook. The lubrication of rollers, I. *Phil. Trans.*, **A250** (1958), 18–34.
3. A. W. Crook. The lubrication of rollers, II and III. *Phil. Trans.*, **A254** (1961), 141–50.
4. A. W. Crook. The lubrication of rollers, IV. *Phil. Trans.*, **A255** (1963), 261–9.
5. B. Jacobson. On the lubrication of heavily loaded cylindrical surfaces considering surface deformations and solidification of the lubricant. *Trans. ASME, J. Lub. Technol.*, **95** (1973), 176–84.
6. D. Dowson and G. R. Higginson. *Elastohydrodynamic Lubrication*. New York: Pergamon Press, 1966.

7 Rolling-contact bearings

7.1. Introduction

In contrast with hydrodynamically lubricated journal bearings, which, for their low friction characteristics depend on a fluid film between the journal and the bearing surfaces, rolling-contact bearings employ a number of balls and rollers that roll, nominally, in an annular space. To some extent, these rolling elements help to avoid gross sliding and the high coefficient of friction that is associated with sliding. The mechanism of rolling friction is, therefore, discussed first followed by the review of the factors affecting the frictional losses.

As a matter of fact, the contact between the rolling elements and the races or rings consists more of sliding than of actual rolling. The condition of no interfacial slip is seldom maintained because of material elasticity and geometric factors. It is natural then that the contact stress and the kinematics of the rolling-element bearing are presented in some detail in order to stress their importance in the service life of this type of bearing. The advantages and disadvantages of rolling-contact bearings when they are compared with hydrodynamic bearings are well known and shall not be discussed here. Instead, more attention is given to the lubrication techniques and the function of the lubricant in bearing operation. Finally, vibration and acoustic emission in rolling-element bearings are discussed as they are inherently associated with the running of the bearing. Some methods to combat the excessive noise emission are also suggested.

7.2. Analysis of friction in rolling-contact bearings

The resistance to relative motion in rolling-contact bearings is due to many factors, the basic one being rolling friction. This was long assumed to be the only resistance to motion in this type of bearing. It was established, however, that the contribution of rolling friction is small though its effect on wear and tear and operating temperature is important. These factors are especially important for miniature instrument rolling-contact bearings operating in the very accurate mechanisms of servo-systems, magnetic recorder mechanisms and other precision parts of instruments.

Previously, rolling friction was treated as a mechanical process, i.e. the interaction of rough surfaces of absolutely rigid bodies. In 1876 Reynolds put forward the hypothesis according to which the frictional force due to the rolling motion of a perfectly elastic body along a perfectly elastic substrate was a result of relative slip between the contacting surfaces resulting directly from their deformations. Figure 7.1 provides a graphic

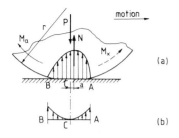

Figure 7.1

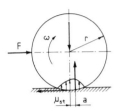

Figure 7.2

explanation of the Reynolds hypothesis. In the region AC surface layers of the rolling body are compressed in a direction along the area of contact and elongated in the plane of the figure. In the region CB of the contact area these deformations take place in the opposite direction, as a result of which micro-slip occurs. Later on, this hypothesis was supplemented by experimental findings which showed that slip in the contact zone is not the only source of frictional losses during rolling. From a practical point of view, the hypothesis that rolling friction results from the imperfect elastic properties of engineering materials, was a significant step forward. Figure 7.2 illustrates the rationale behind this hypothesis. When a perfectly hard roller, rolls along a yielding surface the load distribution on the roller is unsymmetric and produces a force resisting the motion. Modern approach to the rolling friction recognizes the fact that many factors contribute to the total friction torque in rolling-contact bearings. Friction torque can be expressed as follows

$$M = (M_{ds} + M_{gr} + M_{hs} + M_{de} + M_c + M_e + M_m + M_T)K, \qquad (7.1)$$

where M_{ds} is the friction torque due to the differential slip of the rolling element on the contact surface, M_{gr} is the friction torque arising from gyroscopic spin or deviation of the axis of rotation of the rolling elements, M_{hs} is the friction torque due to losses on elastic hysteresis in the material of the bodies in contact, M_{de} is the friction torque resulting from the deviation of the bearing elements from the correct geometric shape and the micro-roughness of contacting surfaces, M_c is the friction torque due to sliding taking place along the guide edges of the raceway and the torque arising from the contact of the rollers with the raceway housing, M_e is the friction torque due to the shearing of a lubricant, M_m is the friction torque resulting from the working medium of the bearing (gas, liquid, air, vacuum), M_T represents a complex increase in friction torque due to an increase in temperature and K is a correction factor taking into account complex changes in the friction torque due to the action of forces not taken into account when computing individual components, for example, the action of axial and radial forces, vibrational effects, etc.

7.2.1. Friction torque due to differential sliding

Let us consider the friction torque due to differential sliding, M_{ds}, for the case where the ball rolls along a groove with a radius of curvature R in a plane perpendicular to the direction of rolling. Pure rolling will occur along two lines (see Fig. 7.3), located on an ellipse of contact, at a distance $2a_c$ apart. In other parts of the ellipse there will be sliding because of the unequal distance of contact points from the axis of rotation. Friction torque due to differential sliding can be expressed in terms of work done, A, by the bearing in a unit time as a result of differential sliding

$$A = (F_i l_i + F_o l_o)z, \qquad (7.2)$$

where F_i, F_o are the frictional forces resulting from the differential sliding

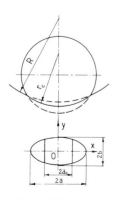

Figure 7.3

during the rolling of the ball along the surface of the inner and outer races respectively, l_i, l_o are the distances moved in unit time by the point of contact of the ball with the inner and outer races respectively and z is the number of balls. The distances moved in unit time are given by the formulae

$$l_i = 2\pi R_{l_i} n_b = \frac{\pi}{2} d_b \left(1 - \frac{d_b^2}{D_0^2} \cos^2 \alpha \right) n, \tag{7.3}$$

$$l_o = 2\pi R_{l_o}(n - n_b) = \frac{\pi}{2} d_b \left(1 - \frac{d_b^2}{d_o^2} \cos^2 \alpha \right) n, \tag{7.4}$$

where n is the rotational velocity of the inner race, n_b is the rotational velocity of the ball, d_b is the diameter of the ball, D_o is the diameter of the pitch circle, R_{l_i}, R_{l_o} are the radii of curvature of the roller groove of the inner and outer races respectively and α is the contact angle.

Equation (7.2) is valid for the case where the inner race rotates and the outer one is stationary. The friction torque due to differential sliding is given by

$$M_{ds} = \frac{A}{2\pi n} \tag{7.5}$$

or after expansion

$$M_{ds} = \frac{P D_0 z}{4 d_b} \left(1 - \frac{d_b^2}{D_0^2} \cos^2 \alpha \right) (r_o B_o - r_i B_i) f, \tag{7.6}$$

where f is the coefficient of sliding friction, B_i, B_o are the factors determined by the geometric conditions of deformations and r_i, r_0 are the respective radii of deformed raceway surfaces of the inner and outer rings.

7.2.2. Friction torque due to gyroscopic spin

Friction torque arising from gyroscopic spin appears in the presence of the contact angle α. To find M_{gr} it is necessary to determine the moment of inertia of the ball, I, obtained from the expression

$$I = \frac{\pi}{60} d_b^2 \frac{\gamma}{g}, \tag{7.7}$$

where $\gamma = 7.8$ gm cm^{-3} is the density of the steel ball. Friction torque due to gyroscopic spin equals

$$M_{gr} = I \omega_b z \sin \alpha, \tag{7.8}$$

where $\omega_b = \dfrac{\pi n_b}{30}$ is the angular velocity of the ball and

$$n_b = n_{sh} \frac{D_o^2 - d_b^2 \cos^2 \alpha}{2 D_0 d_b}, \tag{7.9}$$

where n_{sh} denotes the rotational speed of the shaft supported by the bearing.

Torque, M_{gr}, increases with an increase in contact angle α and attains a maximum value when $\alpha = \pi/2$. Thrust bearings, for example, have such an angle. Radial bearings are characterized by $\alpha = 0$, so that $M_{gr} = 0$. In order to avoid gyroscopic spin of the balls it is necessary to satisfy the inequality

$$M_{gr} < Pd_b fz, \tag{7.10}$$

where P is the axial load. The effect of gyroscopic spin is especially noticeable in radial thrust bearings.

7.2.3. Friction torque due to elastic hysteresis

The energy losses due to elastic hysteresis in the material of the bearing can be determined, assuming that during the loading cycle a specific quantity of kinetic energy is expended. Friction torque caused by elastic hysteresis has been sufficiently studied for the case of slow-running bearings. In the case of high-speed running, however, large centrifugal forces arise and they create an additional load. Hence, in all the expressions given below, friction torque M must be treated as a function of rotational speed.

When the ball rolls along the raceway, the Hertz contact ellipse is formed between them as shown in Fig. 7.4. The energy developed during the rolling of the body about one point of the raceway is proportional to the load and to the magnitude of deformation. Contact time, t, of the rolling body with the raceway is directly proportional to the extent of the contact surface in the direction of motion $2b$ and inversely proportional to the circumferential velocity, $r\omega$ (see Fig. 7.5), thus

$$t \approx \frac{2b}{r\omega}. \tag{7.11}$$

The energy losses during the contact of the rolling body with one point of the raceway are

$$M_{hs}\omega t = P\delta, \tag{7.12}$$

where P is the load and δ denotes the associated deformation.

According to the Hertz theory, the deformation for point contact is given by

$$\delta = K(P^3 \Sigma \rho)^{\frac{1}{3}}, \tag{7.13}$$

where $\Sigma \rho$ is the sum of the reciprocal of the principal radii of curvature, i.e.

$$(\rho_1 + \rho_2)_{\text{I}}/(\rho_1 + \rho_2)_{\text{II}}$$

here ρ_1 and ρ_2 are the reciprocals of the principal radii of curvature of the bodies at the point of contact and K is a coefficient depending on the function $F(\rho)$ defined as

$$F(\rho) = \left| \frac{(\rho_1 - \rho_2)_{\text{I}} + (\rho_1 - \rho_2)_{\text{II}}}{\Sigma \rho} \right|$$

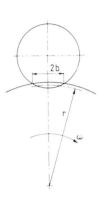

Figure 7.4

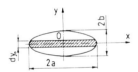

Figure 7.5

and is usually determined from tables contained in the textbooks on elasticity. The expression for the loss caused by elastic hysteresis can be written as

$$M_{hs} = 1.25 \times 10^{-4} D_0 d_b^{-\frac{2}{3}} \Sigma p_i^{\frac{4}{3}}, \tag{7.14}$$

where p_i is the load on the ith ball. Hysteresis losses are practically independent of the shape of the raceway.

7.2.4. Friction torque due to geometric errors

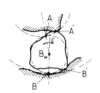

Figure 7.6

The main cause of changes in the friction torque are errors in the geometric shape of the bearing elements. The process leading to the formation of the friction torque is shown in Fig. 7.6. The line of action is formed at every given instant. In an ideal bearing this line passes through the centre of the loaded ball. In a real bearing the line of action does not coincide with the direction of the theoretical line of action.

In a real radial bearing, centre O of the raceway circle of the moving inner ring does not coincide with the theoretical centre O (see Fig. 7.7). Total errors Δz_1 and Δz_2 arise due to geometric imperfections. While determining the centre O_1 it is assumed that the radial bearing is loaded purely by radial load R applied at the point O_1. Here the radial load is taken by two balls which roll without slip. Then

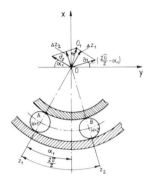

$$M_{de} = \left(R / \sin \frac{\alpha \pi}{z} \right) \left[\frac{dz_1}{dk} \sin \left(\frac{2\pi}{z} - \alpha_1 \right) \right.$$

$$\left. - \Delta z_1 \frac{d_i}{2D_0} \cos \left(\frac{2\pi}{z} - \alpha_1 \right) + \frac{dz_2}{dk} + \Delta z_2 \frac{d_i}{2D_0} \cos \alpha_1 \right], \tag{7.15}$$

Figure 7.7

where d_i is the diameter of the inner ring raceway and k is the turning angle of the moving ring of the bearing, relative to the stationary one.

7.2.5. Friction torque due to the effect of the raceway

If it is assumed that the inner ring of the bearing rotates about a vertical axis and the raceway under the action of its own weight touches the ball only at one point, then the friction torque due to the contact of the rolling elements with the raceway housing is expressed in the form

$$M_c = \frac{D_0}{4} \left(1 - \frac{d_b^2}{D_0^2} \cos^2 \alpha \right) \sin \left[\alpha + \tan^{-1} \left(\frac{d_b \sin \alpha}{2R_r} \right) \right] G_c f_c, \tag{7.16}$$

where G_c is the raceway housing mass, R_r is the radius of the raceway of the inner ring and f_c is the friction coefficient between the rolling elements and the raceway housing.

7.2.6. Friction torque due to shearing of the lubricant

The presence of a lubricant between the rolling bodies in a bearing leads to additional energy losses due to the viscosity of the lubricant, its physical characteristics, contact pressure, relative velocity of the oil flow, temperature range and the design features of the bearing assembly.

The lubricant film formed in the bearing prevents direct contact of the rolling elements, thereby reducing wear, tear and stress in the material at the points of contact and at the same time, increasing the resistance to motion. Power spent on overcoming friction in the lubricant film per unit length of a cylindrical roller bearing is given by

$$W_0^c = \frac{6.5v^2\mu}{\sqrt{h_0\alpha}} \qquad \text{when} \quad \alpha l_2^2 \gg h_0, \tag{7.17}$$

where $v = \partial x/\partial t$ is the longitudinal velocity of the point of contact during the rolling of the cylinder, μ is the viscosity of the lubricant, h_0 is the clearance filled with the lubricant and

$$\alpha = \tfrac{1}{2}\left(\frac{1}{R_x} \pm \frac{1}{R_1}\right),$$

where R_x is the radius of curvature of the cylinder surface at point O and R_1 is the radius of curvature of the raceway. The geometry of the contact with a viscous layer of lubricant is shown in Fig. 7.8.

The case of a rolling ball can be expressed as

$$W_0^b \approx \frac{6\pi v^2\mu\sqrt{\alpha}}{(2\beta/3\alpha)\sqrt{\beta}}\ln\left(\frac{4m^2\beta}{h_0}\right) \qquad \text{for} \quad 4m^2\beta \gg h_0, \tag{7.18}$$

where $2m$ represents the thickness of the oil film

$$\alpha = \tfrac{1}{2}\left(\frac{1}{R_x} \pm \frac{1}{R_1}\right) \qquad \beta = \tfrac{1}{2}\left(\frac{1}{R_y} \pm \frac{1}{R_2}\right).$$

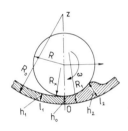

Figure 7.8

Here, R_y and R_2 are the radii of curvature perpendicular to the direction of rolling. Power expended on overcoming friction in this case is sensitive to changes in the value of m and h_0. The resistance moment due to the presence of the lubricant can be approximately expressed as

$$M_e = 2z W_0^b. \tag{7.19}$$

7.2.7. Friction torque caused by the working medium

The progress of space research has made a series of specific demands on the operation of rolling-contact bearings. In outer space conditions, there are many factors, such as the low pressure of the surrounding medium, ionizing radiation, absence of gravitation, presence of substances in the state of plasma, that appreciably modify the nature of the friction process.

It is well known that under low pressure conditions, surface films evaporate from contacting elements, and as a result of that, the friction between clean surfaces reaches a large value due to strong adhesion. Therefore, to create a protective surface film under low pressure conditions, helium is sometimes used. It is characterized by low vapour pressure and it can remain in the liquid state for a wide range of temperatures, providing efficient wetting.

7.2.8. Friction torque caused by temperature increase

Temperature does not uniformly affect friction torque. With an increase in temperature to 100 °C–120 °C, friction torque decreases, which is explained by the decrease in the viscosity of the lubricant. An increase in temperature beyond 100° C–140 °C causes an appreciable increase in the contact component of the friction torque as a result of changes in the geometric dimensions of individual components of the bearing. There are no analytical models which can be used to estimate the effect of temperature on friction torque. The majority of the expressions mentioned in this section contain a well-founded constant friction coefficient, varying over large intervals depending on the working conditions, though for approximate analysis of less important bearing units it is probably sufficient to use an average value chosen for a particular load and rotational speed. Thus, the friction coefficient, f, for self-aligning ball-bearings is usually taken as 0.001, for cylindrical roller-bearings 0.0011, for thrust ball-bearings 0.0013, for deep groove ball-bearings 0.0015, for tapered roller-bearings 0.0018, and needle roller-bearings 0.0045.

7.3. Deformations in rolling-contact bearings

In addition to knowing the stresses set up within the components of rolling bearings by the external loads (for information on contact stress please refer to Chapter 3), knowing the amount that the bearing components will elastically deform under a given load is important.

The normal approach, δ, of two bodies in point contact is

$$\delta = \frac{3P(\Theta a + \Theta_b)}{8\pi a} K(\alpha), \tag{7.20}$$

where $K(\alpha)$ is a complete elliptic integral of the first order, Θ_a and Θ_b are the elastic constants for the two bodies in contact and are the functions of the modulus of elasticity E and the Poisson ratio and a is the one of the contact area semiaxes. Using known relationships from contact mechanics (see Chapter 3) eqn (7.20) can be rearranged to yield

$$\delta = \frac{0.192417(\Theta_a + \Theta_b)^{\frac{2}{3}} K(\alpha) \cos^{\frac{2}{3}} \alpha}{[E(\alpha)]^{\frac{1}{3}}} \left[\Sigma \left(\frac{1}{R} \right) \right]^{\frac{1}{3}} P^{\frac{2}{3}}, \tag{7.21}$$

where $E(\alpha)$ is a complete elliptic integral of the second order and R is the equivalent radius of the contacting bodies.

The compressive deformation in line contact cannot be simply expressed. Palmgren gives an expression for the approach between the axis of a finite-length cylinder compressed between two infinite flat bodies and a distant

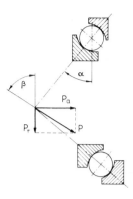

Figure 7.9

point in either of the support bodies

$$\delta = 0.39(\Theta_a + \Theta_b)^{0.9} \frac{P^{0.9}}{L^{0.8}}. \qquad (7.22)$$

Equation (7.22) may also be applied to roller-bearings. It is seen from eqn (7.22), for line contact, that the radius of curvature of the roller does not affect the deflection.

From the equations describing the elastic deflection of two bodies in contact it is apparent that their normal approach depends on the normal load, the geometry, and certain material constants. The calculation of deflections in a complete bearing requires a knowledge of its geometry, material and the radial and axial components of the load. From the load components, the load on the most heavily loaded rolling element must be calculated. Exact calculations of deflection are complex and tedious.

In general, when both radial and axial loads are applied to a bearing inner ring, the ring will be displaced both axially and radially. The direction of the resultant displacement, however, may not coincide with the direction of the load vector. If α is the bearing contact angle defined as the angle between a line drawn through the ball contact points and the radial plane of the bearing, Fig. 7.9, and β is the angle between the load vector and the radial plane, then the relation between the radial displacement δ_r and the axial displacement δ_a is given in the form of a graph as shown in Fig. 7.10. With the thrust load, $\tan\alpha / \tan\beta = 0$ and the resulting displacement is in the axial direction ($\delta_r = 0$). For a radial displacement ($\delta_a = 0$), $\tan\alpha / \tan\beta = 0.823$ for point contact and 0.785 for line contact. A radial load can be applied to single-row bearings only when $\alpha = 0$ in which case the displacement is also radial.

Palmgren gives the deflection formulae which are approximately true for standard bearings under load conditions that give radial deflection. For self-aligning ball-bearings

$$\delta_r = \frac{6.858 \times 10^{-8}}{\cos\alpha} \left(\frac{P^2}{d}\right)^{\frac{1}{3}}. \qquad (7.23)$$

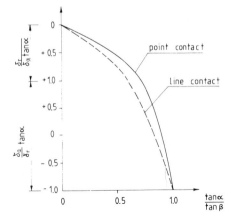

Figure 7.10

For deep-groove and angular-contact bearings

$$\delta_r = \frac{4.3 \times 10^{-8}}{\cos \alpha} \left(\frac{P^2}{d}\right)^{\frac{1}{3}}. \tag{7.24}$$

For roller-bearings with point contact at one race and line contact at the other

$$\delta_r = \frac{5.61 \times 10^{-9}}{\cos \alpha} \frac{(P^3)^{\frac{1}{4}}}{l^{\frac{1}{2}}}, \tag{7.25}$$

where l is the length of the contact.
For roller-bearings with line contact at both raceways

$$\delta_r = \frac{3.0 \times 10^{-10}}{\cos \alpha} \frac{P^{0.9}}{l^{0.8}}. \tag{7.26}$$

With axial load, the corresponding equations are

$$\delta_a = \frac{6.86 \times 10^{-8}}{\sin \alpha} \left(\frac{P^2}{d}\right)^{\frac{1}{3}}, \tag{7.23a}$$

$$\delta_a = \frac{4.3 \times 10^{-8}}{\sin \alpha} \left(\frac{P^2}{d}\right)^{\frac{1}{3}}, \tag{7.24a}$$

$$\delta_a = \frac{5.61 \times 10^{-9}}{\sin \alpha} \frac{(P^3)^{\frac{1}{4}}}{l^{\frac{1}{2}}}, \tag{7.25a}$$

$$\delta_a = \frac{3.0 \times 10^{-10}}{\sin \alpha} \frac{P^{0.9}}{l^{0.8}}. \tag{7.26a}$$

All deflections are in metres and are for zero-clearance bearings. The load on the most heavily loaded rolling element, P, is in Newtons, and the rolling element diameter, d, is in metres.

7.4. Kinematics of rolling-contact bearings

7.4.1. Normal speeds

The relative motion of the separator, the balls, and the races of a ball-bearing is an important study towards the understanding of the performance of this type of bearing. If ω_i is the angular velocity of the inner race, then the velocity of point A, Fig. 7.11, is

$$V_A = \omega_i \overline{AB}. \tag{7.27}$$

Similarly, if ω_o is the angular velocity of the outer race, then the velocity of point C is

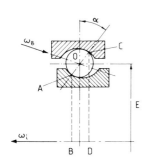

Figure 7.11

$$V_c = \omega_o \overline{CD}. \tag{7.28}$$

From the geometry shown in Fig. 7.11

$$\overline{AB} = \frac{E}{2} - \frac{d}{2}\cos\alpha, \tag{7.29}$$

$$\overline{CD} = \frac{E}{2} + \frac{d}{2}\cos\alpha, \tag{7.30}$$

where d is diameter of the ball. The velocity of the ball centre V_0 is

$$V_0 = \tfrac{1}{2}(V_A + V_C). \tag{7.31}$$

Substituting eqns (7.27) and (7.28) into eqn (7.31) gives

$$V_0 = \tfrac{1}{2}\left[\frac{\omega_i}{2}(E - d\cos\alpha) + \frac{\omega_o}{2}(E + d\cos\alpha)\right]. \tag{7.32}$$

The angular velocity of the separator or ball set ω_c about the shaft axis is

$$\omega_c = \frac{V_0}{E/2}. \tag{7.33}$$

Then

$$\omega_c = \tfrac{1}{2}\left[\omega_i\left(1 - \frac{d\cos\alpha}{E}\right) + \omega_o\left(1 + \frac{d\cos\alpha}{E}\right)\right]. \tag{7.34}$$

The speed of the separator when the outer race is fixed is

$$\omega_c = \frac{\omega_i}{2}\left(1 - \frac{d\cos\alpha}{E}\right). \tag{7.35}$$

For $\alpha < 90°$, the separator is always less than half the shaft speed. When both races rotate, the speed of the inner race relative to the separator is

$$\omega_{i/c} = \omega_i - \omega_c = \tfrac{1}{2}(\omega_i - \omega_o)\left(1 + \frac{d\cos\alpha}{E}\right). \tag{7.36}$$

The speed of the outer race relative to the separator is

$$\omega_{o/c} = \omega_o - \omega_c = \tfrac{1}{2}(\omega_o - \omega_i)\left(1 - \frac{d\cos\alpha}{E}\right). \tag{7.37}$$

From eqns (7.36) and (7.37) the speed of the inner race relative to the separator is always greater than that of the outer race relative to the separator. A point on the inner-race ball track will receive a greater number of stress cycles per unit time than will a point on the outer race. For ball-bearings that operate at nominal speeds, the centrifugal force on the balls is so negligible that the only forces that keep the ball in equilibrium are the two contact forces. For such conditions, the contact forces are equal and opposite, and the inner- and the outer-race contact angles are approximately equal. Even when the two contact angles are equal, true rolling of the ball can occur at only one contact. As shown in Fig. 7.12, the ball is rolling about axis OA, and true rolling occurs at the inner-race contact B

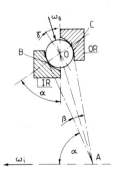

Figure 7.12

Figure 7.13

but not at the outer-race contact C. If the ball has an angular velocity ω_B about the axis OA, then it has a rolling component ω_r and a spin component $\omega_{s,o}$ relative to the outer race as shown in Fig. 7.13. The frictional heat generated at the ball-race contact, where slip takes place, is

$$H_f = \omega_s M_s, \tag{7.38}$$

where M_s is the twisting moment required to cause slip. Integrating the frictional force over the contact ellipse gives

$$M_s = \tfrac{3}{8} f P a E(\alpha) \tag{7.39}$$

when $b/a = 1$; $\alpha = 0°$ and $E = \pi/2$, but when $b/a = 0$; $\alpha = 90°$ and $E = 1$. For the same P, M_s will be greater for the ellipse with the greater eccentricity because the increase in a is greater than the decrease in E. In a given ball-bearing that operates under a given speed and load, rolling will always take place at one race and spinning at the other.

Rolling will take place at the race where M_s is greater because of the greater gripping action. This action is referred to as ball control. If a bearing is designed with equal race curvatures (race curvature is defined as the ratio of the race groove radius in a plane normal to the rolling direction to the ball diameter) and the operating speed is such that centrifugal forces are negligible, spinning will usually occur at the outer race. This spinning results from the fact that the inner-race contact ellipse has a greater eccentricity than the outer-race contact ellipse. The frictional heat generated at the ball-race contact where spinning takes place accounts for a significant portion of the total bearing friction losses. The closer the race curvatures, the greater the frictional heat developed. On the other hand, open race curvatures, which reduce friction, also increase the maximum contact stress and, consequently, reduce the bearing fatigue life.

7.4.2. High speeds

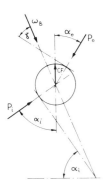

Figure 7.14

At high speeds, the centrifugal force developed on the balls becomes significant, and the contact angles at the inner and the outer races are no longer equal. The divergence of contact angles at high speeds tends to increase the angular velocity of spin between the ball and the slipping race and to aggravate the problem of heat generation. Figure 7.14 illustrates contact geometry at high speed in a ball-bearing with ball control at the inner race. The velocity diagram of the ball relative to the outer race remains the same as in the previous case (normal speed) except that γ has become greater and the magnitude of $\omega_{s,o}$ has increased. As the magnitude of P becomes greater with increasing centrifugal force, ball control probably will be shifted to the outer race unless the race curvatures are adjusted to prevent this occurring. Figure 7.15 illustrates ball control at the outer race. The velocity of the ball relative to the inner race is shown in Fig. 7.16. The inner-race angular velocity ω_i must be subtracted from the angular velocity of the ball ω_B to obtain the velocity of the ball relative to the inner race $\omega_{B,i}$.

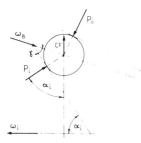

Figure 7.15

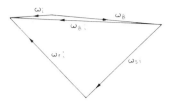

Figure 7.16

The spin component of the ball relative to the inner race is then $\omega_{s,i}$. In most instances, $\omega_{s,i}$ will be greater than $\omega_{s,o}$ so that great care must be taken in designing a ball-bearing for a high-speed application where heat generation is critical.

The spinning moments given by eqn (7.39) can be calculated to determine which race will have ball control. The heat generated because of ball spin can be calculated by solving for the value of ω_x in velocity diagrams similar to those presented earlier.

A further cause of possible ball skidding in lightly loaded ball-bearings that operate at high speed is the gyroscopic moment that acts on each ball. If the contact angle α is other than zero, there will be a component of spin about the axis through O normal to the plane of Fig. 7.12. A gyroscopic couple will also develop. The magnitude of this moment is

$$M_{gr}' = I\omega_B\omega_C \sin(\alpha + \beta_1), \tag{7.40}$$

where I is the moment of inertia of a ball about the axis through O and is given by eqn (7.7).

Gyroscopic moment will tend to rotate the ball clockwise in the plane of the figure. Rotation will be resisted by the friction forces at the inner- and the outer-race contacts, which are fP_i and fP_o, respectively. Whether slip takes place depends on the magnitude of the bearing load. In lightly loaded bearings that operate at high speeds, slippage is a possibility.

7.5. Lubrication of rolling-contact bearings

7.5.1. Function of a lubricant

A liquid or a grease lubricant in a rolling-element bearing provides several functions. One of the major functions is to separate the surfaces of the raceways and the rolling elements with an elastohydrodynamic film. The formation of the elastohydrodynamic film depends on the elastic deformation of the contacting surfaces and the hydrodynamic properties of the lubricant. The magnitude of the elastohydrodynamic film is dependent mainly on the viscosity of the lubricant and the speed and load conditions on the bearing. For normal bearing geometries, the magnitude of the elastohydrodynamic film thickness is of the order of 0.1 to 1.0 μm. In many applications, conditions are such that total separation of the surfaces is not attained, which means that some contact of the asperities occurs. Since the surfaces of the raceways are not ideally smooth and perfect, the existing asperities may have greater height than the generated elastohydrodynamic film and penetrate the film to contact the opposing surface. When this happens, it is a second function of the lubricant to prevent or minimize surface damage from this contact. Action of additives in the lubricants, aid in protecting the surfaces by reacting with the surfaces and forming films which prevent excessive damage. Contacts between the cage and the rolling elements and the cage and guiding loads on the race may also be lubricated by this means.

If the operating conditions are such that the asperity contacts are frequent and sustained, significant surface damage can occur when the lubricant no longer provides sufficient protection. The lubricant film parameter λ is a measure of the adequacy of the lubricant film to separate the bearing surfaces. In order for the frequency of asperity contacts between the rolling surfaces to be negligible, λ must be greater than 3. When λ is much less than 1, we can expect significant surface damage and a short service life of the bearing. When λ is between approximately 1.5 and 3, some asperity contact occurs, but satisfactory bearing operation and life can be obtained due to the protection provided by the lubricant.

Predicting the range of λ for a given application is dependent on knowing the magnitude of the elastohydrodynamic film thickness to a fair degree of accuracy. Surface roughness can be measured but may be modified somewhat during the running-in process. The film thickness can be evaluated using one of several equations available in the literature. Some of them are presented and discussed in Chapter 6.

Liquid lubricants also serve other functions in rolling-element bearings. The heat generated in a bearing can be removed if the lubricant is circulated through the bearing either to an external heat exchanger or simply brought into contact with the system casing or housing. Other cooling techniques with recirculating lubricant systems will be discussed later. Circulating lubricant also flushes out wear debris from intermittent contact in the bearing. Liquid lubricant can act as a rust and corrosion preventer and help to seal out dirt, dust and moisture. This is especially true in the case of grease.

7.5.2. Solid film lubrication

When operation of rolling-element bearings is required at extreme temperatures, either very high or very low, or at low pressure (vacuum), normal liquid lubricants or greases are not usually suitable. High-temperature limits are due to thermal or oxidative instability of the lubricant.

At low temperatures, such as in cryogenic systems, the lubricant's viscosity is so high that pumping losses and bearing torque are unacceptably high. In high-vacuum systems or space applications, rapid evaporation limits the usefulness of liquid lubricants and greases.

For the unusual environment, rolling-element bearings can be lubricated by solid films. The use of solid film lubrication generally limits bearing life to considerably less than the full fatigue life potential available with proper oil lubrication. Solid lubricants may be used as bonded films, transfer films or loose powder applications. Transfer film lubrication is employed in cryogenic systems such as rocket engine turbopumps. The cage of the ball- or roller-bearing is typically fabricated from a material containing PTFE. Lubricating films are formed in the raceway contacts by PTFE transferred from the balls or rollers which have rubbed the cage pocket surfaces and picked up a film of PTFE. Cooling of bearings in these applications is readily accomplished since they are usually operating in the cryogenic working fluid. In cryogenic systems where radiation may also be present,

PTFE-filled materials are not suitable, but lead and lead-alloy coated cages can supply satisfactory transfer film lubrication.

In very high temperature applications, lubrication with loose powders or bonded films has provided some degree of success. Powders such as molybdenum disulphide, lead monoxide and graphite have been tested up to 650 °C. However, neither loose powders nor bonded films have seen much use in high-temperature rolling-element bearing lubrication. Primary use of bonded films and composites containing solid film lubricants occurs in plain bearings and bushing in the aerospace industry.

7.5.3. Grease lubrication

Perhaps the most commonplace, widely used, most simple and most inexpensive mode of lubrication for rolling-element bearings is grease lubrication. Lubricating greases consist of a fluid phase of either a petroleum oil or a synthetic oil and a thickener. Additives similar to those in oils are used, but generally in larger quantities.

The lubricating process of a grease in a rolling-element bearing is such that the thickener phase acts essentially as a sponge or reservoir to hold the lubricating fluid. In an operating bearing, the grease generally channels or is moved out of the path of the rolling balls or rollers, and a portion of the fluid phase bleeds into the raceways and provides the lubricating function. However, it was found that the fluid in the contact areas of the balls or rollers and the raceways, appears to be grease in which the thickener has broken down in structure, due to its being severely worked. This fluid does not resemble the lubricating fluid described above. Also, when using grease, the elastohydrodynamic film thickness does not react to change with speed, as would be expected from the lubricating fluid alone, which indicates a more complicated lubrication mechanism. Grease lubrication is generally used in the more moderate rolling-element bearing applications, although some of the more recent grease compositions are finding a use in severe aerospace environments such as high temperature and vacuum conditions. The major advantages of a grease lubricated rolling-element bearing are simplicity of design, ease of maintenance, and minimal weight and space requirements.

Greases are retained within the bearing, thus they do not remove wear debris and degradation products from the bearing. The grease is retained either by shields or seals depending on the design of the housing. Positive contact seals can add to the heat generated in the bearing. Greases do not remove heat from a bearing as a circulating liquid lubrication system does.

The speed limitations of grease lubricated bearings are due mainly to a limited capacity to dissipate heat, but are also affected by bearing type and cage type. Standard quality ball and cylindrical roller-bearings with stamped steel cages are generally limited to 0.2 to 0.3 × 10⁶ DN, where DN is a speed parameter which is the bore in millimetres multiplied by the speed in r.p.m. Precision bearings with machined metallic or phenolic cages may be operated at speeds as high as 0.4 to 0.6 × 10⁶ DN. Grease lubricated tapered roller-bearings and spherical roller-bearings are generally limited to less than 0.2×10^6 DN and 0.1×10^6 DN respectively. These limits are basically those stated in bearing manufacturers' catalogues.

The selection of a type or a classification of grease (by both consistency and type of thickener) is based on the temperatures, speeds and pressures to which the bearings are to be exposed. For most applications, the rolling element bearing manufacturer can recommend the type of grease, and in some cases can supply bearings prelubricated with the recommended grease. Although in many cases, a piece of equipment with grease lubricated ball- or roller-bearings may be described as sealed for life, or lubricated for life, it should not be assumed that grease lubricated bearings have infinite grease life. It may only imply that that piece of equipment has a useful life, less than that of the grease lubricated bearing. On the contrary, grease in an operating bearing has a finite life which may be less than the calculated fatigue life of the bearing. Grease life is limited by evaporation, degradation, and leakage of the fluid from the grease. To eliminate failure of the bearing due to inadequate lubrication or a lack of grease, periodic relubrication should take place. The period of relubrication is generally based on experience with known or similar system. An equation estimating grease life in ball-bearings in electric motors, is based on the compilation of life tests on many sizes of bearings. Factors in the equation usually account for the type of grease, size of bearing, temperature, speed and load. For more information on grease life estimation the reader is referred to ESDU–78032.

7.5.4. Jet lubrication

For rolling-element bearing applications, where speeds are too high for grease or simple splash lubrication, jet lubrication is frequently used to lubricate and control bearing temperature by removing generated heat. In jet lubrication, the placement of the nozzles, the number of nozzles, jet velocity, lubricant flow rates, and the removal of lubricant from the bearing and immediate vicinity are all very important for satisfactory operation. Even the internal bearing design is a factor to be considered. Thus, it is obvious that some care must be taken in designing a jet-lubricated bearing system. The proper placement of jets should take advantage of any natural pumping ability of the bearing. This is illustrated in Fig. 7.17.

Centrifugal forces aid in moving the oil through the bearing to cool and lubricate the elements. Directing jets into the radial gaps between the rings and the cage is beneficial. The design of the cage and the lubrication of its surfaces sliding on the rings greatly effects the high-speed performance of jet-lubricated bearings. The cage is usually the first element to fail in a high-speed bearing with improper lubrication. With jet lubrication outer-ring riding cages give lower bearing temperatures and allow higher speed capability than inner-ring riding cages. It is expected that with outer-ring riding cages, where the larger radial gap is between the inner ring and the cage, better penetration and thus better cooling of the bearing is obtained.

Lubricant jet velocity is, of course, dependent on the flow rate and the

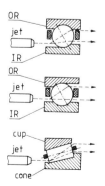

Figure 7.17

nozzle size. Jet velocity in turn has a significant effect on the bearing temperature. With proper bearing and cage design, placement of nozzles and jet velocities, jet lubrication can be successfully used for small bore ball-bearings with speeds of up to 3.0×10^6 DN. Likewise for large bore ball-bearings, speeds to 2.5×10^6 DN are attainable.

7.5.5. Lubrication utilizing under-race passages

During the mid 1960s as speeds of the main shaft of turbojet engines were pushed upwards, a more effective and efficient means of lubricating rolling-element bearings was developed. Conventional jet lubrication had failed to adequately cool and lubricate the inner-race contact as the lubricant was thrown outwards due to centrifugal effects. Increased flow rates only added to heat generation from the churning of the oil. Figure 7.18 shows the technique used to direct the lubricant under and centrifically out, through holes in the inner race, to cool and lubricate the bearing. Some lubricant may pass completely through and under the bearing for cooling only as shown in Fig. 7.18. Although not shown in the figure, some radial holes may be used to supply lubricant to the cage rigid lands. Under-race lubricated ball-bearings run significantly cooler than identical bearings with jet lubrication. Applying under-race lubrication to small bore bearings (<40 mm bore) is more difficult because of the limited space available for the grooves and radial holes, and the means to get the lubricant under the race. For a given DN value, centrifugal effects are more severe with small bearings since centrifugal forces vary with DN^2. The heat generated, per unit of surface area, is also much higher, and the heat removal is more difficult in smaller bearings. Tapered roller-bearings have been restricted to lower speed applications relative to ball-bearings and cylindrical roller-bearings. The speed limitation is primarily due to the cone-rib/roller-end contact which requires very special and careful lubrication and cooling consideration at higher speeds. The speed of tapered roller-bearings is limited to that which results in a DN value of approximately 0.5×10^6 DN (a cone-rib tangential velocity of approximately $36 \, \mathrm{m \, s^{-1}}$) unless special attention is given to the design and the lubrication of this very troublesome

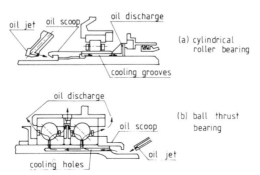

Figure 7.18

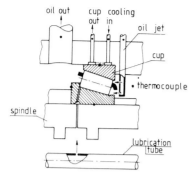

Figure 7.19

contact. At higher speeds, centrifugal effects starve this critical contact of lubricant.

In the late 1960s, the technique of under-race lubrication was applied to tapered roller-bearings, that is, to lubricate and cool the critical cone-rib/roller-end contact. A tapered roller-bearing with cone-rib and jet lubrication, is shown schematically in Fig. 7.19. Under-race lubrication is quite successful in reducing inner-race temperatures. However, at the same time, outer-race temperatures either remain high or are higher than those with jet lubrication. The use of outer-race cooling can be used to reduce the outer-race temperature to a level at or near the inner-race temperature. This would further add to the speed capability of under-race lubricated bearings and avoid large differentials in the bearing temperature that could cause excessive internal clearance. Under-race lubrication has been well developed for larger bore bearings and is currently being used with many aircraft turbine engine mainshaft bearings. Because of the added difficulty of applying it, the use of under-race lubrication with small bore bearings has been minimal, but the benefits are clear. It appears that the application at higher speeds of tapered roller-bearings using cone-rib lubrication is imminent, but the experience to date has been primarily in laboratory test rigs.

The use of under-race lubrication requires holes through the rotating inner race. It must be recognized that these holes weaken the inner-race structure and could contribute to the possibility of inner-race fracture at extremely high speeds. However, the fracture problem exists even without the lubrication holes in the inner races.

7.5.6. Mist lubrication

Air-oil mist or aerosol lubrication is a commonly used lubrication method for rolling-element bearings. This method of lubrication uses a suspension of fine oil particles in air as a fog or mist to transport oil to the bearing. The fog is then condensed at the bearing so that the oil particles will wet the bearing surfaces. Reclassification is extremely important, since the small oil particles in the fog do not readily wet the bearing surfaces. The reclassifier generally is a nozzle that accelerates the fog, forming larger oil particles that more readily wet the bearing surfaces.

Air-oil mist lubrication is non-recirculating; the oil is passed through the bearing once and then discarded. Very low oil-flow rates are sufficient for the lubrication of rolling-element bearings, exclusive of the cooling function. This type of lubrication has been used in industrial machinery for over fifty years. It is used very effectively in high-speed, high-precision machine tool spindles. A recent application of an air-oil mist lubrication system is in an emergency lubrication system for the mainshaft bearings in helicopter turbine engines. Air-oil mist lubrication systems are commercially available and can be tailored to supply lubricant from a central source for a large number of bearings.

7.5.7. Surface failure modes related to lubrication

As discussed earlier, the elastohydrodynamic film parameter, λ, has a significant effect on whether satisfactory bearing operation is attained. It has been observed that surface failure modes in rolling-element bearings can generally be categorized by the value of λ. The film parameter has been shown to be related to the time percentage during which the contacting surfaces are fully separated by an oil film. The practical meaning of magnitude for lubricated contact operations is discussed in detail in Chapter 2. Here it is sufficient to say that a λ range of between 1 and 3 is where many rolling element bearings usually operate. For this range, successful operation depends on additional factors such as lubricant/ material interactions, lubricant additive effects, the degree of sliding or spinning in the contact, and surface texture other than surface finish measured in terms of root mean square (r.m.s.). Surface glazing or deformation of the asperity peaks may occur, or in the case of more severe distress superficial pitting occurs. This distress generally occurs when there is more sliding or spinning in the contact such as in angular contact ballbearings and when the lubricant/material and surface texture effects are less favourable.

Another type of surface damage related to the film parameter λ, is peeling, which has been seen in tapered roller-bearing raceways. Peeling is a very shallow area, uniform in depth and usually less than 0.013 mm. Usually this form of distress could be eliminated by increasing the λ value. In practical terms it means the improvement in surface finish and the lowering of the operating temperature. To preclude surface distress and possible early rolling-element bearing failure, λ values less than 3 should be avoided.

7.5.8. Lubrication effects on fatigue life

The elastohydrodynamic film parameter, λ, plays an important role in the fatigue life of rolling element bearings. Generally, this can be represented in the form of the curve shown in Fig. 7.20. It is worth noting that the curve extends to values of less than 1. This implies that even though λ is such that significant surface distress could occur, continued operation would result in surface-initiated spalling fatigue. The effects of lubrication on fatigue life have been extensively studied. Life-correction factors for the lubricant effects are now being used in sophisticated computer programs for analysis of the rolling-element bearing performance. In such programs, the lubricant film parameter is calculated, and a life-correction factor is used in bearing-life calculations. Up to now, research efforts have concentrated on the physical factors involved to explain the greater scatter in life-results at low λ values. Material/lubricant chemical interactions, however, have not been adequately studied. From decades of boundary lubrication studies, however, it is apparent that chemical effects must play a significant role where there is appreciable asperity interaction.

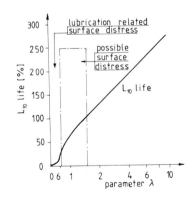

Figure 7.20

7.5.9. Lubricant contamination and filtration

It is well recognized that fatigue failures which occur on rolling-element bearings are a consequence of competitive failure modes developing primarily from either surface or subsurface defects. Subsurface initiated fatigue, that which originates slightly below the surface in a region of high shearing stress, is generally the mode of failure for properly designed, well lubricated, and well-maintained rolling-element bearings. Surface initiated fatigue, often originating at the trailing edge of a localized surface defect, is the most prevalent mode of fatigue failure in machinery where strict lubricant cleanliness and sufficient elastohydrodynamic film thickness are difficult to maintain. The presence of contaminants in rolling-element systems will not only increase the likelihood of surface-initiated fatigue, but can lead to a significant degree of component surface distress. Usually the wear rate increases as the contaminant particle size is increased. Furthermore, the wear process will continue for as long as the contaminant particle size exceeds the thickness of the elastohydrodynamic film separating the bearing surfaces. Since this film thickness is rarely greater than 3 microns for a rolling contact component, even extremely fine contaminant particles can cause some damage. There is experimental evidence showing that 80 to 90 per cent reduction in ball-bearing fatigue life could occur when contaminant particles were continuously fed into the recirculation lubrication system. There has been a reluctance to use fine filters because of the concern that fine lubricant filtration would not sufficiently improve component reliability to justify the possible increase in the system cost, weight and complexity. In addition it is usually presumed that fine filters will clog more quickly, have a higher pressure drop and generally require more maintenance than currently used filters.

7.5.10. Elastohydrodynamic lubrication in design practice

Advances in the theory of elastohydrodynamic lubrication have provided the designer with a better understanding of the mechanics of rolling contact. There are procedures based on scientific foundations which make possible the elimination of subjective experience from design decisions. However, it is important to know both the advantages and the limitations of elastohydrodynamic lubrication theory in a practical design context.

There are a number of design procedures and they are summarized in Fig. 7.21. A simple load capacity in a function of fatigue life approach is used by the designers to solve a majority of bearing application problems. The lubricant is selected on the basis of past experience and the expected operating temperature. Elastohydrodynamic lubrication principles are not commonly utilized in design procedures. However, in special non-standard cases, design procedures based on the ISO life-adjustment factors are used. These procedures allow the standard estimated life to be corrected to take into account special reliability, material or environmental requirements. Occasionally, a full elastohydrodynamic lubrication analysis coupled with

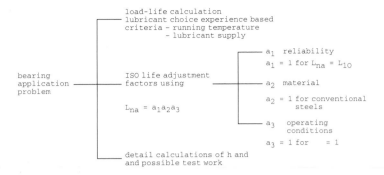

Figure 7.21

experimental investigation is undertaken as, for instance, in the case of very low or very high speeds or particularly demanding conditions. In this section only a brief outline of the ISO design procedures is given. If required, the reader is referred to the ISO Draft International Standard 281–Part 1 (1975) for further details.

An adjusted rating life L is given as

$$L_{na} = a_1 a_2 a_3 \left(\frac{C}{P}\right)^p \tag{7.41}$$

or

$$L_{na} = a_1 a_2 a_3 L_{10} \tag{7.42}$$

where a_1 is the life-correction factor for reliability, a_2 is the life-correction factor for material and a_3 is the life-correction factor for operating conditions.

The reliability factor has been used in life estimation procedures for a number of years as a separate calculation when other than 90 per cent reliability was required. The ISO procedure uses a_1 in the context of material and environmental factors. Therefore, when $L_{na} = L_{10}$, $a_1 = 1$, which means the life of the bearing with 90 per cent probability of survival and 10 per cent probability of failure.

Factors accounting for the operating conditions and material are very specific conceptually but dependent in practice. The material factor takes account of the improvements made in bearing steels since the time when the original ISO life equation was set up. The operating condition factor refers to the lubrication conditions of the bearing which are expressed in terms of the ratio of minimum film thickness to composite surface roughness. In this way the conditions under which the bearing operates and their effect on the bearing's life are described. In effect, it is an elastohydrodynamic lubrication factor with a number of silent assumptions such as; that operating temperatures are not excessive, that cleanliness conditions are such as would normally apply in a properly sealed bearing and that there is no serious misalignment. Both factors, however, are, to a certain extent, interdependent variables which means that it is not possible to compensate for poor operating conditions merely by using an improved material or vice versa. Because of this interrelation, some rolling-contact bearing manufacturers have employed a combined factor a_{23}, to account for both the material and the operating condition effects.

It has been found that the DN term (D is the bearing bore and N is the rotational speed) has a dominating effect on the viscosity required to give a specified film thickness. In a physical sense this can be regarded as being a shear velocity across the oil film. Before the introduction of elastohydrodynamic lubrication there was a DN range outside which special care in bearing selection had to be taken. This is still true, although the insight provided by elastohydrodynamic analysis makes the task of the designer much easier. The DN values in the range of 10 000 and 500 000 may be regarded as permitting the use of the standard life calculation procedures where the adjustment factor for operating conditions works satisfactorily. It should be remembered that the standard life calculations mean a clean running environment and no serious misalignment. In practice, these requirements are not often met and additional experimental data are needed. However, it can be said that elastohydrodynamic lubrication theory has confirmed the use of the DN parameter in rolling contact bearing design.

7.6. Acoustic emission in rolling-contact bearings

Noise produced by rolling-element bearings may usually be traced back to the poor condition of the critical rolling surfaces or occasionally to an unstable cage. Both of these parameters are dependent upon a sequence of events which start with the design and manufacture of the bearing components and ends with the construction and methods of assembly of the machine itself.

The relative importance of the various causes of noise is a function of machine design and manufacturing route so that each type of machine is prone to a few major causes. For example, on high-speed machines, noise levels will mostly depend on basic running errors, and parameters such as bearing seating alignment will be of primary importance. Causes of bearing noise are categorized in terms of:

(i) inherent sources of noise;
(ii) external influences.

Inherent sources include the design and manufacturing quality of the bearings, whereas external influences include distortion and damage, parameters which are mostly dependent on the machine design and the method of assembly. Among the ways used to control bearing noise we can distinguish:

(i) bearing and machine design;
(ii) precision;
(iii) absorption and isolation.

7.6.1. Inherent sources of noise

Inherent noise is the noise produced by bearings under radial or misaligning loads and occurs even if the rolling surfaces are perfect. Under

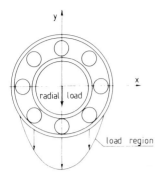

Figure 7.22

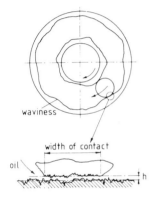

Figure 7.23

these conditions applied loads are supported by a few rolling elements confined to a narrow load region (Fig. 7.22). The radial position of the inner ring with respect to the outer ring depends on the elastic deflections at the rolling-element raceway contacts. As the position of the rolling elements change with respect to the applied load vector, the load distribution changes and produces a relative movement between the inner and outer rings. The movements take the form of a locus, which under radial load is two-dimensional and contained in a radial plane; whilst under misalignment, it is three-dimensional. The movement is also periodic with a base frequency equal to the rate at which the rolling elements pass through the load region. Frequency analysis of the movement yields a basic frequency and a series of harmonics. For a single-row radial ball-bearing with an inner-ring speed of 1800 r.p.m., a typical ball pass rate is 100 Hz and significant harmonics to more than 500 Hz can be generated.

7.6.2. Distributed defects on rolling surfaces

The term, distributed defects, is used here to describe the finish and form of the surfaces produced by manufacturing processes and such defects constitute a measure of the bearing quality. It is convenient to consider surface features in terms of wavelength compared to the Hertzian contact width of the rolling element-raceway contacts. It is usual to form surface features of wavelength of the order of the contact width or less roughness whereas longer-wavelength features waviness. Both these terms are illustrated in Fig. 7.23.

7.6.3. Surface geometry and roughness

The mechanism by which short-wavelength features produce significant levels of vibration in the audible range is as follows. Under normal conditions of load, speed and lubrication the rolling contacts deform elastically to produce a small finite contact area and a lubricating film is generated between the surfaces. Contacts widths are typically 50–500 μm depending on the bearing load and size, whereas lubricating film thicknesses are between 0.1 and 0.4 μm for a practical range of operating conditions. Roughness is only likely to be a significant factor and a source of vibration when the asperities break through the lubricating film and contact the opposing surface. The resulting vibration consists of a random sequence of small impulses which excite all natural modes of the bearing and supporting structure. Natural frequencies which correlate with the mean impulse rise time or the mean interval between impulses are more strongly excited than others. The effects of surface roughness are predominant at frequencies above the audible range but are significant at frequencies as low as sixty times the rotational speed of the bearing.

The ratio of lubricant film thickness to composite r.m.s. surface roughness is a key parameter which indicates the degree of asperity interaction. If it is assumed that the peak height of the asperities is only three times the r.m.s. level, then for a typical lubricant film thickness of $0.3\,\mu m$, surface finishes better than $0.05\,\mu m$ are required to achieve a low probability of surface–surface interaction.

Waviness

For the longer-wavelength surface features, peak curvatures are low compared to that of the Hertzian contacts and hence rolling motion is continuous with the rolling elements following the surface contours. The relationship between the surface geometry and vibration level is complex, being dependent upon bearing and contact geometry as well as the conditions of load and speed. The published theoretical models aimed at predicting bearing vibration levels from the surface waviness measurements have been successful only on a limited scale. Waviness produces vibration at frequencies up to approximately 300 times rotational speed but is predominant at frequencies below about 60 times rotational speed. The upper limit is attributed to the finite area of the Hertzian contacts which average out the shorter-wavelength features. In the case of two discs in rolling contact, the deformation at the contact averages out the simple harmonic waveforms over the contact width.

Bearing quality levels

The finish and form of the rolling surfaces, largely determine the bearing quality but there are no universally accepted standards for their control. Individual bearing manufacturers set their own standards and these vary widely. Vibration testing is an effective method of checking the quality of the rolling surfaces but again there is no universal standard for either the test method or the vibration limits. At present there are a number of basic tests in use for measuring bearing vibration, of these the method referred to by the American Military Specification MIL-B-17913D is perhaps the most widely used.

7.6.4.   External influences on noise generation

There are a number of external factors responsible for noise generation. Discrete defects usually refer to a wide range of faults, examples of which are scores of indentations, corrosion pits and contamination. Although these factors are commonplace, they only occur through neglect and, as a consequence, are usually large in amplitude compared to inherent rolling surface features. Another frequent source of noise is ring distortion. Mismatch in the precision between the bearing and the machine to which it is fitted, is a fundamental problem in achieving quiet running. Bearings are precision components, roundnesses of $2\,\mu m$ are common and unless the bearing seatings on the machines are manufactured to a similar precision, low frequency vibration levels will be determined more by ring distortion, after fitting, than by the inherent waviness of the rolling surfaces.

Bearings which are too lightly loaded can produce high vibration levels.

A typical example is the sliding fit, spring preloaded bearing in an electric motor where spring loads can barely be sufficient to overcome normal levels of friction between the outer ring and the housing. A certain preload is necessary to seat all of the balls and to ensure firm rolling contact, unless this level of preload is applied, balls will intermittently skid and roll and produce a cage-ball instability. When this occurs, vibration levels may be one or even two orders of magnitude higher than that normally associated with the bearing. Manufacturers catalogues usually give the values of the minimum required preload for single radial ball-bearings.

7.6.5. Noise reduction and vibration control methods

Noise reduction and vibration control problems can be addressed first by giving some consideration to the bearing type and the arrangement. The most important factors are skidding of the rolling elements and vibration due to variable compliance. These two factors are avoided by using single row radial ball-bearings in a fixed-free arrangement with the recommended level of preload applied through a spring washer. When this arrangement is already used, secondary improvements in the source of vibration levels may be achieved by the selection of bearing designs which are insensitive to distortion and internal form errors. The benefit of this is clearly seen at frequencies below sixty times the rotational speed. The ball load variation within the bearing is a key issue and the problem of low-frequency vibration generation would disappear if at all times all ball loads were equal. There are many reasons for the variation in ball loads, for instance, bearing ring distortion, misalignment, waviness errors of rolling surfaces all contribute to load fluctuation. Design studies have shown that for given levels of distortion or misalignment, ball load variation is a minimum in bearings having a minimum contact angle under thrust load. Significant reduction in low-frequency vibration levels can be achieved by selecting the clearance band to give a low-running clearance when the bearing is fitted to a machine. However, it is important to bear in mind that running a bearing with no internal clearance at all can lead to thermal instability and premature bearing failure. Thus, the minimum clearance selection should therefore be compatible with other design requirements. Another important factor influencing the noise and the vibration of rolling-contact bearings is precision. Rolling-element bearings are available in a range of precision grades defined by ISO R492. Although only the external dimensions and running errors are required to satisfy the ISO specification and finish of the rolling surfaces is not affected it should be noted, however, that the manufacturing equipment and methods required to produce bearings to higher standards of precision generally result in a higher standard of finish. The main advantage of using precision bearings is clearly seen at frequencies below sixty times rotational speed where improvements in basic running errors and the form of the rolling surfaces have a significant effect. It is important to match the level of precision of the machine to the bearing, although it presents difficulties and is a common cause of noise.

Accumulation of tolerances which is quite usual when a machine is built up from a number of parts can result in large misalignments between housing bores.

The level of noise and vibration produced by a rolling-contact bearing is an extremely good indicator of its quality and condition. Rolling bearings are available in a range of precision grades and the selection of higher grades of precision is an effective way to obtain low vibration levels, particularly in the low-frequency range. It should be remembered, however, that the machine to which the bearing is going to be fitted should be manufactured to a similar level of precision.

References to Chapter 7

1. W. K. Bolton. *Elastohydrodynamics in Practice*; Rolling contact fatigue performance testing of lubricants. London: Institute of Petroleum, 1977.
2. T. A. Harris. *Rolling Bearing Analysis.* New York: Wiley, 1966.
3. A. Fogg and J. S. Webber. The lubrication of ball bearings and roller bearings at high speed. *Proc. Instn Mech. Engrs*, **169** (1953), 87–93.
4. J. H. Harris. The lubrication of roller bearings. London: Shell Max and B.P., 1966.
5. F. Hirano. Motion of a ball in an angular contact bearing. *Trans. ASLE*, **8** (1965), 101–8.
6. F. Hirano and H. Tanon. Motion of a ball in a ball bearing. *Wear*, **4** (1961), 324–32.
7. C. T. Walters. The dynamics of ball bearings. *Trans. ASME*, **93** (1971), 167–72.

8 Lubrication and efficiency of involute gears

8.1. Introduction

Because it is assumed that the reader already has an understanding of the kinematics, stress analysis and the design of gearing, no further presentation of these topics will be given in this chapter. Instead, prominent attention will be given to lubrication and wear problems, because the successful operation of gears requires not only that the teeth will not break, but also that they will keep their precise geometry for many hours, even years of running. The second topic covered in this chapter is the efficiency of gears. It is customary to express the efficiencies of many power transmitting elements in terms of a coefficient of friction. A similar approach has been adopted here. In order to arrive at sensible solutions a number of simplifying assumptions are made. They are:

    (i) perfectly shaped and equally spaced involute teeth;

    (ii) a constant normal pressure at all times between the teeth in engagement;

    (iii) when two or more pairs of teeth carry the load simultaneously, the normal pressure is shared equally between them.

8.2. Generalities of gear tribodesign

If two parallel curved surfaces, such as the profiles of meshing spur gear-teeth, made of a truly rigid material, were pressed together they would make contact along a line, which implies that the area of contact would be zero, and the pressure infinite. No materials are rigid, however, so deformation of an elastic nature occurs, and a finite, though small area, carries the load. The case of two cylinders of uniform radii R_1 and R_2 was solved by Hertz. If we take the case of two steel cylinders for which $v = 0.286$ then the maximum compressive stress is given by

$$p_{max} = 0.416 \sqrt{\left[PE\left(\frac{1}{R_1} + \frac{1}{R_2}\right)\right]},$$

where P is the compressive load per unit length of the cylinders and E is the equivalent Young modulus. If the radius of relative curvature R of the cylinders is defined as $1/R_1 + 1/R_2$ then

$$p_{max} = 0.416 \sqrt{\frac{PE}{R}}.$$

It should be noted that this stress is one of the three compressive stresses, and as such is unlikely to be an important factor in the failure of the material. The maximum shear stress occurs at a small depth inside the material, and has a value of $0.3p_{max}$; at the surface, the maximum shear stress is $0.25p_{max}$. However, when sliding is introduced, a tangential stress field due to friction is added to the normal load. As the friction increases, the region of maximum shear stress (located at half the contact area radius beneath the surface), moves upwards whilst simultaneously a second region of high yield stress develops on the surface behind the circle of contact. The shear stress at the surface is sufficient to cause flow when the coefficient of friction reaches about 0.27. These stresses are much more likely to be responsible for the failure of the gear teeth. The important point for the designer at this stage is that each of these stresses is proportional to p_{max}, and therefore for any given material are proportional to $\sqrt{P/R}$.

For several reasons, however, this result cannot be directly applied to gear teeth. The analysis assumes two surfaces of constant radii of curvature, and an elastic homogeneous isotropic stress-free material. First, a gear tooth profile has a continuously varying radius of curvature, and the importance of this departure from the assumption may be emphasized by considering the case of an involute tooth where the profile starts at the base circle. The radius of curvature, say R_1, is at all times the length of the generating tangent, so at this point it is, from a mathematical point of view, zero; but it remains zero for no finite length of the involute curve, growing rapidly as we go up the tooth and having an unknown value within the base circle. If contact were to occur at this point the stress would not be infinite, as an infinitely small distortion would cause the load to be shared by the adjoining part of the involute profile, so that there would be a finite area of contact. Clearly, the Hertz analysis is rather inapplicable at this point; all that can be said is that the stresses are likely to be extremely high. In the regions where contact between well-designed gear teeth does occur the rate of change of R_1 is much less rapid, and it is not unreasonable to take a mean value at any instant for the short length in which we are interested.

Second, the assumption that the material is elastic will certainly break down if the resulting shear stress exceeds the shear yield strength of the material. The consequences are quite beyond our ability to predict them mathematically. We might manage the calculations if one load application at one instant were all we had to deal with; but the microscopic plastic flow which then occurred would completely upset our calculations for contact at the next point on the tooth profile and so on. The situation when the original contact recurred would be quite different; and we have to deal with millions of load cycles as the gears revolve. All that can be said is that the repeated plastic flow is likely to lead to fatigue failure, but that it will not necessarily do so, since the material may perhaps build up a favourable system of residual stress, and will probably work-harden to some extent. If such a process does go on then there is no longer a homogeneous isotropic stress-free material.

Third, gears which are transmitting more than a nominal power must be lubricated. The introduction of a lubricating film between the surfaces

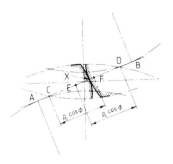

Figure 8.1

might be expected to alter the situation drastically, but it does not, simply because the oil film assumes the form of an extremely thin film of almost constant thickness.

In view of all these qualifying remarks, it is hardly to be expected that we could design gear teeth on the basis that the maximum shear stress is equal to $0.3p_{max}$ and has to equate to the shear strength of the material in fatigue. Nevertheless, the Hertz analysis is of vital qualitative value in indicating the parameter P/R, which, for any given material, can be taken as a criterion of the maximum stress, the actual value to be allowed being determined experimentally.

In order to proceed, we have to determine the minimum value of R when only a single pair of teeth is in contact, as shown in Fig. 8.1. It is known that for involute teeth the radius of curvature is the length of the generating tangent, so, with reference to Fig. 8.1, we can write for contact at X

$$R_1 = AX, \qquad R_2 = BX, \qquad R = \frac{R_1 R_2}{R_1 + R_2} = \frac{AX \, BX}{AX + BX}.$$

R will have a minimum value at either E or F, depending on which is nearer the adjoining base circle. In this case the critical point is E, and R can be calculated. If the permissible surface stress factor determined experimentally is denoted by S_c, then

$$P = S_c R \tag{8.1}$$

and the permissible tangential load at the pitch circle on a unit width of the tooth will be

$$P \cos \phi = S_c R \cos \phi. \tag{8.2}$$

Again, as in the case of bending stresses the need, for the individual designer, to work out each particular case is obviated by the provision of a factor Z which corresponds to $(R \cos \phi)$ for the meshing teeth of module 1, and a speed factor X_c accounting for impact and dynamic loads. In fact, tables containing Z and X_c, given in many books on gears, are based on a slightly more empirical approach, which suggests that $(R)^{0.8}$ gives better agreement with practice than $(R)^{1.0}$. The final simple form of the critical factor S_c for surface wear is

$$S_c = \frac{KP}{X_c Z}, \tag{8.3}$$

where K is not directly proportional to $(1/m)$ but to $(1/m)^{0.8}$, where m denotes module. Values of S_c for commonly used gear materials can be found in books on gears.

8.3. Lubrication regimes

There are three clearly distinguishable regimes of operation for gears with regard to lubrication. They may be described and defined as follows:

    (i) Boundary lubrication. This regime of lubrication is characterized by a velocity so low that virtually no elastohydrodynamic lubricating film is formed between the surfaces in contact. The friction and wear is mainly controlled by the adsorbed surface film, a few Angstroms thick, formed by the lubricant and its additives.

(ii) Mixed lubrication. The mixed lubrication regime is predominant when the velocity of the gears is sufficient to develop a lubricating film but its thickness does not provide full separation of the contacting surfaces. As a result of that, direct contact between the highest asperities takes place which may lead to accelerated running-in. The magnitude of the frictional force and the rate of wear are significantly lower than in the case of boundary lubrication.

(iii) Thick film lubrication. When the speed of the gears attains a sufficiently high value, an elastohydrodynamic film is developed, the thickness of which is adequate to separate completely the surfaces of two teeth in mesh. In principle all the friction resistance comes from the shearing of the elastohydrodynamic film. There is practically no wear if a small amount of initial wear during running-in is ignored. The only potential sources of wear in this lubrication regime are those due to abrasive particles contaminating the oil and the surface fatigue resulting in pitting. Each of the lubrication regimes can be assigned a characteristic value of friction coefficient. In boundary lubrication, a friction coefficient as high as 0.10–0.20 is not unusual. However, when care is taken of the surface finish of the gear teeth and a good boundary lubricant is used, the coefficient of friction can be substantially reduced to, say, the 0.05–0.10 range. Mixed lubrication is characterized by a coefficient of friction in the range of 0.04–0.07. Thick film lubrication produces the lowest friction and a coefficient of friction in the order of 0.01–0.04 can be regarded as typical. The graph in Fig. 8.2 provides an illustration of the three lubrication regimes discussed. They are defined in terms of the load intensity and the velocity measured at the pitch diameter. The load intensity value employed is the Q-factor. The Q-factor represents the average intensity of loading on the surface, whereas conventional stresses used in gear rating formulae represent the worst condition, with allowances made for misalignment, tooth spacing error, etc. This approach seems to be more justifiable as it takes into account the average conditions rather than exceptional conditions resulting from misalignment or spacing

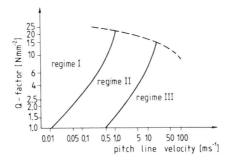

Figure 8.2

errors. Any unusual load concentration will be relieved due to running-in and the average conditions of loading will prevail.

The other important variable is the velocity measured at the pitch diameter. The usual practice is to use the relative velocities of rolling and sliding in any analysis, as they are responsible, among other factors, for developing an oil film. In a first attempt, however, aimed at finding the lubrication regime, the velocity of rolling at the pitch diameter can be used. The upper limit in the Fig. 8.2 represents the approximate highest intensity of tooth loading that case-carburized gears are able to carry. It also represents the surface fatigue strength upper limit for a relatively good design. It is well known that the pitting of gear teeth is markedly influenced by the quality of the lubrication. Under thick film lubrication conditions the S–N curve characterizing the tendency of gear teeth to pit is quite flat. As the lubrication changes from thick film to mixed lubrication and finally boundary regime lubrication the slope of the S–N curve becomes progressively steeper. Figure 8.3 shows typical S–N curves for contact stress expressed as a function of the number of gear tooth contacts. The data are valid for hard case-carburized gears (approximate hardness 60 HRC). During one full revolution each gear tooth is subjected to one load cycle. The contact stress in the gear teeth is proportional to the square root of the tooth load P. The relation between the load on the tooth and the number of cycles is given by

$$\frac{P_a}{P_b} = \left(\frac{N_b}{N_a}\right)^{1/q} \quad \text{or} \quad \frac{N_b}{N_a} = \left(\frac{P_a}{P_b}\right)^{q}, \tag{8.4}$$

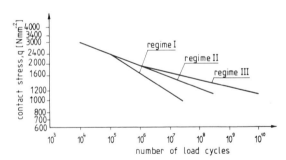

Figure 8.3

where P_a is the tooth load for N_a cycles before pitting, P_b is the tooth load for N_b cycles before pitting and q is the slope exponent for load versus cycle fatigue curve. The slopes of the curves, plotted in Fig. 8.3, are defined by the exponents in the following equations

$$N_b/N_a = \left(\frac{P_a}{P_b}\right)^{3.2} \quad \text{boundary lubrication regime,} \tag{8.5}$$

$$N_b/N_a = \left(\frac{P_a}{P_b}\right)^{5.3} \quad \text{mixed lubrication regime,} \tag{8.6}$$

$$N_b/N_a = \left(\frac{P_a}{P_b}\right)^{8.4} \quad \text{full film lubrication regime.} \tag{8.7}$$

In practice the specified exponent values may be quite different depending on the load and service life required, expressed by the number of load cycles N. Besides, depending on the microstructure of the material, the surface finish, the character of the oil additives and other similar factors, the slope of the S–N curve in a given lubrication regime may change. Thus, in the boundary lubrication regime the slope may vary from an exponent of as low as 2 to as high as 5. A mixed lubrication regime may vary in slope from 4 to 7. The thick film lubrication regime is usually characterized by an exponent in the range of 8 to 16, particularly in the range of 10^7 cycles to 10^{12} cycles.

Figure 8.3 should be considered as representing the average data and the real application conditions may vary considerably from that shown in the figure. In the case of heavy pitting some action should be taken in order to stop or at least slow down the damaging process. Usually an oil with higher viscosity provides the remedy by slowing down the pitting and creating the conditions for the pitted surfaces to recover. Pitting is not particularly dangerous in the case of low-hardness gears and a moderate amount of pitting is usually tolerated in medium-hardness gears. The opposite is true for hard gears where virtually no pitting can be tolerated. Work-hardening of the surface material is taking place during pitting, due to that, the surface is toughened and becomes more resistant to pitting. It is quite often the case that if the lubrication of the gears is efficient, pitting is a transient problem ceasing completely after some time.

8.4. Gear failure due to scuffing

Scuffing is usually defined as excessive damage characterized by the formation of local welds between sliding surfaces. For metallic surfaces to weld together the intervening films on at least one of them must become disrupted and subsequently metal–metal contact must take place through the disrupted film.

When two spheres, modelling the asperities on two flat surfaces, are loaded while in contact, they will at first deform elastically. The region of contact is a circle of radius a, given by the Hertz theory discussed in Chapter 3. When the load is increased, plasticity is first reached at a point beneath the surface, at about $0.5a$ below the centre of the circle of contact. The value of the shear stress depends slightly on the Poisson ratio but for most metals has a value of about $0.47P_m$, where $P_m = (W/\pi a^2)$ is the mean pressure over the circle of contact. At this stage P_m takes the value $1.1Y$, where Y is the yield stress of the softer metal.

As the load is increased, the amount of plastic deformation increases and the mean pressure rises. Eventually the whole of the material in the contact zone is in the plastic state and at this point the mean pressure P_m acquires its maximum value of about $3Y$. The load corresponding to full plasticity is about 150 times that at the onset of plasticity.

There is, therefore, an appreciable range of loads over which plastic flow takes place beneath the surface without it extending to the surface layers themselves. In these conditions, welding does not occur and this possibility of changing the surface profile by plastic flow of the material beneath, gives a means of smoothing out surface irregularities without causing excessive damage. This is one of the mechanisms utilized during running-in.

When sliding is introduced, however, a tangential stress field due to friction is added to the normal load. As the friction increases the region of maximum shear stress moves from $0.5a$ beneath the surface upwards whilst, simultaneously, a second region of high yield stress develops on the surface behind the circle of contact. The shear stress at the surface is sufficient to cause flow when the coefficient of friction reaches about 0.27. With plastic deformation in the surface layer itself, welding becomes possible. For a normal load which just suffices to cause shear at $0.5a$ beneath the surface, an increase in the friction to 0.5 causes shear over the whole area of contact in considerable depth. Also, as the load increases, the coefficient of friction necessary to cause flow in the surface, decreases. Experiments strongly suggest that scuffing originates primarily with an increase in the coefficient of friction. Scuffing is usually associated with poor lubrication.

As scuffing starts, the damage is not great when the oxide films are disrupted and the metals first come into contact. Usually the damage builds up as sliding proceeds. At first it is localized near the individual surface asperities where it is initiated. During further motion the regions of damage grow, and eventually coalesce with a great increase in the scale of the deformation. This could imply that the tendency to scuff depends upon the amount of sliding. In spur gears, for example, the motion is one of rolling at the pitch line and the proportion of sliding increases as the zone of contact moves away from it. It was observed that scuffing occurs away from the pitch line.

Another significant factor to consider is the speed of sliding as it directly influences the surface temperature. The temperature rise is sensitive to the load but varies as the square root of the speed. The rise is usually greater for hard metals than for soft but it is most sensitive of all to the coefficient of friction. Therefore, the maintenance of low friction, through efficient lubrication, is of prime importance in reducing the risk of scuffing.

The risk of scuffing could also be significantly reduced by the proper selection of gear materials. The first rule is that identical materials should not rub together. If for some reason the pair of metals must be chemically similar, their hardness should be made different so that the protective surface film on at least one of them remains intact, preventing strong adhesion. Metallic pairs which exhibit negligible solid solubility, are more resistant to welding and subsequently to scuffing, then those which form a continuous series of alloys. The role of the natural surface films is to prevent welding. If they are hard and brittle and the metal beneath is soft, the likelihood of them being broken increases significantly. The best films are ductile but hard enough to compete with the underlying metal.

There are many factors which may initiate gear scuffing but only two of them are really important. The first is the critical temperature in the contact zone and the other is the critical thickness of the film separating the two contacting surfaces.

8.4.1.   Critical temperature factor

The idea that scuffing is triggered when the temperature in the contact zone exceeds a certain critical temperature was first introduced by Blok in 1937. Failure of the lubricant film due to too high a temperature developed at the points of real contact between two teeth in mesh is central to this hypothesis. Contacting surface asperities form instantaneous adhesive junctions which are immediately ruptured because of the rolling and sliding of the meshing gears. This mechanism usually operates with gear teeth running in a thick film lubrication regime.

A severe form of scuffing is usually accompanied by considerable wear and as a result of that the teeth become overloaded around the pitch line. A practical consequence of this is pitting in an accelerated form leading to tooth fracture. One of the objectives the designer of gears must attain is to secure their operation without serious scuffing. It is generally accepted that a mild or light form of scuffing may be tolerated, provided it stops and the gears recover. Simple measures such as changing to a more efficient oil, operating the gears at less than service load until the completion of the running-in of the teeth or even removing bad spots on large teeth by hand can often be very effective in saving the gear drive from serious scuffing problems.

A commonly used design procedure to avoid scuffing because of excessively high temperature in the contact zone depends on the flash temperature estimation which in turn is compared with the maximum allowable temperature for a given oil. The approximate formula used to estimate flash temperature is

$$T_f = T_b + \left[\frac{1.25}{1.25 - R_a} \right] G_c \left[\frac{P_e}{b} \right]^{\frac{3}{4}} \left[(\omega_1)^{0.5} \frac{m^{0.25}}{1.094} \right], \tag{8.8}$$

where T_f is the flash temperature index [°C], T_b is the gear bulk temperature [°C], b is the face width in contact [mm], m is the module [mm], R_a is the surface finish [μm], G_c is the geometry constant (see Table 8.1) and ω_1 is the angular velocity of pinion.

Table 8.1. *Geometry constant G_c for pressure angle $\phi = 20°$*

G_c (at pinion tip)	pinion (number of teeth)	gear (number of teeth)	G_c (at gear tip)
0.0184	18	25	−0.0278
0.0139	18	35	−0.0281
0.0092	18	85	−0.0307
0.0200	25	25	−0.0200
0.0144	25	35	−0.0187
0.0088	25	85	−0.0167
0.0161	12	35	−0.0402
0.0101	35	85	−0.0087

As a first approximation, the value of P_e may be assumed to be equal to the full load on the tooth. When certain conditions are met, that is, spacing accuracy is perfect and the profile of the tooth is very accurate and modified, then P_e may be taken to be equal to 60 per cent of P. Equation (8.8) was formulated with the assumption that the tooth surface roughness is in the range of 0.5 to 0.7 μm.

8.4.2. Minimum film thickness factor

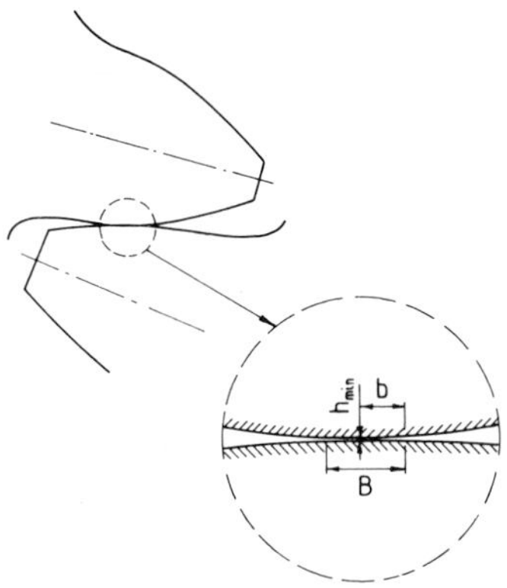

Figure 8.4 illustrates the contact zone conditions on the gear teeth when the running velocity is low and consequently the thickness of an elastohydrodynamic film is not sufficient to completely separate the interacting surfaces. The idea of the minimum thickness of the lubricant film is based on the premise that it should be greater than the average surface roughness to avoid scuffing. Conditions facilitating scuffing are created when the thickness of the lubricant film is equal to or less than the average surface roughness. It is customary to denote the ratio of minimum thickness of the film to surface roughness by

Figure 8.4

$$\lambda = \frac{h_{\min}}{R_a}, \tag{8.9}$$

where $R_a = (R_1 + R_2)/2$, R_1 is the root mean square (r.m.s.) finish of the first gear of a pair and R_2 is the finish in r.m.s. of a second gear of a pair.

The minimum thickness of the lubricant film created between two teeth in mesh is calculated using elastohydrodynamic lubrication theory developed for line contacts (see Chapter 6). Contacting teeth are replaced by equivalent cylinders (see Chapter 3) and the elastic equation together with the hydrodynamic equation are solved simultaneously. Nowadays, it is a rather standard problem which does not present any special difficulties. For spur or helical gears the following approximate formula can be recommended

$$h = 44.6(C_f) \frac{(L_f)(V_f)}{(P_f)}, \tag{8.10}$$

where

$$C_f = \text{curvature factor} = \frac{C \sin \phi}{\cos^2 \psi} \frac{m}{(m+1)^2},$$

$L_f = $ lubricant factor $= (\alpha E')^{0.54}$,
$\alpha = $ lubricant pressure-viscosity coefficient,
$E' = $ effective modulus for a steel gear set,

$$E' = \frac{\pi}{2} \frac{E}{(1-v)^2},$$

$v = $ the Poisson ratio $= 0.3$ for steel,

$$V_f = \text{velocity factor} = \left[\frac{\mu_0 u}{E'A}\right]^{0.70},$$

$\mu_0 = $ lubricant viscosity at operating temperature,

$$u = \text{rolling velocity} = \frac{\pi n_p d_p \sin\phi}{60},$$

$d_p = $ pinion pitch diameter,
$n_p = $ pinion r.p.m.,
$\phi = $ pressure angle,

$$P_f = \text{loading factor} = \left(\frac{P_t}{bE'A}\right)^{0.13},$$

$P_t = $ tangential load on tooth and
$b = $ face width.

Knowing both the thickness of the oil film, eqn (8.10) and the roughness of the gear tooth surfaces, the parameter λ can be determined. It is standard practice to assume a thick film lubrication regime and no danger of scuffing, when λ is greater than 1.2. In the case when λ is less than 1.0, some steps should be taken to secure the gear set against a high probability of scoring. This scoring is a direct result of insufficient thickness of the oil film. The usual remedy is to use an oil containing surface-active additives.

8.5. Gear pitting

It is not absolutely necessary to have contact between interacting gear teeth in order to produce wear, provided the running time is long enough. The Hertzian stresses produced in the contact zone of interacting gear teeth can lead to a fatigue which is regarded as a standard mode of failure. It takes the form of pitting; a pit being a small crater left in the surface as a result of a fragment of metal falling out. The presence of a lubricant does not prevent this, for under elastohydrodynamic conditions the surface pressure distribution is essentially that found by Hertz for unlubricated contacts. It could be argued that pitting is caused by lubrication in the sense that without lubrication the surface would fail long before pitting could appear. However, there are some reasons to believe that the lubricant is forced into the surface cracks by the passage of very high pressure and the lubricant then acts as a wedge to help open up and extend the cracks.

It is known from experiment that smooth surfaces pit less readily. It was found from tests run on a disc machine using a small slide/roll ratio, that increasing the oil film thickness, also reduced the tendency to pit and that the ratio of the surface roughness to the oil film thickness, was the dominant parameter. The correlation between the number of revolutions before pitting occurred, and the surface roughness to oil film thickness ratio, holds over a 500-fold variation in the above-mentioned ratio.

It must be emphasized, however, that the surface roughness measured was the initial value, and that the roughness when pitting occurred was very much less. Fatigue failures can originate either at or beneath the surface and it is clear that there are, at least, two competing failure mechanisms; one associated with inclusions in the material of the gear and the other with surface roughness and lubrication. A material with many internal imperfections is likely to fail by subsurface fatigue, and the life of the gear made of it will show little dependence on the ratio of the oil film thickness to the surface roughness. An inclusion free material will have a much longer life depending strongly on the oil film thickness to surface roughness ratio. Improvements in the manufacture of materials, mean that in practice, surface originated failures dominate, and therefore that surface roughness is now a critical factor.

There are a number of different types of pitting described in the literature on gears, however, all of them stem from two basic mechanisms. The basis used here for the classification of fatigue-related gear teeth failure, commonly called pitting, is the location at which the failure process originates.

8.5.1. Surface originated pitting

It has been found that in the case of through-hardened gears (hardness in the range of 180 to 400 HB) pitting usually originates at the surface of the tooth. Due to the stresses developed in the contact zone, small cracks are created on the surface. These cracks grow inwards and after reaching some depth they eventually turn upwards. As a result of that small metal particles are detached from the bulk and fall out. In most cases pitting is initiated in the vicinity of the pitch line. At the pitch point there is only pure rolling while above and below it there is an increasing amount of sliding along with rolling. Experiments suggest that pitting usually starts at the pitch line; a fact never fully explained, and progresses below the pitch line towards dedendum. It sometimes happens, especially with gears having a small number of teeth (less than twenty), that pitting begins at mid-dedendum or even lower. Usually, the dedendum part of the tooth is the first to undergo pitting, and only in considerably overloaded gears or in gears which have suffered a significant dedendum wear, is pitting attacking the addendum part of the tooth observed. The technique used to measure the wear extent due to pitting consists in taking a replica of the worn tooth profile and then cutting it normal to the tooth surface. Figure 8.5 shows, schematically, a typical worn involute tooth.

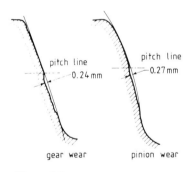

Figure 8.5

8.5.2. Evaluation of surface pitting risk

Only a very approximate evaluation of pitting risk is possible and that is done simply by comparing the contact stress with a certain value characteristic for the material of the gear. If the working contact stress does not exceed that value then the probability that there will be pitting in the design life of the gear is rather small. The practical implementation of this simple rule is somewhat difficult. This is because the evaluation of such things as material quality, tooth accuracy and lubrication efficiency always pose a problem for the designer. For example, the material may have hidden metallurgical defects or its hardness may not be uniform but it still meets the required contact stress value. Similar problems are encountered when tooth accuracy and lubrication efficiency are estimated in quantitative terms. One should then rely on past experience or on similar successful designs already in service.

8.5.3.   Subsurface originated pitting

Subsurface originated cracks are quite usual in the case of surface-hardened teeth. Cracks usually start at the case/core interface or entirely within the case, then coalesce, and as a result of that, rather large areas of the tooth may be damaged. The continuous operation of gears promotes the progress of tooth damage and this cannot be tolerated for long periods of time. A standard engineering practice in fighting against subsurface originated pitting has been to make the case deeper. Other changes in the operating conditions of gears, for instance, better surface finish, use of different types of oil, alterations in pitch line velocity, proved to be rather ineffective. The only factors which have any real influence on this type of tooth failure are the intensity of the load, the case depth and the tooth geometry. It is known from engineering practice that the final drives in slow moving vehicles are exposed to a considerable danger of subsurface originated pitting. The gears used in the final drives are spur, helical, bevel or hypoid. They are usually case-carburized and almost always heavily loaded with contact stresses in the region of 1.8 GPa.

8.5.4.   Evaluation of subsurface pitting risk

Subsurface originated pitting is a serious form of tooth wear and under no circumstances can it be tolerated. Therefore it is vitally important to ensure during the design stage that the gear unit will operate with a minimum risk of subsurface pitting. There are a number of procedures to estimate the risk of subsurface pitting recommended by specialist books on gears and each of them is based on experimental data. Usually the first step in estimation procedure is to determine the half-width of the Hertzian contact zone, a. The next step is to determine the maximum shear stress at a depth $1a, 2a, 3a$ and so on, below the surface, assuming that the friction between the contacting teeth is about 0.1. Proceeding in the way described above, a plot can be produced of maximum shear stress as a function of the depth below the surface of a tooth. Additional curves can be plotted to illustrate the relationship between the shear stress capacity of the material and the case-hardness expected at different depths. In order to minimize the risk of subsurface pitting this second curve should be above the maximum shear stress curve. If we know at which depth the shear strength of the material is equal to or higher than, the maximum shear stress it is possible to evaluate, in an approximate way, the case depth required. For example, if the load on the tooth is below 0.2 GPa then a case of hardness 50 HRC should extend to a depth of at least $2a$. Securing a case depth of sufficient magnitude should minimize the risk of subsurface pitting. This is true for all standard loading conditions with a good quality material, free of inclusions and other structural defects. In the case of shock loading or a material with hidden structural defects, subsurface pitting will appear even though the case is deep enough. There are no standards specifying the minimum case depth to avoid damage. Any departure from established practice regarding case depth should be preceded by thorough testing.

8.6. Assessment of gear wear risk

It is important for the designer to be familiar with the accepted general procedures which are used to assess the risk of gear wear and to be able to decide which of the expected wear rates may be tolerated.

Engineering practice shows that power gear trains, transmitting over 500 kW, run at speeds sufficient to secure thick film lubrication. Consequently, there should be no wear, provided that allowable surface contact stresses are not exceeded and the lubricant is clean. Gears with teeth of low hardness might undergo wear on lower parts of the tooth flanks. This wear is mainly due to pitting, although contribution of a limited plastic flow of the material in highly loaded gears, may be of significance.

When gears run in a mixed lubrication regime the risk of pitting is considerably increased after 10^7 or more contact cycles. This is the case with final drive gears in vehicles, or gears in the last stage of multiple stage electric motor drives. The risk of wear can be significantly reduced by using oils with a relatively high viscosity and containing surface active additives. However, the teeth should be at full hardness and the contact stresses should not exceed the assumed design limits for a given material. There are some highly loaded gear trains which are run at speeds which exclude the formation of a thick lubricating film. The speed is not sufficiently high to secure even mixed lubrication. Under such circumstances the practical remedy, in many cases, is to use very viscous oils containing highly active surface agents. A properly selected oil combined with good worn-in tooth surfaces may move gear operation from a boundary to a mixed lubrication regime.

Scuffing is usually characterized by excessive damage of tooth surfaces and virtually cannot be controlled. Therefore it is extremely important to ensure, at the design stage, that the risk of scuffing is as low as it practically can be. Another important factor in the smooth operation of gears is the cleanliness of the lubricant. When the lubricant is free from any form of contamination the gear train will operate without serious wear problems. It is vitally important to ensure that the new gear units are thoroughly cleaned before they are put into service.

When the power transmitted by the gear train is in the range 1–100 kW and the pitch circle velocity is less than $10 \, \text{m s}^{-1}$, splash lubrication can be quite effective provided that the lubricant is replenished at regular intervals, usually after 6 to 8 months, or when the level of the lubricant is below that recommended. When splash lubrication is used its cooling effectiveness must always be checked. It is especially important in the case of gear units transmitting power in the range of 100–500 kW at a pitch line velocity not exceeding 15 m s^{-1}. The usual procedure is to determine both a thermal rating and a mechanical rating for the unit. The thermal rating tells us how much power can be transmitted by the unit before its steady-state temperature is too high. Steady-state temperature over about 120 °C can be considered as too high and sufficient to induce detrimental and permanent changes in the lubricant. Gears should be run relatively cool to prevent wear in the long term.

Special lubrication problems are presented by large gear trains transmitting 1000 kW or more. It is usual to employ a pressurized lubrication system to lubricate, by means of oil jets, the gear teeth and bearings. Sufficient lubrication is achieved when the oil flow rate is approximately 6.5 × 10^{-5} m^3 s^{-1} per 360 kW transmitted. This is applicable in the case of a single mesh and spur or for helical gears. Other types of gears or different configurations may require higher rates of oil flow. The location of the oil jets depends on the pitch line velocity. For the highest velocities, oil jets should be used on both sides of the mesh; two-thirds of the oil is supplied on the outgoing side and one-third on the incoming side.

Long service life, free from wear problems depends crucially on the lubrication system; its ability to keep gears cool, and to deliver lubricant free from hard particles (filters should filter out particles down to 5 μm).

8.7. Design aspect of gear lubrication

It is accepted in gear design, that with the increase in the pitch line velocity, the lubricant used should be less viscous in order to minimize the power losses. In the case of heavily loaded gears, however, it is recommended that more viscous lubricants are used. Table 8.2 provides some guidance as to what is the recommended lubricant viscosity for a given pitch line velocity. The higher values of viscosity are selected for heavily loaded gears.

The next problem is to determine, more or less precisely, the amount of oil required in the gearbox, as too much oil would cause an increase in power losses due to oil churning and, on the other hand, an insufficient amount of oil would adversely affect heat dissipation. The following formula can be used to find, with sufficient accuracy, the amount of oil required (in cubic metres)

$$V = (0.0035-0.011)N \left(\frac{0.1}{Z_\mathrm{p} \cos \psi} + \frac{0.03}{2+V} \right), \tag{8.11}$$

where N is the power transmitted in kW, Z_p is the number of teeth in the

Table 8.2. *Recommended lubricant viscosity*

Pitch line velocity (m s^{-1})	1	2.5	4.0	10	16	25
Kinematic viscosity (cSt)		180–300	125–200	100–160	70–100	50–80	40–65

pinion, V is the pitch line velocity in m/s and ψ is the helix angle (for spur gears $\psi = 0$).

As a general rule, the lower values of V are used in the case of single-stage gear trains and the higher values of V are recommended for multi-stage gear trains. Slow running gears are usually splash lubricated. The depth to which the gear is immersed in the oil bath is given by

$$h = (1\text{–}6)m, \tag{8.12}$$

where m is the module in mm.

In multi-stage gear trains it is difficult to obtain a proper immersion of all the gearwheels in the oil bath. This is always the case when there are substantial differences in the diameters of the gearwheels. The usual solution is to install an auxiliary oil tank for each gearwheel in order to achieve the required depth of immersion.

Splash lubrication is effective up to a certain clearly defined pitch line velocity. This velocity can be determined in the way illustrated schematically in Fig. 8.6. The centrifugal force, acting on an element of oil having a mass dm will cause the motion of the element in the radial direction. This will be counteracted by the force required to shear the oil film formed on the surface of the tooth. Using symbols from Fig. 8.6 we can write

Figure 8.6

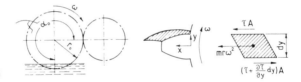

$$\omega^2 r \, dm = A \frac{d\tau}{dy} dy. \tag{8.13}$$

But

$$dm = A\rho \, dy$$

and

$$\tau = \mu \frac{dV}{dy},$$

therefore, eqn (8.13) can be written as

$$\omega^2 r A \rho \, dy = A \frac{d\tau}{dy} dy. \tag{8.14}$$

Assuming that the thickness of the oil film formed on the tooth surface is h_o and integrating, eqn (8.14) gives

$$h_o \rho \omega^2 r = \mu \int_0^{h_o} \frac{d^2 V}{dy^2} dy. \tag{8.15}$$

Assuming further that there is a parabolic distribution of the velocity within the oil film

$$V = V_{max} - (y - h_o)^2 \frac{V_{max}}{h_o^2}$$ (8.16)

the final form of eqn (8.15) is

$$\omega^2 r = 2v \frac{V_{max}}{h_o^2}.$$ (8.17)

In order to derive the expression for the maximum velocity of the oil due to the action of centrifugal force, we assume that the tooth is immersed in the oil bath, $x_e = 0.2m$, below the dedendum and that the time of action of the centrifugal force is defined by the angle of rotating, α, and the angular velocity, ω; thus,

$$t = \frac{\alpha}{\omega}$$

and finally

$$V_{max} = \frac{x_o}{t} = \frac{0.2m\omega}{\alpha}.$$ (8.18)

Substituting eqn (8.18) into eqn (8.17) and rearranging gives

$$V_l = \omega r = \frac{0.4vm}{\alpha h_o^2},$$ (8.19)

where v is the kinematic viscosity in cSt.

According to eqn (8.19) the allowable value of the pitch line velocity, V_l, at which splash lubrication is still effective, is a function of the oil viscosity, the module, the angle between the point of immersion and the point of engagement, α, and, indirectly, the surface roughness of the tooth surface. Gear trains operating at high speeds and also power gear units are jet lubricated. Each pair of meshing gears should receive an amount of oil resulting from the expression

$$Q = 0.6 + 2 \times 10^{-6} mV,$$ (8.20)

where Q is the flow rate of oil per 1 cm of the tooth width, measured in $[m^3/(min\ cm)]$, m is the module in mm and V is the pitch line velocity in $m\ s^{-1}$.

For a very approximate estimate of the oil flow rate required, the following formula can be used:

$$Q = (0.015 \times 10^{-3})N \quad [m^3\ min^{-1}],$$ (8.21)

where N is the power transmitted in kW.

8.8. Efficiency of gears

The power loss in properly lubricated spur, helical or similar types of gearing is usually very low, that is due to the tooth friction being of the order of only 1 per cent or less of the power transmitted at full load. To this, the losses due to oil churning and bearing friction have to be added. In such gears, there is inevitably, sliding at all points in the path of contact, except at the pitch point, and it can be deduced that the coefficient of friction is low and that lubrication must therefore be effective in spite of the extremely high contact pressures.

In the case of skew, and more particularly worm, gearing, sliding occurs not only as in spur gears but, much more importantly, in a direction at right angles to this. In fact, we can obtain a sufficiently close approximation to the situation in a worm gear by ignoring the pressure angle of the thread and thinking of this thread as perpendicular to the axis. We can then regard the thread as an inclined plane which moves relatively to the surface of the worm wheel; and the analogous situation of a block being pushed up an inclined plane by a horizontal force is quite common in mechanics. Thus, the expression for efficiency can be written as

$$\eta = \frac{\tan \alpha}{\tan(\alpha + \phi)}, \tag{8.22}$$

where α is the inclination of the plane, or in this case the pitch angle of the worm, and ϕ is the angle of friction. For the case of the worm-wheel driving the worm the expression for efficiency is

$$\eta' = \frac{\tan(\alpha - \phi)}{\tan \alpha}. \tag{8.23}$$

Now, in the case of a single-tooth worm, α may be only a few degrees, and if the surfaces are dry or poorly lubricated ϕ may well exceed α; in this case η will be less than 0.5 and η' will be negative. In other words, the drive will be irreversible. Such a gear has its uses, but would be unthinkable for power transmission. For multi-start worm gears, however, α can be made of the order of 45°, and if the gears are well lubricated, ϕ, under running conditions, particularly at high speeds, may well be of the order of 1° or less. The efficiency is then of the order 0.97–0.98, i.e. of the same order as that for spur gears. As far as power loss is concerned, the difference is probably negligible but it should be noted that the losses have to be dissipated as heat, and since the amount of heat which has to be expended is almost directly proportional to the effective coefficient of friction, it is vital to ensure that the best possible lubrication is maintained, and in the case of highly loaded gears that sufficient cooling is provided.

8.8.1. Analysis of friction losses

In Fig. 8.7, one gear rotating clockwise drives another. Subscript 1 is used on the symbols for the driver and the subscript 2 is used on those for the driven gear. All the parameters used in the following analysis are clearly defined in Fig. 8.7. Using the assumption that when two or more pairs of teeth carry the load simultaneously, the normal pressure is shared equally

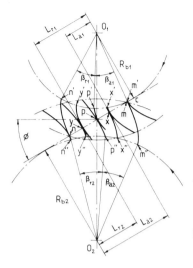

Figure 8.7

between them, it can be shown that the total friction loss and the power input to the driven gear during the engagement of one pair of mating teeth are the same as when one pair of mating teeth carry the entire load throughout their period of engagement.

During approach, considering any position of contact as at x (see Fig. 8.7), the normal force W opposes the rotation of the driver, while the frictional force (fW) assists rotation. The torque exerted by the driver at any approach position is as follows:

$$T_{a1} = W_n R_{b1} - f W_n L_{a1} \tag{8.24}$$

but

$$L_{a1} = R_{b1} \tan \phi - R_{b1} \beta_{x1}$$

thus

$$T_{a1} = W_n R_{b1} [1 - f(\tan \phi - \beta_{x1})]. \tag{8.25}$$

The work output for the driver during approach is as follows

$$W_{a1} = \int T_{a1} \, \mathrm{d}\beta_1 = W_n R_{b1} \int_{-\beta_{a1}}^{0} [1 - f(\tan \phi - \beta_1)] \, \mathrm{d}\beta_1. \tag{8.26}$$

During recess, the direction of sliding between the teeth is reversed, so that

$$T_{r1} = W_n R_{b1} + f W_n L_{r1}$$

and the work output for the driver during recess

$$W_{r1} = \int T_{r1} \, \mathrm{d}\beta_1 = W_n R_{b1} \int_{0}^{\beta_{r1}} [1 + f(\tan \phi + \beta_1)] \, \mathrm{d}\beta_1. \tag{8.27}$$

Now, turning to the driven gear, during approach the normal force and the frictional force oppose one another. Thus

$$T_{a2} = W_n R_{b2} - f W_n L_{a2}. \tag{8.28}$$

Expressing L_{a2} in the form

$$L_{a2} = R_{b2} \tan \phi + R_{b2} \beta_{x2}$$

the work output for the driven gear during approach is given by

$$W_{a2} = W_n R_{b2} \int_{-\beta_{a2}}^{0} [1 - f(\tan \phi + \beta_2)] \, \mathrm{d}\beta_2. \tag{8.29}$$

During recess, both the normal force and the tangential force assist the rotation of the driven gear, therefore

$$T_{r2} = W_n R_{b2} + f W_n L_{r2} \tag{8.30}$$

but

$$L_{r2} = R_{b2} \tan \phi - R_{b2} \beta_{x2}$$

so

$$W_{r2} = W_n R_{b2} \int_0^{-\beta_{r2}} [1 + f(\tan\phi - \beta_2)]\,\mathrm{d}\beta_2 \qquad (8.31)$$

Case I: The coefficient of friction is constant

The first case to be considered is that of the constant friction coefficient throughout the engagement. Integrating the equations describing the work output, gives

$$W_{a1} = W_n R_{b1} \left[\beta_{a1} - f'(\tan\phi)\beta_{a1} + \frac{f'}{2}\beta_{a1}^2 \right], \qquad (8.32)$$

$$W_{r1} = W_n R_{b1} \left[\beta_{r1} + f'(\tan\phi)\beta_{r1} + \frac{f'}{2}\beta_{r1}^2 \right], \qquad (8.33)$$

$$W_{a2} = W_n R_{b2} \left[\beta_{a2} - f'(\tan\phi)\beta_{a2} - \frac{f'}{2}\beta_{a2}^2 \right], \qquad (8.34a)$$

$$W_{a2} = W_n R_{b1} \left[\beta_{a1} - f'(\tan\phi)\beta_{a1} - \frac{f' R_{b1}}{2 R_{b2}}\beta_{a1}^2 \right], \qquad (8.34b)$$

$$W_{r2} = W_n R_{b2} \left[\beta_{r2} + f'(\tan\phi)\beta_{r2} - \frac{f'}{2}\beta_{r2}^2 \right], \qquad (8.35a)$$

$$W_{r2} = W_n R_{b1} \left[\beta_{r1} + f'(\tan\phi)\beta_{r1} - \frac{f' R_{b1}}{2 R_{b2}}\beta_{r1}^2 \right]. \qquad (8.35b)$$

The efficiency of the gears is therefore equal to

$$\frac{W_{a2} + W_{r2}}{W_{a1} + W_{r1}} = \frac{(\beta_{a1} + \beta_{r1}) - f'(\tan\phi)(\beta_{a1} - \beta_{r1}) - (f'/2i)(\beta_{a1}^2 + \beta_{r1}^2)}{(\beta_{a1} + \beta_{r1}) - f'(\tan\phi)(\beta_{a1} - \beta_{r1}) + (f'/2)(\beta_{a1}^2 + \beta_{r1}^2)}, \qquad (8.36)$$

where i is the gear ratio.

The friction losses per minute are equal to

$$W = \frac{W_n \omega_1 R_{b1}}{W_{a1} + W_{r1}} \left[\frac{f'}{2}(\beta_{a1}^2 + \beta_{r1}^2)\left(1 + \frac{1}{i}\right) \right]. \qquad (8.37)$$

The efficiency can be written more simply and almost exactly by considering the work input to be equal to

$$W_{a1} + W_{r1} = W_n \omega_1 R_{b1}$$

hence

$$\text{efficiency} = 1 - \left[\frac{1 + (1/i)}{\beta_{a1} + \beta_{r1}} \right] \frac{f'}{2}(\beta_{a1}^2 + \beta_{r1}^2). \qquad (8.38)$$

Case II. The coefficient of friction considered as variable

As a matter of fact, the friction coefficient is not constant but varies with different loads, speeds, lubricants and gear materials, as well as with different types of types of surface finish and many other factors. Actual tests carried out on gears have revealed that the form of the relationship between the average friction coefficients and the pitch line velocities is very much the same as in the case of journal bearings. At low speeds, the values of the friction coefficient are high, decreasing rapidly to a certain minimum value with increasing velocity, and then rising slowly with further increase in velocity. There is, however, one important difference in the lubrication mechanism operating in plain journal bearings and in gears. In the case of the journal bearings, hydrodynamic lubrication is usually a dominant type of lubrication while in gears, elastohydrodynamic lubrication is the main mechanism. It is known that the nature of sliding between involute gear teeth consists of sliding in one direction during approach, reducing to zero at the pitch point where the direction of sliding changes, and increasing again as the contact progresses through the recess action. This is shown, in a schematic way, in Fig. 8.8.

Since the direction of sliding changes at the pitch point, we may conclude that the coefficient of friction will assume the value characteristic for a thick

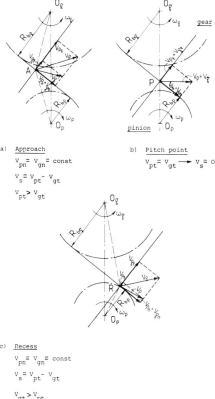

a) Approach
$$v_{pn} = v_{gn} = const$$
$$v_s = v_{pt} - v_{gt}$$
$$v_{pt} > v_{gt}$$

b) Pitch point
$$v_{pt} = v_{gt} \longrightarrow v_s = 0$$

c) Recess
$$v_{pn} = v_{gn} = const$$
$$v_s = v_{pt} - v_{gt}$$
$$v_{gt} > v_{pt}$$

Figure 8.8

film lubrication regime during the period of engagement of a pair of mating teeth. It is possible to set up an efficiency equation in various ways. The chances are, however, that the most that can be determined by experiment is to establish some average values of friction coefficient for the approach action and similarly for the recess action.

Experimental results suggest that at very low pitch line velocities (up to $1.5 \, \text{m min}^{-1}$) the friction of approach period appears to be approximately double that of the recess period on hobbed, milled and shaped gears made of cast-iron, soft-steel, bronze and aluminium. On hardened and ground steel gears, however, the difference between the friction of approach and the friction of recess is almost non-existent. When the contact passes through the pitch point, a significant increase in friction (about 150 per cent) takes place. Thus, introducing different average values for the friction coefficients of the approach and recess, gives

$$W_{a1} = W_n R_{b1} \left[\beta_{a1} - f_a (\tan \phi) \beta_{a1} + \frac{f_a}{2} \beta_{a1}^2 \right], \tag{8.39}$$

$$W_{r1} = W_n R_{b1} \left[\beta_{r1} + f_r (\tan \phi) \beta_{r1} + \frac{f_r}{2} \beta_{r1}^2 \right], \tag{8.40}$$

$$W_{a2} = W_n R_{b1} \left[\beta_{a1} - f_a (\tan \phi) \beta_{a1} - \frac{f_a}{2i} \beta_{a1}^2 \right], \tag{8.41}$$

$$W_{r2} = W_n R_{b1} \left[\beta_{r1} + f_r (\tan \phi) \beta_{r1} - \frac{f_r}{2i} \beta_{r1}^2 \right]. \tag{8.42}$$

The friction losses per minute are given by

$$W_f = \frac{W_n \omega_1 R_{b1}}{\beta_{a1} + \beta_{r1}} \left[\left(1 + \frac{1}{i} \right) \left(\frac{f_a}{2} \beta_{a1}^2 + \frac{f_r}{2} \beta_{r1}^2 \right) \right]. \tag{8.43}$$

Considering the work input to be equal to

$$W_{a1} + W_{r1} = W_n \omega_1 R_{b1}$$

then the efficiency is given by

$$\text{efficiency} = 1 - \left[\frac{1 + (1/i)}{\beta_{a1} + \beta_{r1}} \right] \left(\frac{f_a}{2} \beta_{a1}^2 + \frac{f_r}{2} \beta_{r1}^2 \right). \tag{8.44}$$

8.8.2. Summary of efficiency formulae

In order to collate the material presented in the previous section the following summary is made: when N_1, N_2 is the number of teeth on driver and driven gear, respectively, i is the gear ratio, β_a, β_r is the arc of approach and recess on the driver, respectively, f is the average coefficient of friction, f_a is the average coefficient of friction during the approach period and f_r is the average coefficient of friction during the recess period, then, for the constant friction coefficient

$$\text{efficiency} = 1 - \left[\frac{1 + (1/i)}{\beta_a + \beta_r}\right]\frac{f}{2}(\beta_a^2 + \beta_r^2) \tag{8.45}$$

and for different average friction coefficients during the approach and recess periods

$$\text{efficiency} = 1 - \left[\frac{1 + (1/i)}{\beta_a + \beta_r}\right]\left(\frac{f_a}{2}\beta_a^2 + \frac{f_r}{2}\beta_r^2\right). \tag{8.46}$$

References to Chapter 8

1. H. M. Martin. Lubrication of gear teeth. *Engineering*, **102**, (1916), 16–19.
2. D. W. Dudley. *Practical Gear Design*. New York: McGraw-Hill, 1954.
3. K. F. Martin. The efficiency of involute spur gears. *ASME Technical Paper*, No. 80-C2/DET-16, 1980.
4. D. W. Dudley. *Gear Handbook*. New York: McGraw-Hill, 1962.
5. D. Dowson and G. R. Higginson. *A Theory of Involute Gear Lubrication*. Gear Lubrication Symposium. London: Inst. of Petroleum, 1964.
6. D. W. Dudley. Information sheet – Gear scoring design guide for aerospace spur and helical power gears. Washington, D.C.: AGMA, 1965.
7. H. Blok. The postulate about the constancy of scoring temperature. *Interdisciplinary Approach to the Lubrication of Concentrated Contacts*, NASA SP-237, 1970.

Index

Abrasive wear, 19, 20
Acoustic emission, 268
Addendum, 10
Adhesive interaction, 15
Adhesive junction, 14, 15
Adhesive wear, 19
Adhesive wear equation, 39
Aerosol lubrication, 264
Angle of lap, 129, 133
Apparent area of contact, 14
Asperity, 14
Attitude angle, 192
Attitude of journal, 57
Axially loaded bearing, 123

Ball bearing, 7
Band and block brake, 144
Band brake, 136
Basic dynamic capacity, 7
Bearing clearance, 54
Bearing eccentricity, 54
Bearing materials, 220
Belt drive, 128
Belt power transmission rating, 132
Big-end bearing, 213
Blistering, 167
Blok theory, 75, 280
Boundary lubricated bearing, 121
Brake design, 136
Braking of vehicle, 145
Bulk temperature, 79

Cam, 9
Cam-follower, 9, 246
Centrifugal clutch, 120
Chemical wear, 19
Coefficient of adhesion, 146
Coefficient of viscosity, 48
Collar bearing, 124

Concave surface, 67
Concentrated force, 65
Cone clutch, 114
Conformal surfaces, 2
Conjunction temperature, 75
Connecting-rod bearing, 213
Contact mechanics, 64
Convex surface, 67
Copper-lead alloy, 221
Cornering of tyre, 152
Counterformal surfaces, 2
Crankshaft bearing, 213
Creep of tyre, 152
Critical slope, 188
Critical temperature, 82, 280
Critical temperature hypothesis, 11
Curvature factor, 281
Curved brake block, 138
Cylinder liner, 8

Dedendum, 10
Deformations in rolling-contact
    bearing, 254
Diametral clearance, 190, 195
Differential sliding, 249
Distributed force, 65
Driven rolling, 156
Dynamic hydroplaning, 158
Dynamically loaded journal bearing,
    212

Eccentricity ratio, 190, 192, 203
Efficiency of involute gears, 273, 288
Elastic contact, 14
Elastic extension of belt, 131
Elastic hysteresis, 251
Elasticity parameter, 241
Elastohydrodynamic lubrication, 3
Elliptical bearing, 206

Energy dissipation, 18
Engineering design, 1
Equivalent cylinder, 95
Equivalent speed method, 214
Extreme pressure oil, 11
Externally pressurized bearing, 181

Fatigue wear equation, 40
Film lubrication, 48
Flash temperature, 75, 83, 280
Flat pivot, 184
Fleming-Suh model, 45
Fluid film, 3, 6, 210
Four lobe bearing, 206
Fractional film defect, 34
Fracture mechanics and wear, 45
Fracture of adhesive junction, 16
Fracture toughness, 17
Free rolling, 156
Friction angle, 98
Friction circle, 122
Friction coefficient, 13
Friction drive, 10, 127
Friction due to adhesion, 15
Friction due to deformation, 17
Friction due to ploughing, 16
Friction in slideways, 98
Friction losses, 289
Friction stability, 100
Friction torque, 249
Frictional force, 13
Frictional traction, 10

Gas bearing, 210
Gear lubrication, 286
Gear tribodesign, 273
Gear wear, 285
Grease lubrication, 261
Grubin approximation, 245
Gyroscopic spin, 250

Heat of adsorption of lubricant, 35
Helical seal, 163
Hertzian area, 2
Hertzian stress, 9
Higher kinematic pair, 232
Hydrodynamically lubricated bearing, 174, 204
Hydrostatic bearing, 178
Hydrostatic thrust bearing, 225
Hypoid gears, 11

Hysteresis losses, 234
Hysteresis loss factor, 234

Inlet zone temperature, 244
Interfacial adhesive bonds, 15
Interfacial shear strength, 16
Involute gears, 10, 273

Jet lubrication, 262
Journal bearing, 189, 204
Journal bearing with:
  fixed non-preloaded pads, 205
  fixed preloaded pads, 205
  movable pads, 207
  special geometric features, 207
Junction growth, 15

Kinematics of rolling-contact bearing, 256
Kinetic friction, 98
Kingsbury, 186

Labyrinth seals, 164
Lambda ratio, 26, 29, 260, 265, 281
Line contact, 242
L life, 7, 267
Load bearing capacity, 196
Load number, 197, 215
Load sharing, 37
Load transmission, 1
Loading factor, 282
Lower kinematic pair, 97
Lubricant contamination, 266
Lubricant factor, 281
Lubricant filtration, 266
Lubricant viscosity, 33
Lubricated contact, 31
Lubrication effect on fatigue life, 265
Lubrication of cylinders, 238
Lubrication of rolling-contact bearings, 259
Lubrication of seals, 172
Lubrication of involute gears, 273
Lubrication regimes, 275

Marangoni effect, 162
Mechanical seal, 160
Michell, 186
Michell bearing, 223
Micro-slip, 236
Misalignment, 6
Mist lubrication, 264

Nonconforming contact, 22

Ocvirk number, 197
Ocvirk solution, 191
Offset factor, 206, 207
Oil film thickness, 214
Oil flow, 194

Pad pivot, 207
Palmgren, 255
Peclet number, 76
Petroff law, 50
Piston, 8
Piston ring, 8
Pitting, 10, 12, 72, 282
Pivot bearing, 124
Plastic deformation, 15
Plasticity index, 14, 30
Plate clutch, 111
Ploughing, 14
Pneumatic tyres, 151
Point-contact lubrication, 245
Preload factor, 206, 208
Pressure gradient, 175
Pressure-viscosity coefficient, 33, 241,
    242, 281
Propulsion of vehicle, 145
Protective layer, 4
PV limit, 230

Radial clearance, 190
Rayleigh step, 163
Real area of contact, 14
Recess, 180
Reynolds equation, 174, 177, 239
Reynolds hypothesis, 249
Reynolds number, 163, 238
Reynolds theory, 51
Rim clutch, 116
Roller bearing, 7
Rolling contact bearing, 7, 248
Rolling friction, 235, 248
Rolling of tyre, 155
Rope drive, 134
Run-in, 11

Screw jack, 105
Scuffing, 9, 11, 12, 278
Self-lubricating bearing, 226
Short-bearing approximation, 191
Short-bearing theory, 203
Sliding bearing, 6, 174

Sliding of tyre, 155
Solid film lubrication, 260
Sommerfeld diagram, 60
Sommerfeld number, 215
Sommerfeld solution, 190
Square thread, 103
Squeeze-film lubrication, 181
Standard deviation, 27
Static load rating, 7
Subcase fatigue, 73
Subsurface fatigue, 11
Surface active additives, 12
Surface failure, 71, 265
Surface fatigue, 73
Surface fatigue wear, 19, 21
Surface finish, 9
Surface peak, 90
Surface roughness, 2
Surface temperature, 74
Surface tension, 162, 240
Surface topography, 14
Surface traction, 233

Taper rolling bearing, 7
Thermal bulging, 80
Thermal correction factor, 244
Thermal effects, 74
Thermal loading factor, 244
Three lobe bearing, 206
Thrust bearing, 183, 221
Tilting pad bearing, 207, 223
Tin-aluminium alloy, 220
Traction effort, 146
Tractive resistance, 150
Tread pattern, 155
Triangular thread, 109
Tribodesign, 1
Tribology, 1
Tyre performance, 157
Tyre surface, 154

Unloaded bearing, 53

Variance, 27
V-belt drive, 134
Velocity factor, 282
Virtual coefficient of friction, 59
Viscosity, 180, 202, 204, 215
Viscosity parameter, 241
Viscosity-temperature coefficient, 245
Viscous flow, 50

Viscous hydroplaning, 158
Viscous shear, 210

Wear in mechanical seals, 164

Wear rate, 13
Worm gears, 12

Yield strength, 14, 21